TEUBNER-TEXTE zur Physik Band 28

P. C. Bosetti (Hrsg.)

Trends in Astroparticle-Physics

TEUBNER-TEXTE zur Physik

This regular series includes the presentation of recent research developments of strong interest as well as comprehensive treatments of important selected topics of physics. One of the aims is to make new results of research available to graduate students and younger scientists, and moreover to all people who like to widen their scope and inform themselves about new developments and trends.

A larger part of physics and applications of physics and also its application in neighbouring sciences such as chemistry, biology and technology is covered. Examples for typical topics are: Statistical physics, physics of condensed matter, interaction of light with matter, mesoscopic physics, physics of surfaces and interfaces, laser physics, nonlinear processes and selforganization, ultrafast dynamics, chemical and biological physics, quantum measuring devices with ultimately high resolution and sensitivity, and finally applications of physics in interdisciplinary fields.

Trends in Astroparticle-Physics

Edited by

Prof. Dr. Peter Christian Bosetti

Institute for Physics, Vijlen

Springer Fachmedien Wiesbaden GmbH

Prof. Dr. Peter Christian Bosetti

Born in 1948 in Hückinghausen. Studied physics in Aachen from 1967 to 1972. Dr. rer nat. (1975), Privatdozent (1982), C2 Professor (1987) at RWTH Aachen, Direktor VIP (Vijlen) since 1991, Apl. Professor for Physics at RWTH Aachen since 1993.
Fields of interest: particle physics, astrophysics, cosmology, biophysics.

Die Deutsche Bibliothek – CIP-Einheitsaufnahme

Trends in astroparticle physics /
ed. by Peter Christian Bosetti.
- Stuttgart ; Leipzig ; Teubner, 1994
(Teubner-Texte zur Physik ; 28)
ISBN 978-3-663-01467-6 ISBN 978-3-663-01466-9 (eBook)
DOI 10.1007/978-3-663-01466-9
NE: Bosetti, Peter C. [Hrsg.] ; GT

Ursprünglich erschienen bei B. G. Teubner Verlagsgesellschaft Leipzig 1994
Sofatcover reprint of the hardcover 1st edition 1994

Umschlaggestaltung: E. Kretschmer, Leipzig

Preface

From October 10 to 12, 1991 the Second International Conference on Trends in Astroparticle-Physics took place at Aachen, Germany. The meeting was the second in a series that started in 1990 at UCLA. It was attended by about 100 physicists from all over the world and covered the interface of elementary particles and astrophysics.

Topics covered included Neutrino-Telescopes in preparation, high energy gamma ray detectors, cosmology and particle physics, but also new ideas for future detectors.

Many people have worked hard on the preparation for the meeting and deserve many thanks, in particular D. Rein, L. Sehgal, U. Berson, G. Wurm, C. Wiebusch, C. Ley, and U. Braun. Representing the conference secretariat, I. Goidie, U. Packbier, and L. Jenckins were of great help. We also would like to thank the Stadtsparkasse Aachen, in particular Mr. Fischer, Mrs Cremers and Direktor Schwind for their support and providing the meeting facilities as well as making possible in parallel a public exhibition on the topic of the conference. Very special thanks goes to H. Geller, without whom this conference could not have taken place. Last but not least we wish to thank the speakers for taking their time in preparing their talks and presenting their results, making this publication possible.

Amongst the local participants of the conference was also Prof. Helmuth Faissner, not missing one of the talks. Helmuth Faissner has dedicated his scientific life to fundamental research keeping always unconventional experiments in mind. With his great understanding and intuition on elementary particle physics he has influenced a great number of young (and older) scientists and initiated many important experiments.

We would like to dedicate these proceedings to Professor Helmuth Faissner to his 65th birthday.

Vijlen, October 1993 Peter Christian Bosetti

Contents

Summaries **Chair: P. C. Bosetti**

DUMAND and the search for high energy neutrino point sources

D. Samm

III. Physikalisches Institut, Technische Hochschule Aachen, D-5100 Aachen, Germany

presented for the DUMAND Collaboration*

Abstract

DUMAND is a project to build a deep underwater laboratory. The main goal is to detect and study sources of high energy neutrinos. DUMAND Stage II will consist of 216 optical modules arranged in nine strings 230 m high in an octagon 105 m in diameter. The array is to be located approximately 30 km offshore, west of Keahole Point off the Big Island of Hawaii at an ocean depth of 4.8 km. The array will have an effective area of 20 000 m^2 for throughgoing muons from $> 100\ GeV$ neutrinos of atmospheric and extraterrestrial origin. The angular resolution for reconstructed muons is 1°. The minimum detectable flux from a point source is $10^{-10} cm^{-2} s^{-1}$ above 1 TeV.

1. Concept of DUMAND

DUMAND (Deep Underwater Muon And Neutrino Detector) is a particle astrophysics experiment, whose major goal is to detect and study sources of high energy neutrinos. The detection of these high energy neutrinos is based on the fact, that muons produced via charged current reactions of neutrinos will emit Cherenkov light while passing through water. This light will be detected by an array of large photomultipliers with hemispherical photocathodes deployed in the deep ocean.

The signature of a neutrino interaction within or near the DUMAND detector is a long straight muon from a direction anywhere in the range of 20° above the equator to straight up through the center of the earth. The direction of the muon track will be reconstructed from the arrival time of the photomultiplier hits and measured charge.

To search for point sources the experiment must be able to distinguish between upgoing muons due to the neutrinos searched for and downgoing cosmic ray muons of much more intense flux. The most physical background comes from upward muons induced by atmospheric neutrinos. Due to the large area and the good time resolution of the photomultipliers, which leads to reconstructed muon tracks with high accuracy, none of these backgrounds should be a problem.

The detector is situated in the deep ocean (4.8 km) to shield it from cosmic ray backgrounds and utilize seewater as both target and detection medium. Fig. 1 illustrates the concept of DUMAND. An adequate site to deploy a deep underwater Cherenkov detector was found in the pacific ocean, 30 km off the Big Island of Hawaii

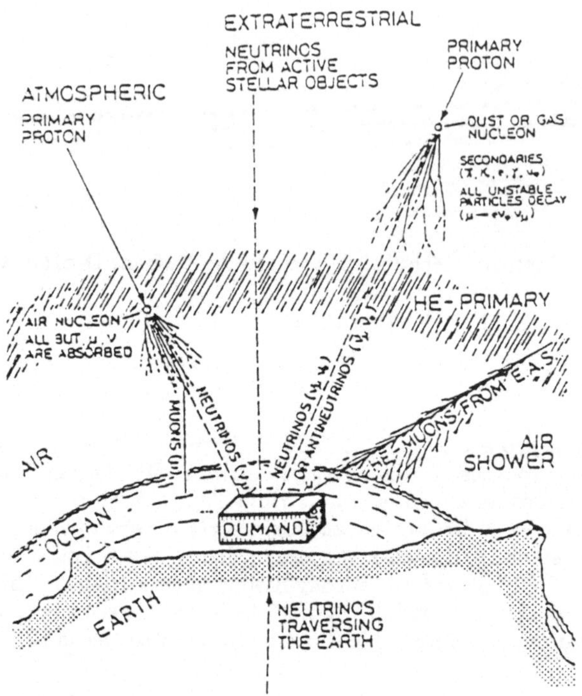

Figure 1. The concept of the DUMAND experiment.

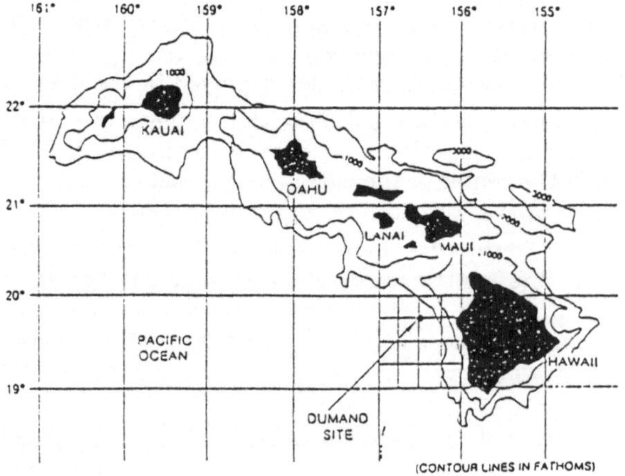

Figure 2. DUMAND deployment site.

(Fig. 2). The coordinates of the location are: 19° 44′N, 156° 19′ W. The location has been chosen after a long series of investigations, as it offers several important advantages:

- water clearity
  At this location the absorption length for photons in the blue-green "window" is about 40 m [1,2]
- low current
  The water currents have been measured and found to be about 2 $cm\ s^{-1}$ [3] on the area
- flat ground
  The ocean ground is flat to about 1 m over more than one kilometer, thus future expansions are possible
- great depth
  At the depth of 4.8 km only the high energy part of the cosmic ray spectrum penetrates and thus the background from cosmic rays is low
- near to shore.

In addition to opening a new window for observations in astrophysics and astronomy, DUMAND will complement acceleration-based high energy physics research. With the detector it is possible to study the neutrino cross section at high energies from the ratio of observed numbers of upward and downward muons [3]. Because of the exceptionally good angular resolution of DUMAND it will be possible to search for neutrino oscillations by using the variation in neutrino path lengths through the earth, which depends on the zenith angle [3]. There exists also the possibility to study neutrino oscillations with the help of the Fermilab beam [4,5]. Monopoles are detectable via their slow velocity and, of course, one can study the physics of cosmic ray muons. Finally one can do studies in Ocean and Earth sience, and if the Stecker et al. model of neutrinos from active galactic nuclei [6] is not to optimistic, one has the chance to measure the core density of the earth with high precision ($\approx 5\%$) [7].

2. Why high energy neutrino astronomy?

There is no doubt that neutrinos play a fundamental and unique role in elementary particle physics. Neutrinos are also important in cosmology, since they might provide a key to the formation of galaxies and the large scale structure of the universe. The emission of neutrinos often dominates the cooling of optically thick objects and thus determines their evolution and their age. Since matter is essentially transparent for neutrinos, they provide a good probe to explore the interior of optically thick objects, like stars. The best known example is given by the flux of MeV solar neutrinos that has been measured. The observation for neutrino bursts associated with the supernova SN1987A has proven the importance of neutrino observation for eluding the dynamics of stellar core collapse. So one might attempt to "see" the universe in neutrinos produced by nuclear processes. But although the fluxes due to stellar processes are high, it was very hard to detect even the 10 MeV neutrinos from our own sun. Thus, the detection of those low energy neutrinos from point sources seems impossible in the forseeable future, except in the case of tremendous bursts of supernovae explosions.

Unfortunately the rate of such spectacular events is not very high. For a supernova of type II in our galaxy the expected rate in one is 10 or 50 years [8,9]. So it seems that the beginning for neutrino astronomy is likely to be in the TeV region for four reasons:

- At high energies, as Fig. 3 indicates, the cross section for charged current weak interactions becomes larger [10] and one has a better chance to detect the neutrinos.

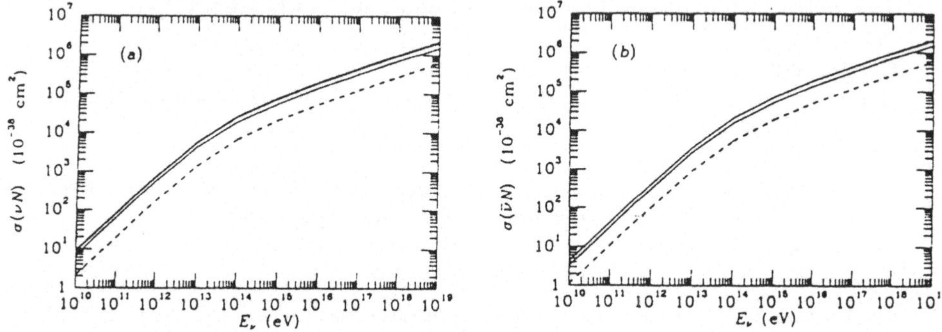

Figure 3. Cross section for charged current reactions at high energies; (a) for νN (b) for $\bar{\nu}N$. The dashed line is $\sigma(\nu N \rightarrow \nu X)$, the lower solid line is $\sigma(\nu N \rightarrow \mu X)$, the upper solid line is the total.

- The angle $\theta_{\mu\nu}$ between the neutrino with an energy E_ν and the muon is given approximately by:

  $\theta_{\mu\nu} \simeq 2.6\sqrt{\frac{100GeV}{E_\nu}}$

  At high energies the angle $\theta_{\mu\nu}$ gets smaller and the muon track orientation provides an excellent correlation with the source location. Thus the background from atmospheric neutrinos gets smaller.
- The neutrino spectrum from the atmosphere can approximately be described by a power law in the energy ($\sim E_\nu^{-\alpha}$). A differential spectral index of $\alpha = 3.8$ corresponds to atmospheric neutrinos while extraterrestrial neutrinos have a differential spectral index in the range of 2.0 to 2.6. Thus the background from atmospheric neutrinos is more steep than that one expects from point sources. In Fig. 4 the integral muon spectrum is shown for a point source that produces 20 events per year in the DUMAND detector with $\alpha = 2$. The atmospheric neutrinos produced within 1° are also shown.
- The range of the muon and thus the effective target volume increase with energy.

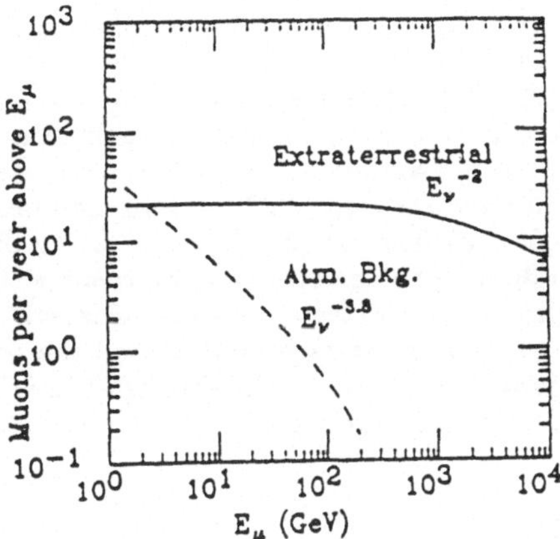

Figure 4. Integral muon spectrum with a E^{-2} differential neutrino spectrum from point sources. The spectrum of muons from atmospheric neutrinos within an angle of 1° is also shown.

There is another good reason to go to high energies: the old question of the origin of high energy cosmic rays. Until now cosmic rays have been observed up to energies of $10^{20}eV$. In particular, there is still uncertainty concerning the source of cosmic rays and the acceleration mechanism. The majority of cosmic rays consists of charged particles, but it is not possible to point these particles back to their source. All directional information has been lost due to interactions with the galactic magnetic field.

To track back to point sources in principle two kinds of known particles can be used: photons and neutrinos. Photons are much more easier to detect and a lot of large surface detectors exist to measure high energy photons from point like sources [11]. Why are neutrinos of special interest? Neutrinos are a unique tracer of energetic hadronic acceleration in sources. Whereas photons can be produced by hadronic and electromagnetic processes, the neutrinos – essentially – can only be produced by hadronic interactions. For this reason neutrinos can be used to understand the acceleration mechanism. Even the lack of them would give important hints of the origin of high energy cosmic rays. Furthermore, neutrinos carry more information than gamma-rays, because they have larger absorption lengths. Neutrinos are generated inside regions of dense matter and their flux should be much larger than the photon flux, because only photons which come from outer shells of the stellar atmosphere ("low density region") are not absorbed. Consequently with neutrinos it is possible to look into regions which are not observable with photons (e.g. core of galaxies).

3. The octagon array

The DUMAND detector will consist of a total of 216 optical modules mounted on nine vertical strings each with 24 modules spaced 10 m apart along the string vertically and with the string spaced 40 m horizontally (Fig. 5). This geometry allows a significant effective detector volume to be achieved with a modest number of photomultiplier tubes. The precise location (within 20 - 30 cm) of each module will be continually determined by acoustic ranging. The strings are located in an octagonal configuration with the nineth string in the center of the equilateral octagon. Beside the optical modules each string consists of three hydrophones, laser calibration units and environmental instruments. The array will have a height of 230 m and a diameter of 105 m and thus an effective muon area of $\approx$ 20 000 m^2 [2]. The lowest module will be approximately 100 m above the ocean floor. The detector spacing ensures that muons with energies $> 50\ GeV$, which pass through the detector from outside, will be detected with high efficiency. Given the spatial location of each module and the time (within 1 ns) of each photon induced photomultiplier pulse, tracks through the array can be reconstructed with an angular resolution of the order of 1°.

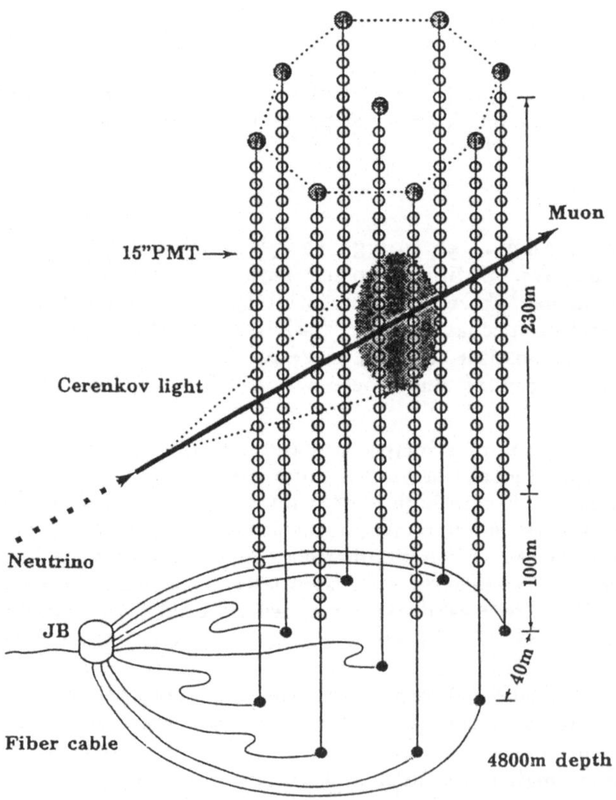

Figure 5. DUMAND II array, string configuration.

The basic DUMAND detector unit - the optical module - is composed of large photomultipliers (Hamamatsu R2018C and Philips XP2600) (Fig. 6) placed in a 1 cm thick pressure glass housing. The photomultipliers (PMT) used have the capability of wide area detection for maximum response to faint Cherenkov light with good energy resolution and a small time jitter on the one photoelectron level [12,13]. A flexible silicon layer mechanically buffers and optically couples the PMTs to the sphere. Beside the PMT, each optical module contains the high voltage power supply for the PMT and the integrated circuit electronics. The optical modules are mounted with the PMTs facing downward, optimizing the sensitivity for muons from upward going neutrinos.

The PMT signal is converted to an optical signal whose leading edge specifies the arrival time of the hit. The optical pulse is sent from the module via a multi-mode fiber optic cable through an optical feed through to a string bottom digitization and multiplexing package. Electric power and communications pass through a second penetrator. The signals, including the data from the hydrophones and the calibration modules will be multiplexed and sent shoreward in two colours on single-mode fibers (one for each string).

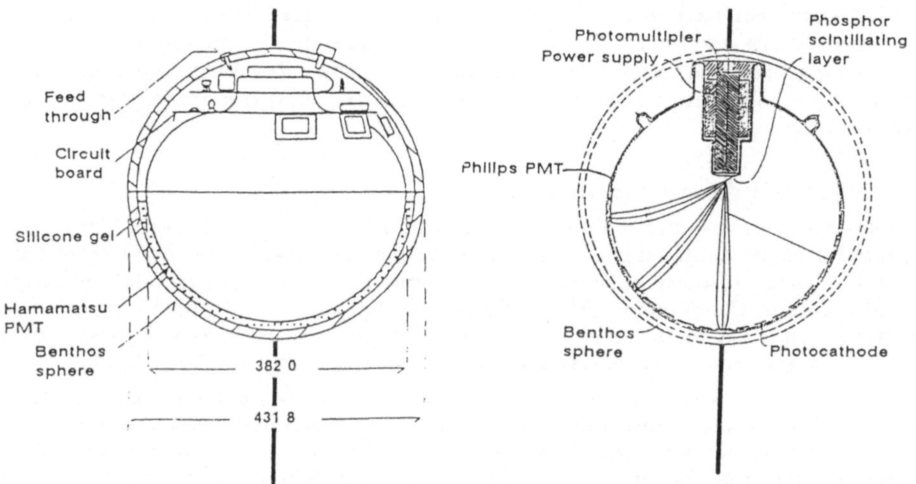

Figure 6. Cross sectional view of the optical modules (taken from Ref. 14).

4. Backgrounds and minmum detectable flux for point sources

For a deep ocean array, the backgrounds will be generally of two types:

- Natural optical background due to bioluminescence and Cherenkov radiation from the decay products of radioactive elements present in the water.
- Background due to atmospheric neutrinos and to cosmic ray muons, which could be falsely reconstructed as upward going and misinterpreted as neutrino induced.

4.1 Optical background

The intensity of surface light, which penetrates the 4.8 *km* of water is orders of magnitudes too small to be measurable. The optical background at this depth is produced by Cherenkov radiation of electrons from natural radioactivity and from bioluminescence. It is important to notice that these phenomena produce uncorrelated background and thus only increase the single rates and not the true coincidences!

A detailed study of the decay modes and concentrations of the isotopes present in the seewater shows that Cherenkov radiation from the electron emitted in the decay of ^{40}K is the dominant contribution to optical background from natural radioactivity. The optical attenuation length of water – e.g. 40 m in the blue window – determines the effective volume which contributes to the ^{40}K light flux. The calculation gives a flux of 120 photons $cm^{-2}s^{-1}$ [15]. These photons give signals at the one photoelectron level.

It is well known that many of the organisms in the ocean do emit light. The DUMAND collaboration undertook a series of in situ measurements to determine this optical background. The light observed consists of rare, but comparatively intense flashes over a much lower and relatively constant background of 218^{+20}_{-60} photons $cm^{-2}s^{-1}$ [16]. This is somewhat less than twice the rate expected from ^{40}K alone. Flashes which exceed the ambient background by a factor of three or greater occur at the rate of approximately one per hour. The observation of pulses in the bioluminescent background shows that it is important to monitor the single rate. With this precaution, both the false coincidence rates and the deadtime ($< 10\%$) are acceptable for rare neutrino events.

4.2 Background due to cosmic rays

The flux of cosmic ray muons at a depth of 4.7 km is $F_\mu = 2.1 * 10^{-5} cm^{-2}s^{-1}$ integrated over all zenith angles which leads to a single cosmic ray muon event rate of 3 min^{-1}. Some of these downward going cosmic ray muons may simulate neutrino induced events in the detector. Monte Carlo calculations showed that the fraction of downward cosmic ray muons, which could be falsely reconstructed as an upward going muon from a neutrino event, is less than 0.0014 with 90% [2] confidence level.

The DUMAND array will detect about 3500 atmospheric neutrino induced muons over all angles per year. But only those within a specified angle contribute as background for the search of neutrino point sources. With the given angular resolution of 1° the atmospheric background to a point source is less than one event per year above 50 GeV in a 1° circle on the sky.

In conclusions all known backgrounds are below one event per year and DUMAND will be largely signal limited, rather than background limited in the search for extraterrestrial point sources.

4.3 Minimum detectable flux

The sensitivity of DUMAND to extraterrestrial neutrinos can be expressed in terms of the minimum detectable flux (MDF). This will depend on the size of the detector, the source declination efficiency of the detector and the number of background events.

Due to the great depth DUMAND has the important and unique feature of essentially 100% sky coverage. The annual exposure of the detector as a function of declination of the source is shown in Fig. 7. Several potential sources are indicated.

We estimate that DUMAND is almost three orders of magnitude more sensitive than any previous detector. The minimum detectable flux for point sources as a function of the spectral index in shown in Fig. 8. The calculations are based on data from

Ref. 16. For convenience the MDF is normalized at the flux above 1 TeV. This does not mean that we are only detecting events above this energy. Because we can detect muons down to energies of 50 GeV, we are sensitive to neutrinos down to roughly 100 GeV.

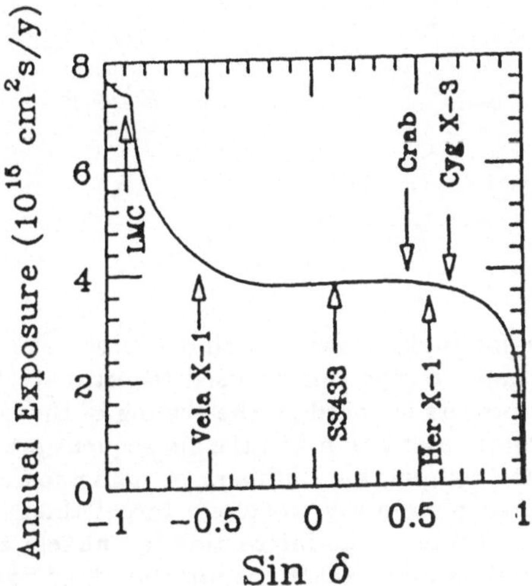

Figure 7. Flux sensitivity of the DUMAND array to neutrino induced muons, for one year as a function of the sine of the source declination.

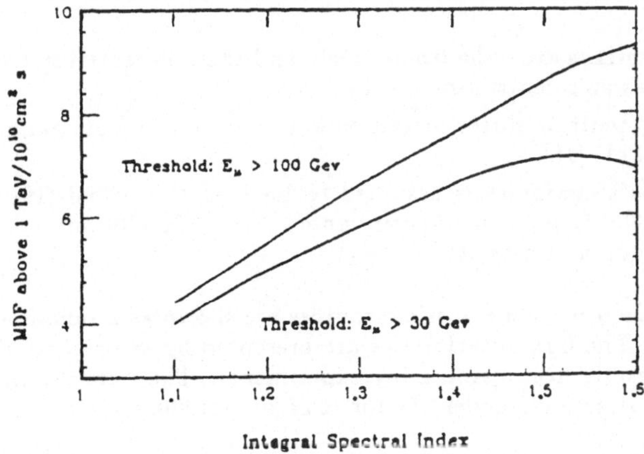

Figure 8. The minimum detectable flux for point sources as a function of the intgral spectral index.

5. Likely sources and neutrino fluxes expected

It is anticipated that high energy cosmic rays may be produced in many different types of objects like:

- X-ray binaries
- Active radio galaxies
- Seyfert galaxies
- Quasars
- Newborn pulsars
- Expanding supernova remnants
- Galactic disk

Although there is strong justification for looking for neutrino sources even if gamma-ray sources are not observed, the gamma-rays can be used to estimate the required neutrino detector sensitivity. This process is based on the scaling of the neutrino flux from the gamma-ray flux. The proton interaction and the subsequent meson (essentially pions and kaons) decay produce comparable numbers of neutrinos and gamma-rays, but in order to escape the source gamma-rays require a target thickness between 10 and 300 $g\ cm^{-2}$. If the target is thinner, the interaction is unlikely and if the target is thicker, the gamma ray cannot escape. Neutrinos, on the other hand, only begin to be absorbed if their path takes them directly through the center of large companion stars. Thus it is not unreasonable to expect neutrino fluxes greater than those measured for gammas [17]. But nevertheless to make predictions of neutrino fluxes is difficult. There exists a high degree of uncertainty about the basic facts. So, in the absence of a solid observational foundation, theoretical estimates of neutrino fluxes must be regarded as speculations.

To estimate the expected flux from neutrino point sources the following assumptions were made:

- The production of gamma-rays are of hadronic origin and the differential spectral index is equal for gammas and neutrinos.
- As a basis for estimating neutrino fluxes measured data were taken from gamma-ray observations at $\geq 1\ TeV$ [18].
- The ratio R of neutrinos to gammas is assumed to be $1 \leq R \leq 1000$ [19] for objects with a large variability in their intensity and $1 \leq R \leq 30$ [20] for sources with roughly constant gamma luminosity.

In Fig. 9 the fluxes expected for some possible sources are shown as a function of the distance from the source. The flux uncertainties are presented by error bars. The lower limit is based on the results from gamma ray experiments. The upper limit of the error bars is given by the source dependent factor R of 30 or 1000.

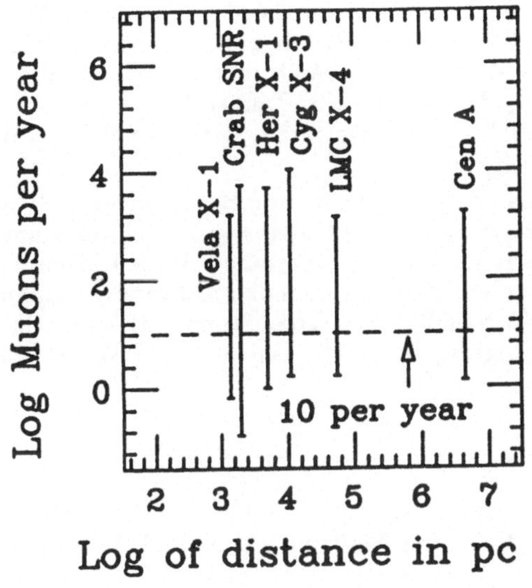

Figure 9. Flux expected for possible sources as a function from the distance of the source.

6. DUMAND - past, present and future

Thirty years ago M.H. Markov formulated the idea to build a large Cherenkov detector for neutrino astrophysics research in lake- or sea-water. In 1973, an international collaboration was proposed by F. Reines to work out a project for a large deep underwater detector for muons and neutrinos. In 1980 first feasibility studies were started to investigate the concepts of the DUMAND detector as well as survey the oceanographic parameters. In 1982 the first trial was made to measure muons with a short string in the pacific ocean. This experiment failed due to the breaking of a support cable. In 1983 and 1984 the bioluminescense background was measured [16]. In 1987 the Short Prototype String (SPS)[12] was successfully completed at the forseen DUMAND site. In 1990 final DOE approval was achieved. In 1992 the deployment operation will begin with the installation of the shore cable and the junction box. At the end of 1992 it is planned to place the first three strings on the ocean bottom and the final six strings will be connected at the end of 1993.

* Members of the DUMAND Collaboration

T. Aoki[11)], R. Becker-Szendy[4)], P. Bosetti[1)], J. Bolesta[4)], P.E. Boynton[14)], H. Bradner[9)], U. Camerini[15)], C. Converse[4)], S. Dye[3)], P.K.F. Grieder[2)], D. Harris[4)], T. Hayashino[10)], M. Ito[10)], M Jaworski[15)], H. Kawamoto[10)], T. Kitamura[7)], K.

Kobayakawa[6] S. Kondo[4], P. Koske[5], J.G. Learned[4], J.J. Lord[14], R March[15], S. Matsuno[4], M. Mignard[4], K. Miller[13], P. Minkowski[2], R. Mitiguy[4], K. Mitsui[11], D. O'Connor[4], Y. Ohashi[11], A. Okada[11], V.Z. Peterson[4], A. Roberts[4], C.E. Roos[13], M. Sakuda[12], D. Samm[1], V.J. Stenger[4], H. Suzuki[10], S. Tanaka[10], S. Uehara[12], C. Wiebusch[1], M. Webster[13], R.J. Wilkes[14], A. Yamaguchi[10], I. Yamamoto[8], K.K. Young[14],

1) Technische Hochschule Aachen, Germany; 2) University of Bern, Sitzerland; 3) Boston University, USA; 4) University of Hawaii, USA; 5) University of Kiel, Germany; 6) Kobe University, Japan; 7) Kinki University, Japan; 8) Okayama Science University, Japan; 9) Scripps Institute of Oceanography, USA; 10) Tohoku University, Japan; 11) ICRR, University of Tokyo, Japan; 12) NLHEP, Tsukuba, Japan; 13) Vanderbilt University, USA; 14) University of Washington, USA; 15) University of Wisconsin, USA.

References

1 V.J. Stenger, Proc. of the 13th Int. Conf. on Neutrino Phys. and Astrophys., Boston, (1988) 344.
2 J.Babson et al., Nucl. Phys. B (Proc.Sup.), 14A, (1990) 157.
3 DUMAND II Proposal, HDC-2-88 (1988).
4 J.G. Learned and V.Z. Peterson, Proc. of the Workshop on Physics at the Main Injector, Fermilab (1989), Hawaii preprint HDC-5-89 (1989).
5 V.J. Stenger, DUMAND External Report HDC-4-91 (1991).
6 F.W. Stecker, C. Done, M.H. Salamon and P. Sommers, Phys. Rev. Let. 21 (1991) 2697.
7 J.G. Learned, Proc. of the 3rd Workshop on Neutrino Telescopes, Venice (1991).
8 S.T. Dye, Hawaii preprint UH-511-667-89 (1988).
9 G.A. Tammann, Proc. of the DUMAND Summer Workshop, Hawaii (1976).
10 C. Quigg, M.H. Reno and T.P. Walker, Fermilab Pub-86/50-AT.
11 H. Meyer, Contribution to this Conference.
12 S. Matsuno et al., Nucl. Inst. and Meth. A276 (1989) 359.
13 D. Samm, Proc. of the Int. Workshop on Neutrino Telescopes, Venice (1988)
14 C. Ley, Master Thesis, RWTH Aachen (1990).
15 H. Bradner et al., Deep-Sea Research 34 (1987) 1831.
16 T. Aoki et al., Nuovo Cimento 9C (1986) 642.
17 V.S. Berezinsky et al., Astron. and Astrophys. 189 (1988) 306.
18 T.C. Weekes, Phys. Rep. 160 (1988) 1.
19 E.W. Kolb, M.S. Turner and T.P. Walker, Phys. Rev. D32 (1985) 1145.
20 T.K. Gaisser and A.F. Grillo, Phys. Rev. D36 (1987) 2753.

Physics Capabilities of the Second-Stage Baikal Detector NT-200

presented by CH. SPIERING

for the BAIKAL-Collaboration:

S.D.Alatin[5], I.A.Belolaptikov[1], L.B.Bezrukov[1], B.A.Borisovets[1], N.M.Budnev[2], E.V.Bugaev[1], A.G.Chensky[2], Zh.A.M.Djilkibaev[1], V.I.Dobrynin[2], G.V.Domogatsky[1], L.A.Donskich[1], A.A.Doroshenko[1], G.N.Dudkin[4], V.Yu.Egorov[4], S.V.Fialkovsky[5], M.D.Galperin[1], A.V.Golikov[3], O.A.Gress[2], M.N.Gushtan[1], R.Heller[7], H.Heukenkamp[7], L.Jenek[8], V.B.Kabikov[3], D.Kiss[8], A.M.Klabukov[1], S.I.Klimushin[1], A.P.Koshechkin[2], J.Krabi[7], V.F.Kulepov[5], Yu.S.Kusner[9], L.A.Kuzmichov[3], J.B.Lanin[2], O.J.Lanin[1], A.A.Levin[6], G.A.Litunenko[2], A.L.Lopin[2], B.K.Lubsandorzhiev[1], A.A.Lukanin[4], M.B.Milenin[5], T.Mikolajski[7], V.A.Naumov[2], M.I.Nemchenko[2], A.I.Nikiforov[6], N.V.Ogievietzky[1], E.A.Osipova[3], A.M.Ovcharov[4], V.M.Padalko[4], A.H.Padusenko[4], A.I.Panfilov[1], Yu.V.Parfenov[2], A.A.Pavlov[2], O.P.Pokalev[2], V.A.Poleschuk[9], V.A.Primin[2], M.I.Rosanov[6], P.P.Sherstyankin[9], I.A.Sokalsky[1], Ch.Spiering[7], A.A.Sumanov[2], L.Tanko[8], V.A.Tarashansky[2], T.Thon[7], I.I.Trofimenko[1], R.Wischnewski[7], E.S.Zaslavskaya[3], V.L.Zurbanov[2]

1 Institute for Nuclear Research, Moscow, Russia
2 Irkutsk State University, Irkutsk, Russia
3 Moscow State University, Moscow, Russia
4 Tomsk Polytechnical Institute, Tomsk, Russia
5 Polytechnical Institute, Nizhni Novgorod, Russia
6 Marine Technical University, St.Petersburg, Russia
7 Institute for High Energy Physics, Zeuthen, Germany
8 Central Research Institute of Fundamental Physics, Budapest, Hungary and Joint Institute of Nuclear Research, Dubna, Russia
9 Limnological Institute, Irkutsk, Russia

Abstract

We describe the lake Baikal deep underwater detector "NT-200" and discuss its physics capabilities to investigate problems in the field of neutrino astrophysics, cosmic ray physics and particle physics.

1 Introduction

One of the most challenging questions of astrophysics is the origin of very high energy particles in cosmic rays. Up to now, the sources powering the high-energy universe are not identified. *Suspected* sources for particles in the VHE (Very High Energy) region, nominally 10^{11} to 10^{14} eV, are - for instance - binary neutron star systems or young pulsars powering supernova shells, whereas objects as Active Galactic Nuclei (AGN) could produce UHE (Ultra High Energy) particles with energies up to $10^{17} eV$ and beyond [1,2].

As far as hadronic processes underly the particle production, the sources should also be powerful emitters of neutrinos. Neutrinos can escape regions of the universe invisible via γ-rays. In this way, neutrino astronomy complements the astronomy of electromagnetic radiation and yields informations, hardly or even not at all accessible by visible light, X-rays, γ-rays etc.

Up to now, the observational window for extraterrestrial neutrinos of high energy has not yet been opened. None of the existing underground detectors has detected point sources of VHE or UHE neutrinos. Obviously, effective area, angular resolution and energy determination for future detectors have to be improved in comparison with underground devices in order to detect the feeble fluxes of VHE neutrinos. The most promising way to do this seems the deployment of the detectors deep underwater (DUMAND, Baikal), deep under ice (AMANDA) or in shallow depths of water (GRANDE-type projects) [3].

There is a wide spectrum of questions to be investigated with deep underwater detectors. Beside the search for point sources of VHE neutrinos, these detectors could be used for testing the hypothesis, that the universe is filled with a diffuse flux of extremely energetic neutrinos from Active Galactic Nuclei [4]. They can be used to search for dark matter candidates as quark nuggets, magnetic monopoles [5] or supersymmetric particles. Furthermore, by registrating muons generated in primary cosmic ray interactions above the detector, questions of standard cosmic ray physics can be tackled. Deep underwater detectors could be made sensitive for low energy collapse neutrinos from nearby supernovae [6]. Last not least, there are proposals to direct accelerator neutrinos to underwater detectors to investigate neutrino oscillations [7,19].

However, the most exiting results might be obtained from *unexpected phenomena.* The great decade of astrophysics, when new and mostly absolutely unexpected phenomena as quasars (1960-63), X-ray stars (1962), the microwave relic radiation (1965), X-ray galaxies (1966), pulsars (1968) and gamma bursters (1973) were detected, was due to the operation of new sensitive devices as radio detectors, X-ray satellites etc., most of which opened a new observational window. The first successes of low energy neutrino astrophysics were connected with objects already known: the sun and a supernova (SN1987A). Having in mind the discoveries mentioned above, this probably will not be the end of the story. In fact, the hope to discover non-anticipated phenomena is one of the main motivations to build big underwater detectors!

In what follows, we describe the lake Baikal second-stage detector NT-200, which is scheduled to start data taking in 1993/94, present its basic parameters with respect to the registration of muons emerging from neutrino interactions and discuss its physics capabilities.

2 The NT-200 Detector

The detector will be deployed in the southern part of lake Baikal, about 4.5 km from shore at a depth of 1 km (see fig.1). At this depth, the water has minimal natural luminescence (from 5 to 20 counts s^{-1} cm^{-2} on a PMT photocatode). The attenuation lenght is about 20 m at $\lambda = 480$ nm (fig.2). The coordinates of the location are 51°50′N and 104°20′E.

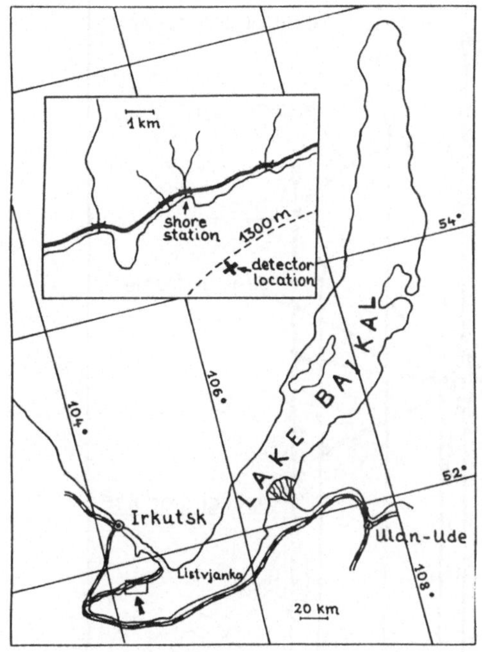

Fig.1: Site of the Baikal experiment

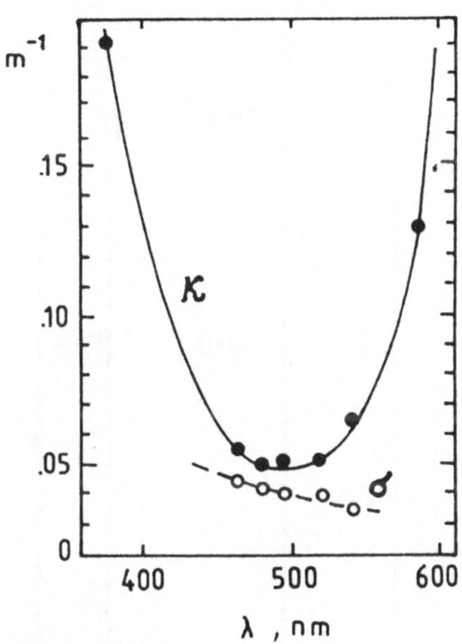

Fig.2: Absorbtion coefficient κ (solid line) and scattering coefficient σ (dashed line) for Baikal water at 1 km depth as a function of wave length λ.

Fig.3 shows a schematic view of the NT-200 array. It consists of 8 strings - one central string surrounded by seven outer strings. The strings are attached to a rigid frame, which is positioned 300 m above the bottom. This umbrella-like frame, consisting of 7 arms each 21.5 m long, is connected to the upper buoy 20 m below surface by a single cable. From that buoy, via a separated string, another cable connects the detector site with the shore station.

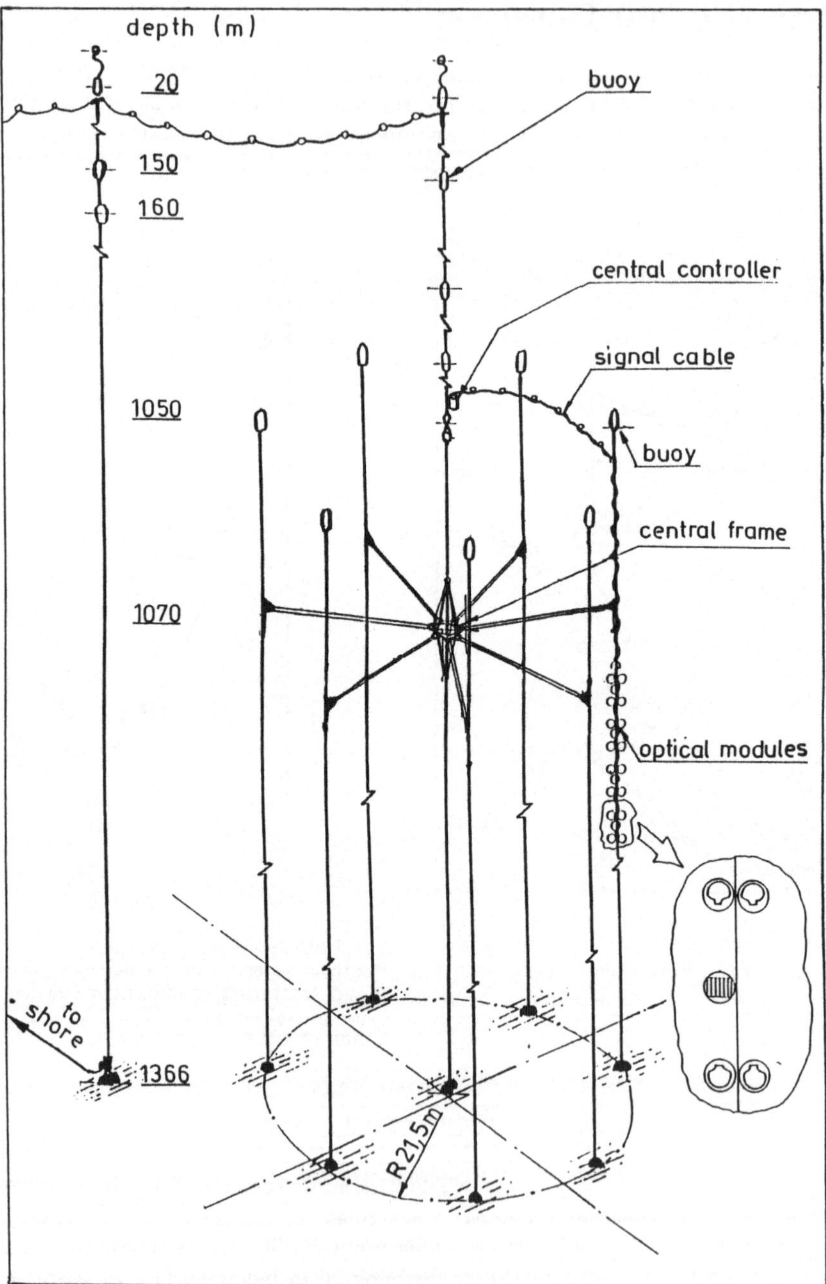

Fig.3: Schematical view of the NT-200 array. Optical modules are indicated for only one of the 8 strings. Part of the complicated buoyancy is omitted in the figure.

In the insert of fig.3, the optical modules are shown. The modules are grouped in pairs looking alternatively upward and downward. The distance of the pairs looking face to face is 7.5 m, the distance of pairs looking back to back is 5 m. The two PMT of a pair are switched in coincidence. In the most simple concept, their signals are transmitted to shore only if at least a minimal number of PMT pairs have given signals within a time window of 600 $nsec$ (this is about two times the time a relativistic particle needs to cross the whole array). The array trigger on the minimal number of fired pairs (3 to 5) is formed in the central controler.

The fast data processing in the shore station is managed by a network of transputers.

The main element of the array is an underwater optical module containing a 37-cm diameter, highly sensitive phototube with excellent resolution in time (1.8 $nsec$ FWHM) and amplitude. The tube (named "QUASAR") is an improved version of the PHILIPS "smart" phototube XP2600 and is produced in Novosibirsk [8]. Both XP2600 and QUASAR have been tested extensively in the lake, the last time with a small 8-PMT prototype-string which was operated in March/April 1991.

Fig.4 gives a sketch of the installations after the winter expedition march/april 1991. Beside two electrical cables connecting detector site and shore, and various test strings, the most remarkable object is the septagonal frame. It was deployed to train the handling of the structure (including its complicatéd buoyancy) and to check its long-term rigidity over one year in 1 km depth.

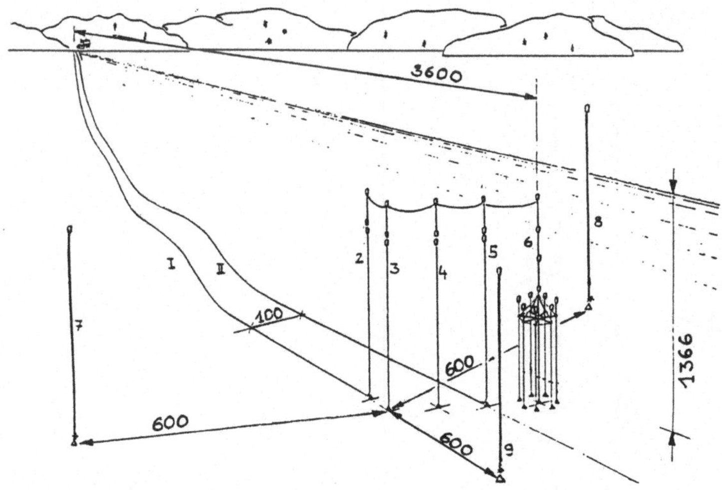

Fig.4: Sketch of the present installations at the Baikal site. I,II: electrical cables to shore. At string 3 and at the strings 7-9 (centered around string 3 at a distance of 600 m) the ultrasonic system for monitoring the space coordinates of the array is tested. String 3 carries ultrasonic receivers, strings 7-9 carry ultrasonic emitters. A optical fiber cable is under long-term test at string 4. The mechanical frame which will hold the NT-200 was deployed for a one year test in april 1991 at string 6.

3 Basic parameters of the array

The crucial question in running an underwater detector as a neutrino telescope is whether one can reject the background of downward atmospheric muons. In 1 *km* depth, these background events exceed the upward moving muons from interactions of atmospheric neutrinos by a factor of 10^6. Sophisticated analysis criteria are necessary to reject wrongly reconstructed downward muons which could fake upward muons.

With the reconstruction algorithms developed up to now, the median angular resolution for single muons is $1^o - 1.5^o$. However, the distribution of the mismatch angle is characterized by a long tail, extending up to angles greater than 40 degrees. Following only the result of the fit procedure, down to zenith angles of 140° fake events would exceed "true" neutrino events by more than ten times (see fig.5). This result can be improved by applying two additional criteria [9]: Firstly, one can check wether all PMT which should have fired (provided the reconstruction was correct) indeed *have* fired. Secondly, one can apply additional cuts on the difference between measured and reconstructed time at each individual PMT. The corrresponding criteria reduce the fake events by an order of magnitude (full circles in fig.5) . Assuming that the background contribution from muon bundles does not exceed the single muon background, one concludes that the NT-200 can work as a neutrino telescope with an aperture of $60^o - 80^o$ around the opposite zenith.

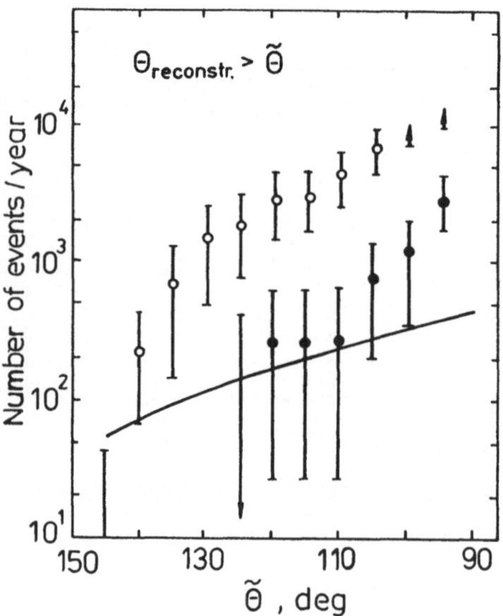

Fig.5: Number of downward going single muons per year reconstructed with zenith angles greater than 90°. Open circles: result of straightforward fit procedure. Full circles: result after application of more sophisticated criteria. Line: "true" upward muons from interactions of atmospheric neutrinos.

The effective area of the array depends on the choosen trigger condition for event selection [10]. From our Monte Carlo calculations we conclude that the trigger condition "6/3" (i.e. signals from $\geq$ 6 PMT pairs on $\geq$ 3 strings) is suitable to select essentially muons which are reconstructable. Fig.6 shows the effective area as a function of the muon energy [11]. Requesting a certain accuracy of the angular reconstruction, the effective area decreases. Note, that for high muon energies the effective area exceeds the geometrical area (dashed line in fig.6)!

The dependence of the effective area on the zenith angle θ is demonstrated in fig.7 for 1-TeV muons. Shown is the effective area for various limits on the accepted mismatch angle ψ. Due to the relatively small diameter of the array, horizontal tracks ($\cos\theta = 0.$) are badly reconstructed. That leads to a strong decrease of the effective area if one requests an angular reconstruction accuracy of better than 3°. With increasing energy, however, this effect is washed out. Note, that NT-200 has symmetrical downward and upward sensitivities and that fig.7 covers only one hemisphere!

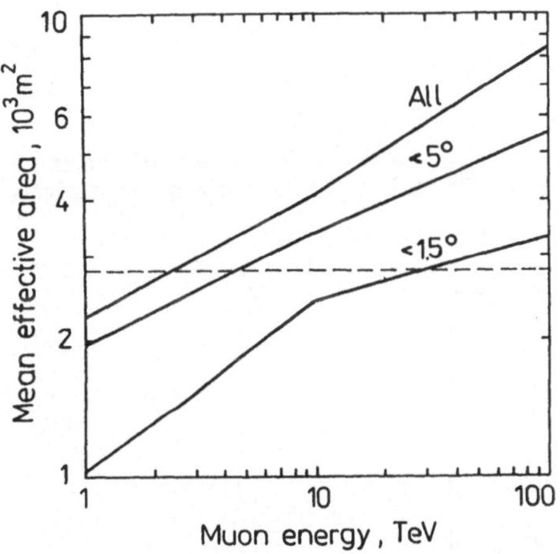

Fig.6: NT-200 effective area as a function of energy. Shown are results for all muons fulfilling the trigger condition (see text) as well as for muons being reconstructed with mismatch angles smaller than 5° and 1.5° respectively. Dashed line: mean geometric area.

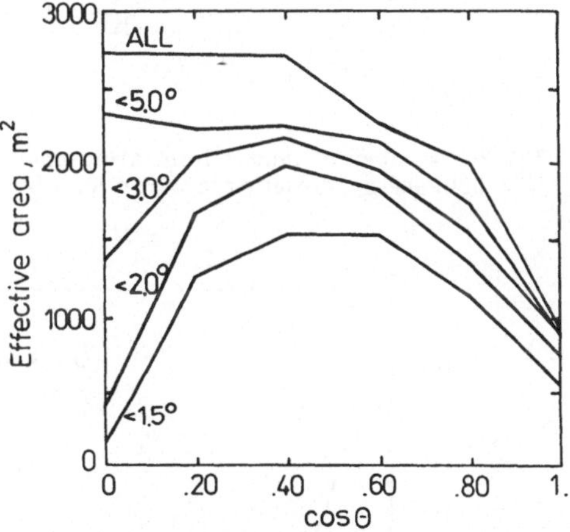

Fig.7: NT-200 effective area as a function of zenith angle θ for 1TeV muons. Shown are results for different reconstruction mismatch angle limits.

Fig.8 shows the number muons from atmoshperic neutrinos per year, as a function of the muon threshold energy. With a typical threshold of 20 GeV, we expect 200-400 events for horizon cuts at 120° to 100°. The number of events per 2° half cone is typically about 0.1 per year. The dependence of this number on muon threshold energy and on the declination is shown in figs. 9 and 10, respectively [12].

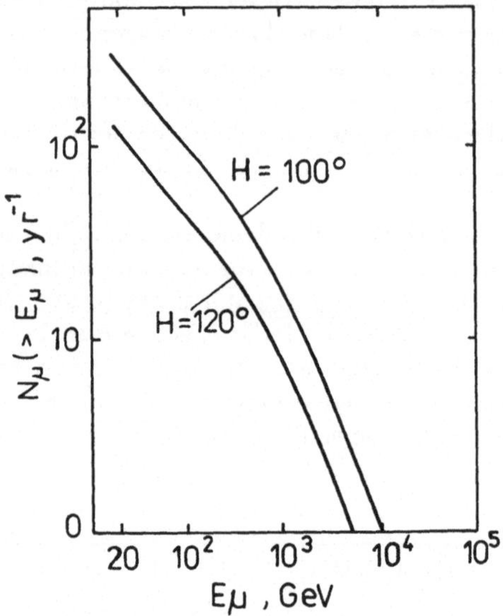

Fig.8: Number of muons from atmospheric neutrinos per year, as a function of the muon threshold energy. Results are shown for 2 horizon cuts: 10° and 30° below horizon, respectively.

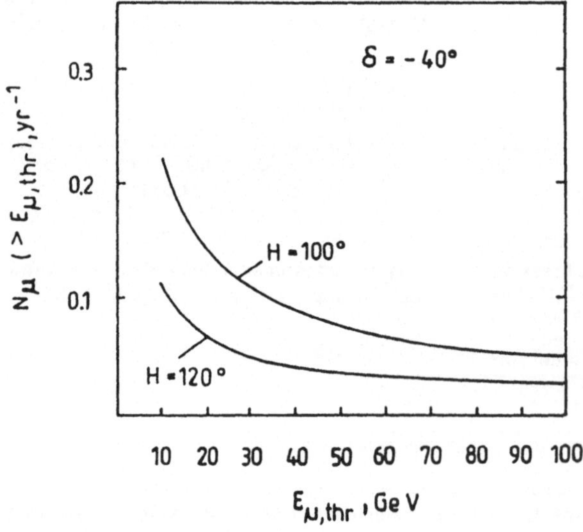

Fig.9: Number of atmospheric neutrino-induced muons in an angular bin of 2° half-cone as a function of threshold energy of muon detection. The angular bin is centered at a declination $\delta = -40^o$. H is the horizon cutoff angle.

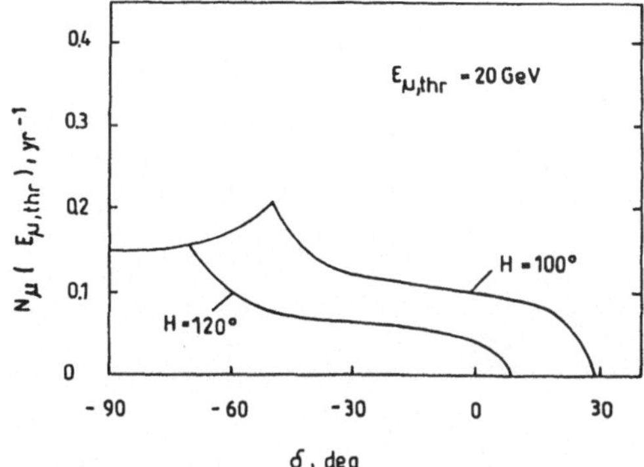

Fig.10: Number of atmospheric neutrino-induced muons in an angular bin of 2° half-cone as a function of declination δ.

Fig.11 illustrates the sky coverage of the array. It shows the mean effective area (multiplied with the time duty factor for the source to be below the horizon-cut angle) as a function of the declination. Shown are the results for cut angles of 120°, 100° and (dashed) 90° and requesting a mismatch angle smaller than 2°. A source with energy spectral index $\gamma = 2.0$ was assumed. Sources permanently below horizon (flat part of the curves), are seen with an effective area of about 2000 m^2 [12].

The minimal detectable flux (MDF) of neutrinos with energies greater than 1 TeV is shown in fig.12, in dependence on the spectral index of the source. It is of the order of $10^{-9}\, s^{-1}\, cm^{-2}$ [12].

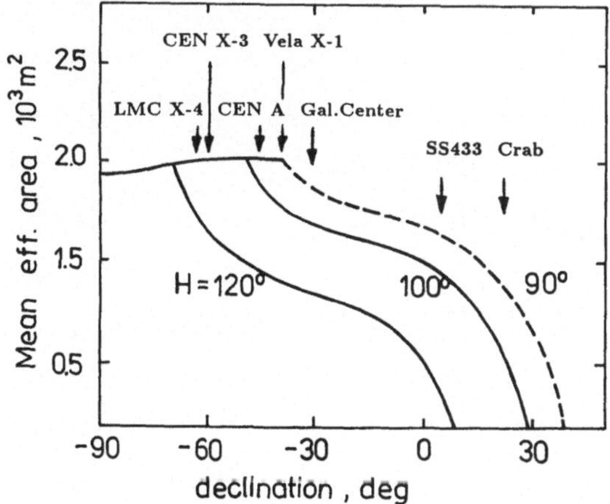

Fig.11: Sky coverage of the array for different horizon cuts and a source with spectral index $\gamma = 2.0$. Accepted events are assumed to be reconstructed with a mismatch angle smaller than 2°.

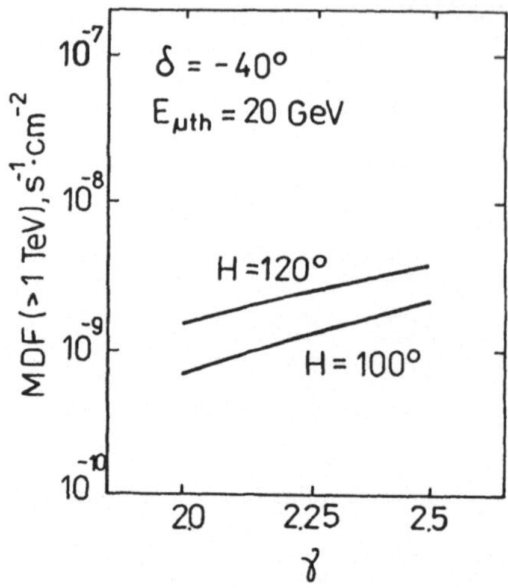

Fig.12: Minimal Detectable Flux of neutrinos above 1 TeV as a function of neutrino differential spectrum index γ.

Stochastic processes as pair production, bremsstrahlung and nuclear reactions increase the energy loss of muons with inceasing muon energy. A major advantage of the huge dimensions of underwater detectors compared to underground detectors is the possibility to determine the energy of muons due to the larger sampling lengthes. Fig. 13 shows the "measured" energy for muons of 1, 10 and 100 TeV, respectively [13]. At muon energies greater than 2-3 TeV, the logarithm of the energy can be estimated with an accuracy of 40-50 %, thus allowing to determine at least the order of magnitude of the muon energy.

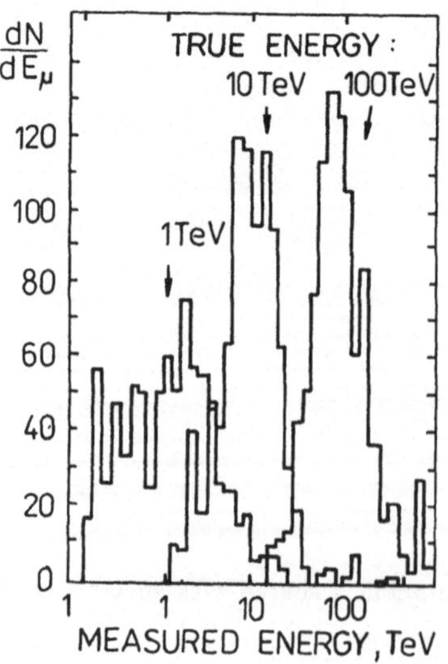

Fig.13: Distribution of the measured energies for 1-, 10- and 100-TeV muons.

4 Physics items

The long-term perspective of the second-stage deep underwater detectors - be it the OCTAGON array of the DUMAND collaboration [14] or the NT-200 of the Baikal collaboration - is their extension to full scale detectors with effective areas of $> 10^5\,m^2$. However, beside testing the feasability of a full-scale detector, they can already tackle a lot of problems in the fields of astrophysics, particle physics and cosmic ray physics, some of them in a way superior to underground detectors.

Steady point sources of high energy neutrinos are expected to provide fluxes of the order $10^{-11}\ s^{-1}\ cm^{-2}$ with $E > 1\,TeV$ for a typical distance of 10 kpc [1]. So, NT-200 could detect them only under very favourable conditions: The sources have to be rather near (1-2 kpc) and should be completely hidden when watched via electromagnetic radiation - otherwise they would have been detected via VHE gamma rays. Another possiblity would be a non-isotropic emission, directed by chance preferently to the earth.

In contrast to steady sources, young pulsars, supernova shells and possibly even other short term sources might produce fluxes of $10^{-9}\ s^{-1}\ cm^{-2}$. In the NT-200 array, these fluxes could yield a signal of about ten events within some monthes.

A recent model [4] of particle production in Active Galactic Nuclei predicts diffuse fluxes of extremely high energetic neutrinos. Following [4], we would expect some hundred events with energies greater than $10\,TeV$ per year. Berezinsky [15] has presented objections against the model and concludes that the prediction is too high by at least a factor of 30. But even for a 30 times lower flux, a douzen events with $E > 10TeV$ might be identified against the high energy atmospheric neutrinos in the NT-200 array.

Another field of research will be the search for dark matter candidates, DMC. For the moment, one of the mostly favoured DMC is the neutralino. The neutralino is assumed to be the lightest SUSY particle in the framework of the Minimal Standard Supersymmetric Model (MSSM). Constraints on the mass parameters of neutralinos have been established recently by measurements at LEP [16].

These particles could accumulate in the sun or the earth and annihilate with their antiparticles into neutrino - antineutrino pairs. In the case of electron neutrinos reacting via charged current just within the array, one would observe monoenergetic showers with energies corresponding to the mass of the neutralino. Under*water* detectors, due to their bigger volume, can search for fully contained, monoenergetic showers up to higher energies than under*ground* Cherenkov detectors (which have established limits up to masses of 30-90 GeV [17]). The total mass of NT-200 is 84.4 *ktons*. The energy of fully contained TeV-showers can be determined with 15% accuracy or better.

Neutralinos annihilating in the earth could yield an enhanced flux of upward muons coming dominantly from a cone of 30° around the opposite zenith.

A recent hypothesis [18] proposes R-parity violation in the MSSM, causing decays of the Lightest Supersymmetric Particle, e.g. into a neutrino and a massless Goldstone boson. Again, the signature would be monoenergetic showers within the array.

There are other exotic particles to be searched for by NT-200. The trigger described in section 2 (at least 3 firing PMT pairs within 600 $nsec$) is tailored to relativistic particles. In fact, this trigger responds to particles moving with velocities down to $\beta \approx 10^{-1}$. Such particles could be (non-GUT) monopoles, which for $\beta \geq 0.6$ might generate much more

Cherenkov light than naked muons [19].

With a modified trigger, the array is able to respond to slowly moving ($10^{-5} < \beta < 10^{-2}$), bright particles, as "nuclearities" or magnetic monopoles catalyzing proton decays via the Rubakov effect. Corresponding results have been published by the Baikal collaboration on the basis of single string results collected from 1984 to 1988 [5]. Using a modified electronics, the NT-200 could improve the upper flux limits by nearly 2 orders of magnitude. There are different approaches to the necessary modifications, starting with the registration of enhanced counting rates due to slowly moving bright particles [19] and ending with the idea to send all informations without application of an underwater trigger to shore - for further, flexible aquisition by fast processors.

The array will registrate $\approx$ 15 muons per second; about 2000 muons with energies greater than 100 TeV will be collected during one year. The energy spectrum of atmospheric muons could be measured, provided, muon bundles can be separated from single muons in 98 % of the cases. Presently, this question is under study. Also, one could try to search for anisotropies of muons at highest energies.This feeble effect, however, needs excellent calibration of the detector.

Finally, we only shortly want to mention further items: The detection of collapse neutrinos from supernovae via enhanced counting rates of all the PMT pairs within some seconds, and the combined work of the underwater detector with a top array of Cherenkov EAS counters on the ice cover. If in the future a neutrino beam could be directed from the UNK to lake Baikal [20], one could also search for oscillations of accelerator generated neutrinos.

References

1. V.S.Berezinsky, Nucl.Phys.B (Proc.Suppl.), V19,375,1991
2. T.K.Gaisser,"Cosmic rays and Particle Physics",Cambridge University Press,1990
3. see, for a review of the different projects: H.Sobel, Proc. III. Int. Workshop on Neutrino telescopes, Venice 1991, p.189
4. F.W. Stecker et al., Phys.Rev.Lett. 66 (1991) 2697
5. see, e.g., L.B.Bezrukov et al., Soviet Journ. of Nucl. Phys. 52 (1990) 54
6. J.G.Learned, Proc. 1990 DUMAND trigger workshop, Seattle 1990
7. see e.g. R.H.Bernstein and S.J.Parke, FERMILAB-91/59T and ref. therein
8. see the talk of B.K.Lubsandorshiev at this conference
9. I.A.Belolaptikov, Zh.A.M.Djilkibaev, Baikal Internal Note, Moscow 1990
10. I.A.Belolaptikov and Ch.Spiering,Preprint PHE 90-2, Zeuthen 1990
11. Zh.A.M.Djilkibaev,O.Lanin,Baikal Internal Note, Moscow 1990
12. J. Krabi et al., Preprint PHE 91-12, Zeuthen 1991
13. E.A.Osipova, talk at the Baikal meeting dec.1990
14. see the talk of D.Samm at this conference
15. V.S.Berezinsky, contribution this conference
16. D.Dechamp et al. (ALEPH Coll.), Phys.Lett. B244 (1990) 541.
17. Y.Suzuki, ICRR-Report-234-91-3, Tokyo 1991
18. V.S.Berezinsky, contribution to this conference
19. Zh.A.M.Djilkibaev,A.A.Doroshenko,I.A.Sokalsky,Baikal Int. Note,Moscow 1991
20. S.S.Gershtein et al., Preprint Lebedev Phys. Inst. FIAN-LECH-87, Moscow 1989

CHARACTERISTICS OF THE GALLEX SPECTROMETER

G. Heusser*

Max-Planck-Institut für Kernphysik, P.O. Box 103980, D-6900 Heidelberg, Germany

*For the GALLEX-COLLABORATION: ***Heidelberg**, Max-Planck-Institut für Kernphysik:* P. Anselmann, W. Hampel, G. Heusser, J. Kiko, T. Kirsten, G. Monninger, E. Pernicka, R. Plaga, B. Povh, U. Roenn, M. Sann, C. Schlosser, H. Völk, R. Wink, M. Wójcik[1]; ***Karlsruhe**, Kernforschungszentrum - KfK:* R. v. Ammon, K. Ebert, T. Fritsch, K. Hellriegel, E. Henrich, L. Stieglitz; ***L'Aquila**, Laboratori Nazionali del Gran Sasso:* M. Balata, E. Bellotti, C. Cattadori, N. Ferraris, H. Lalla, S. Pezzoni, T. Stolarczyk; ***Milano**, Dip. di Fisica dell'Università - INFN*: O. Cremonesi, E. Fiorini, S. Ragazzi, L. Zanotti; ***München**, Physik Dept. E15 - Technische Universität:* F. v.Feilitzsch, R. Mößbauer, U. Schanda; ***Nice**, Université de Nice Observatoire:* G. Berthomieu, E. Schatzmann; ***Rehovot**, Weizmann Institute of Sciences:* I. Carmi, I. Dostrovsky; ***Roma**, II-Università - INFN:* S. d'Angelo, C. Bacci, P. Belli, R. Bernabei, L. Paoluzi; ***Saclay**, DPhPE CEN:* S. Charbit, M. Cribier, G. Dupont, L. Gosset, J. Rich, M. Spiro, C. Chao, D. Vignaud; ***Upton**, N.Y., Brookhaven Nat. Laboratory:* R.L. Hahn, F.X. Hartmann, J.K. Rowley, R.W. Stoenner, J. Weneser

Abstract

A description is given of the spectrometer used for the detection of ^{71}Ge in the Solar Neutrino experiment GALLEX being performed in the Gran Sasso Underground Laboratory. The spectrometer consists of miniaturized proportional counters and a shield with a large well-type NaI pair (Tl) detector (active side) and an inner pure copper shield (passive side). Very careful material selection for the proportional counter- and shield-construction and radon suppression resulted in total background rates (>0.5 keV) between 0.4 and 1 count per day for many proportional counters. With energy and rise time cuts, the average rates for the relevant L- and K-peak of the ^{71}Ge spectrum are 0.1 cpd and 0.03 cpd, respectively and, thus, are far below the signal predicted by the Standard Solar Model. Eight counter positions within the NaI pair detector have the option to detect also ^{69}Ge and ^{68}Ga (positron emitters) in the coincidence mode, though with slightly higher background for the ^{71}Ge decay mode. An analysis of the different background components cannot fully account for the measured background of the proportional counters so that presumably a part of it is due to contamination during the assembling process. Here is a potential for further background reduction. After introduction, the basic concept of the experiment and the present status as of December 1991 are briefly outlined.

---

[1] Institute of Physics, Jagellonian University, Kraków

1. INTRODUCTION

Gallium as the target material for a radiochemical solar neutrino experiment has a low energy threshold of 233 keV and a favourably high cross section [1]. It is therefore well suited to study the solar neutrino problem with its deficit in high energy electron neutrinos by measuring the flux of the low energy neutrinos from the basic energy producing p-p reactions.

The long standing contradiction between the high energy ^{8}B neutrino-flux as experimentally determined in the radiochemical chlorine experiment [2] and the theoretical predictions of the Standard Solar Model [3,4] has recently been confirmed by the water Čerenkov detector of Kamiokande [5].

The rate of the p-p reaction, contrary to the rate of the ^{8}B-neutrino producing branch, hardly depends on the uncertainties of the input parameters and physical assumptions of the solar models. Its determination will allow to decide for one of the two explanations for the solar neutrino problem, commonly discussed to be due either to astrophysical causes (e.g. non-standard model [6]) or, to particle-physical causes like matter-enhanced neutrino oscillation [7].

Two experiments, one performed by the Soviet-American SAGE collaboration at Baksan Laboratory and the other by the European GALLEX collaboration at Gran Sasso Laboratory (LNGS), are using a gallium target to measure the low energy neutrino flux applying the radiochemical technique. In SAGE metallic gallium is used, whereas the GALLEX target is an aqueous $GaCl_3$-solution. The various aspects of GALLEX have been repeatedly reviewed [8-12], therefore this report, after a brief description of the basic concept and the present status (Dec. 91), mainly concentrates on the counting facility operating at LNGS.

1.1. Concept of the Experiment

According to the Standard Solar Model [3], about 1.2 electron neutrinos are captured per day by the reaction $^{71}Ga(\nu_e,e^-)^{71}Ga$ in the 101 tons of target solution containing 30.3 t of gallium. The pp- and pep-neutrinos from the primary fusion chain contribute 56 % to that rate. After an exposure time of 3 weeks, about 14 atoms of ^{71}Ge have accumulated taking into account the decay by the half-life of 11.43 days. Together with 1 mg inactive Ge-carrier they are extracted by sweeping the tank with 3000 m^3 of nitrogen. In the highly acidic solution $GeCl_4$ is volatile so that 100 percent can be removed. In washer columns the $GeCl_4$ is separated from the nitrogen stream with low acidic water. After being concentrated by a further extraction step from the column water, the $GeCl_4$ is transferred to a small volume of tritium-free water and converted into the gaseous compound germane - GeH_4. Purified by gas chromatography, the germane is finally filled into miniature proportional counters together with 70% Xe, and the back-decay of ^{71}Ge into ^{71}Ga is observed.

1.2. Status of GALLEX

After the final removal of a minute portion of cosmogenic ^{68}Ge, which was bound to a nonvolatile slowly dissolving compound [11], the solar neutrino signal is measured since May 1991. Up to now 9 full runs have been performed, all still being counting. At least

10 months of data acquisition are required to achieve even a preliminary result. For a statistically meaningful result, the experimental concept calls for 4 years of data taking.

The 53.5 m^3 of $GaCl_3$-solution is contained in one of two 70 m^3 special tanks installed in the GALLEX main building in Hall A of the Gran Sasso Laboratory. Both tanks are identical, except that tank A has a more efficient sparging system for the GeH_4-extraction and is engineered to take up the Cr-calibration source. In spring 1992 the solution will be transferred from tank B to tank A.

An overall test and calibration of the experiment with a radioactive ^{51}Cr neutrino source [13] is planned in 1993, the source strength will exceed 1 MCi (37 PBq). All technical provisions for the tank exposure have already been installed.

2. THE GALLEX PROPORTIONAL COUNTER

One major challenge in GALLEX is the detection of about 12 ^{71}Ge-atoms (SSM and 90% overall yield for extraction and conversion), which at the beginning of counting correspond to a decay rate of 0.7 per day. ^{71}Ge decays purely by electron capture and results in the spectrum shown in Fig. 1 when measured in a miniature proportional counter filled with the typical Gallex gas mixture and in sufficiently high statistics (about 22 000 total counts). It features the L-peak at ~ 1.2 keV and the K-peak at ~ 10.4 keV. The gas mixture (70% Xe+30% GeH_4 at 1.05 atm pressure) is a compromise choice between the different counting characteristics like efficiency, energy resolution, pulse form discrimination, background, calibration possibility, etc.

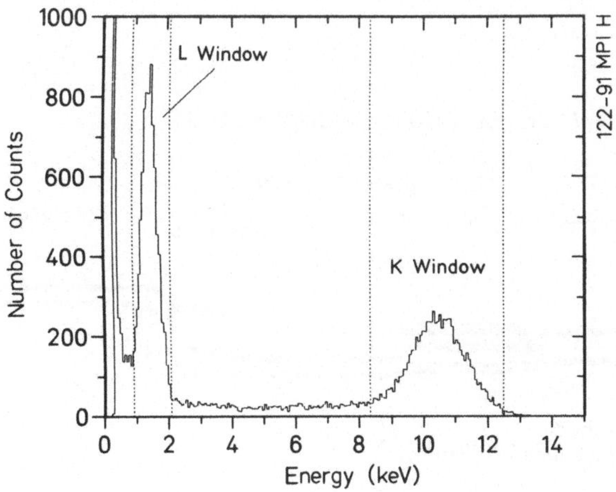

Figure 1. Energy spectrum of ^{71}Ge measured with a Gallex proportional counter and a gas mixture of 70% Xe + 30% GeH_4.

Based on the Davis-type proportional counter [14], many years of development and continuous refinement and optimization of the efficiency and background performance have led to the final design "HD2-Fe" depicted in Fig. 2. The whole body of the proportional counter including the stopcock valve is made from artificial quartz with very low intrinsic activity (see Table 2). The counting gas is pushed into the active volume defined by the iron cathode with mercury through the gas inlet tube. The cathode made from selected low-activity iron (Table 2) is machined to fit precisely the quartz tube to keep the dead volume at its minimum (less than 10% including capillary tubes and plug spacings). For good reproducibility of all counting characteristics the dimensions have been standardized: iron cathode - 6.4 mm outer diameter, 32 mm long, 0.15 - 0.17 mm wall thickness; quartz tube - 6.42 mm inner diameter, thus forming an active volume of nearly 1 cm^3. Both tungsten wires, the cathode wire (spot welded to the iron) and the center wire are directly sealed into the quartz. The normally used graded seals contain radioimpurities like K. A 50 μm thick quartz window is sealed onto the end plug next to the quartz spacer for energy calibration by external sources. Three small holes in the quartz spacer, in addition to the center wire hole, allow easier penetration also of low energy X-rays. An important quality of these counters, like of Davis-type counters, is that they can be baked out at 200 to 300°C to remove volatile impurities, which otherwise might alter the counting gas characteristics during the long counting periods of up to 6 months by outgassing. Also high-purity boron-doped silicon is used as material for the cathodes. Early tests are performed with shaped Si-cathodes which narrow towards the end. Their shape corrects field-strength distortions at the counter ends which are responsible for the so called degraded events between the L- and K-peak (Fig. 1). It has already been demonstrated that by proper shaping part of these events can be shifted back into the K-peak [15].

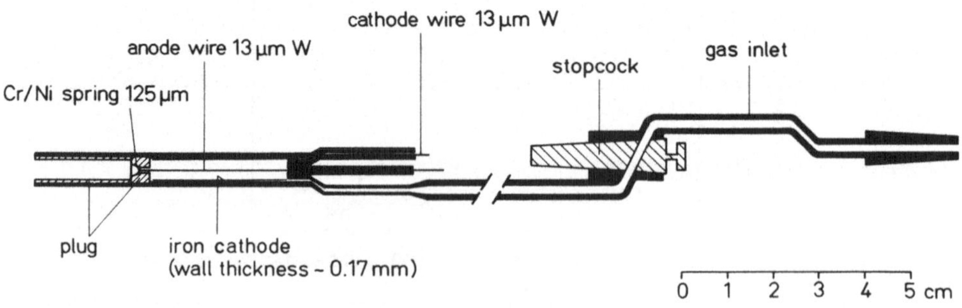

Figure 2. Cross section of a HD2-Fe-type proportional counter.

3. SPECTROMETER SHIELD

Special care has been taken in designing the shield to minimize radioimpurities and to reduce the influence of radon to an insignificant level. Also for economic reasons the activity level of the materials used is decreasing with decreasing distance to the active volume of the proportional counters. The proportional counters are inserted into the shield within specially tailored Cu/Pb housings shown in open state in Fig. 3. Two halves made of low-activity lead (0.3 Bq ^{210}Pb/kg) are machined to match the contours of the proportional counter in order to embed it, with little free space for radon and also to shield the active volume against an U-contamination (about 18 m Bq), in a hybrid preamplifier, which is planned to be implemented into the lower lead half. Both halves are covered with 0.5 mm thick copper to strengthen the soft lead. The copper tube shown on top of Fig. 3 is slid over the closed halves to fix them and also to protect the front part with the active volume of the proportional counter sticking out of the lead halves. It also forms a tight electrostatic shield. For this reason also the front end of the tube is closed with a copper screw when the counter is not being calibrated. Depending on the position within the shield (passive or active side; see below), the tube is either filled with a high-purity copper cylinder fitting over the active volume (for shielding and radon displacement) or, is empty to allow ß-γ coincidence measurements with the NaI(Tl) detector. The rectangularly shaped part (Fig. 3) holds the preamplifier and connectors. The electronic parts have been selected for radiopurity as far as practicable; especially the normally used highly active glass-fibre-reinforced circuit board has been replaced by hard paper. For shielding reasons, the preamplifier, which is not completely free of activity, is positioned out of the line of sight to the proportional counter.

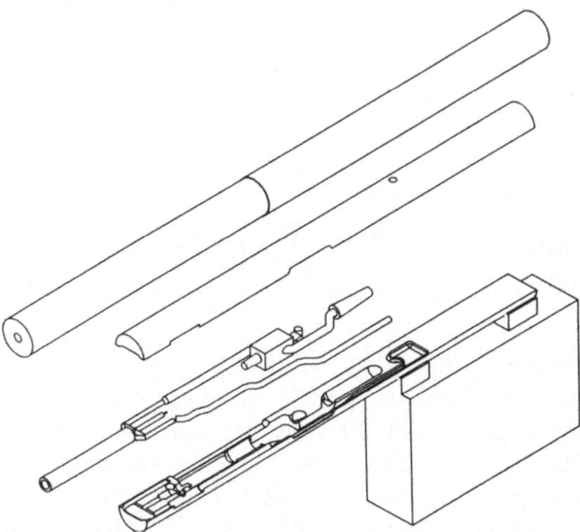

Figure 3. Counter housing in the open state with the proportional counter between the two lead (copper) halves. The tube on top is slid over the two closed halves to fix them.

Figure 4 depicts the Gallex shield tank with the inner dimensions of 760 mm diameter and 1940 mm length. It consists of a cylinder and two sliding doors made of low-activity steel (2 mBq/kg, 20 mm thick) filled with low-activity Boliden lead (30 Bq ^{210}Pb/kg, 155 mm thick). The 3 tons heavy doors are moved on rails by means of a spindle drive along the axis of the 10.5 t weighing cylinder.

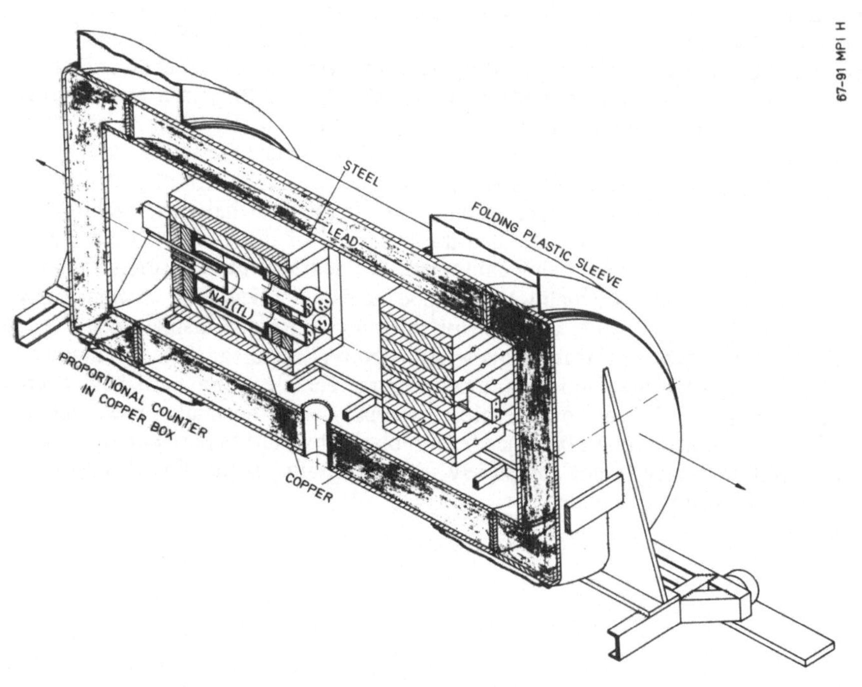

Figure 4. Gallex shield tank with a well-type NaI(Tl)-detector (active side) and a pure copper block (passive side).

The inner shield has an active side with a well type NaI(Tl) pair detector capable of holding up to 8 proportional counters and a passive side made from radiopure (K/Th/U < 2/1/1 mBq/kg) copper with 29 wells for counters. The Ø 254 × 229 mm NaI(Tl) split crystal with the well dimensions Ø 85 × 163 mm is equipped with a Ø 254 mm × 102 mm NaI "pure" light pipe to shield the primordial radioimpurities in the photomultiplier glass. More details, also on the selection of its photomultiplier tubes can be found in [16]. It is shielded on all sides with 10 cm of the same copper used for the passive side. Originally it was tested with a plastic scintillation anticoincidence shield, designed also for operation at the shallower depth (15 m.w.e.) of the MPI Heidelberg Low Level laboratory. At the Gran Sasso Laboratory the muon flux is reduced by 6 orders of magnitude against a factor 2-3 at Heidelberg, compared to sea level. However, a comparison with the copper showed no significant difference in the background spectrum of the NaI(Tl), but a slightly lower total (0.1 - 3.9 MeV) count rate for the passive case: 214.9 cpm against 217.8 cpm. In Fig. 5 two spectra with the copper in place are shown. Here only the lower one is discussed. Its most prominent peak originates from K in the "low potassium" glass of the multipliers [16]. An unvetoed measurement with the plastic scintillators yielded a more pronounced Compton region of ^{40}K (total count rate: 238.5 cmp), thus disclosing that the scintillators with their additional 12 "dirty" photomultipliers add more to the background of the NaI(Tl) than they are able to suppress. Further advantages of a passive shield against an active one are: better control of radioimpurities, almost no construction constraints, and a constant shielding effect, which does not depend on electronic stability. These arguments certainly hold also for the purely passive inner copper shield for the proportional counters shown in Fig. 4 (right side). For services both inner shields can be moved on rails out of the middle cylinder.

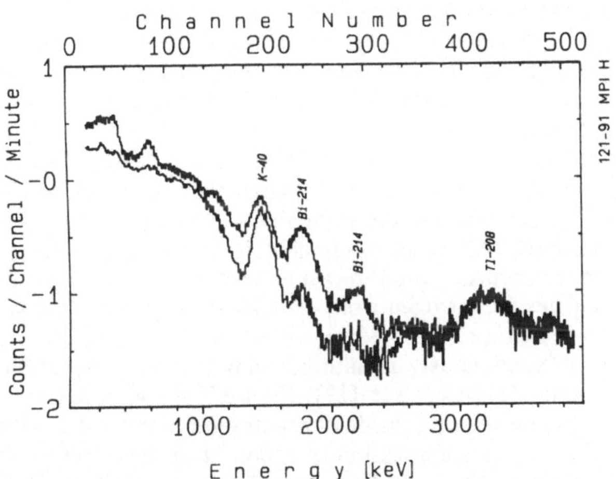

Figure 5. Background spectra of the NaI(Tl) crystal in its copper shield with radon (air) removed from the shield tank (bottom spectrum) and Rn present (top).

4. RADON SUPPRESSION

In this experiment, radon and its progenies can form a dangerous background source in three different ways. 1. The most severe one would occur by admixture to the counting gas. In Gallex this is well controlled by looking for time patterns being most typical of the ^{222}Rn daughter pair ^{214}Bi and ^{214}Po ($T_{1/2}$=162 μsec). 2. The deposition of the daughter pair on the outer surface of the proportional counter is minimized by keeping a narrow quartz cylinder around the active volume during counter handling outside the shield. 3. To prevent Rn from entering the shield and uncontrolled deposition of its progenies, the shield tank has been converted into a kind of glove box with folding plastic foil sleeves, which connect the two doors with the middle cylinder. Fig. 4 depicts the sleeves, their folding structure when the doors are closed, however, is not well presented in the drawing. They prevent the exchange of the radon-free nitrogen inside with the radon-loaded air outside when the doors are slid open. The glove box-system is an additional protection against contamination from the outside. The proportional counters are transferred in their copper box into the shield by means of an air lock. Before the enclosed air in the lock-chamber is flushed out with nitrogen. Handling inside is done with gloves, they and the link-up ring for the air lock are mounted onto the folding plastic sleeves so that they stay outside when the doors are closed. The air lock can be moved on a wheeled rack to the desired link-up position. The extra volume of nitrogen needed when the doors are opened is stored in a balloon. It is pushed back and forth by moving the heavy doors and by means of a blower. Unfortunately, the foil and the glove material have a non-zero permeability against radon [18]. Therefore, the doors are equipped with an O-ring to seal the interior against Rn-influx in the normal, closed position. All cables entering the tank through an aperture in the bottom of the middle cylinder are sealed in vacuum-tight.

The folding sleeves are made of polyurethane soft because of its good transparency (delicate handling of the proportional counters inside), good folding- and welding property, and reasonably low radon permeability [18]. The gloves are made of butyl rubber and the balloon of a multilayer foil.

To even further reduce the radon concentration, a circulation system has been set up which removes the residual radon via a charcoal trap at liquid nitrogen temperature. Figure 6 shows the flow scheme. On the right side the shielding tank, accommodated in a Faraday cage, is indicated. All other equipment is installed outside, except two Lucas chambers for the radon concentration measurement inside and outside the tank. The membrane pump circulates the nitrogen from the tank through the charcoal trap and keeps a slight under-pressure in it to prevent freezing of nitrogen. The radon concentration inside and outside the tank is continuously monitored with 5 hours integration time. For more details on the radon removal system see [19]. Figure 7 shows a one-month example with the outside concentration on top and inside concentration below, also with expanded scale. As demonstrated in Fig. 7, the suppression of radon is up to a factor 100 and more, and also the strong variation on the outside is completely damped on the inside. For low statistic experiments like GALLEX with data evaluation by the maximum likelihood method a constant background is prerequisite. The small increase in the lower plot of Fig. 7 was due to an incomplete venting of the air lock and is still below a critical level. Due to the relatively large free space in the well of the NaI(Tl), proportional counters at the active side in the shield are more sensitive to radon than at the passive side, by about a factor 5. The influence on the background spectrum of the NaI(Tl) by about 80-120 Bq/m^3 inside the shield is rather dramatic as can be seen in the upper spectrum of Fig. 5. The total count rate (0.1-3.9 MeV) increased from 217.8 cpm to 325.9 cpm. Also for the coincidence modes a too high Rn-concentration would be problematic.

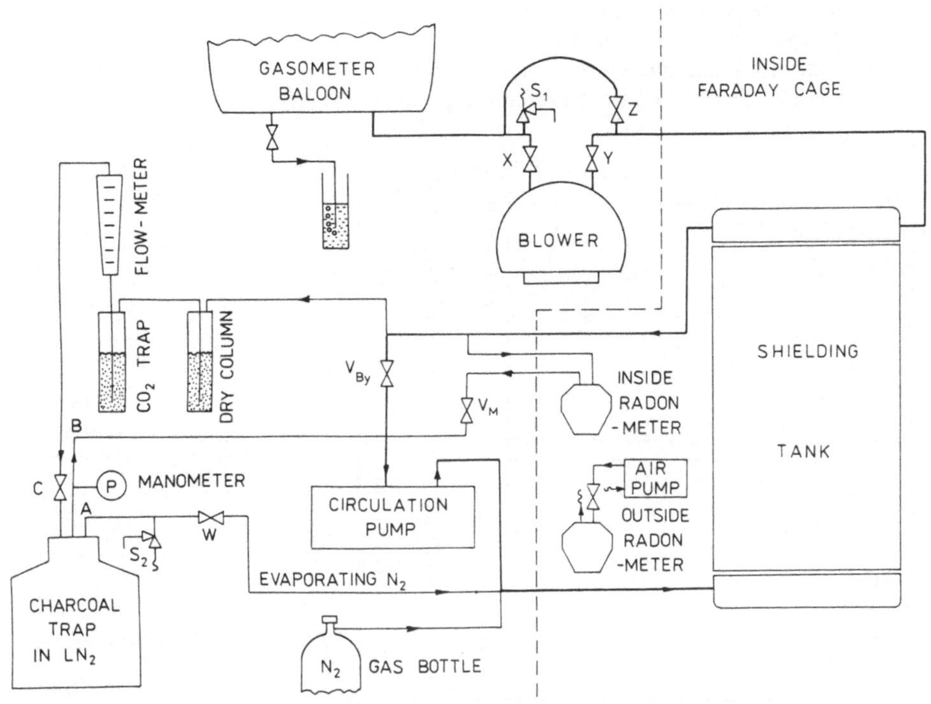

Figure 6. The radon removal line of the shield tank.

5. CALIBRATION

Each pulse of the proportional counters is fully recorded with a fast transient digitizer (0.4 nsec resolution), two slow transient digitizers on different time scales (33 and 333 nsec), and with a CAMAC ADC system. In this way, the full potential of pulse shape (rise time)-, energy-, and time cuts for background reduction is preserved. Also the pulses of both NaI(Tl) halves are separately monitored by a slow transient digitizer and by ADCs in a time-related way, but also independently, to the proportional counter events. The energy- and rise-time calibration of the proportional counters is performed with a cerium X-ray source excited by a ^{153}Gd source [20]. The cerium X-rays create three peaks in the counter, at 1.03, 5.09, 9.75 keV, mainly by conversion with Xe of the gas mixture. Also when shining through the end window, the energy of the Ce X-rays (34-40 keV) is quite uniformly deposited in the active volume so that this calibration (energy and rise time) method is well suited for the internal ^{71}Ge counting.

A comparison is shown in Fig. 8. The concentrated points depict the L- and K-peak of ^{71}Ge (a) and of the energetically similar ones of the calibration source with an additional peak at about 5.09 keV (b). The relations of the energy- and rise time-cuts are defined by intercomparison and can then be applied to identically built counters, which are used for solar neutrino runs and therefore cannot be spoiled by (artificial) ^{71}Ge. The absolute calibration is also determined by the reference proportional counters.

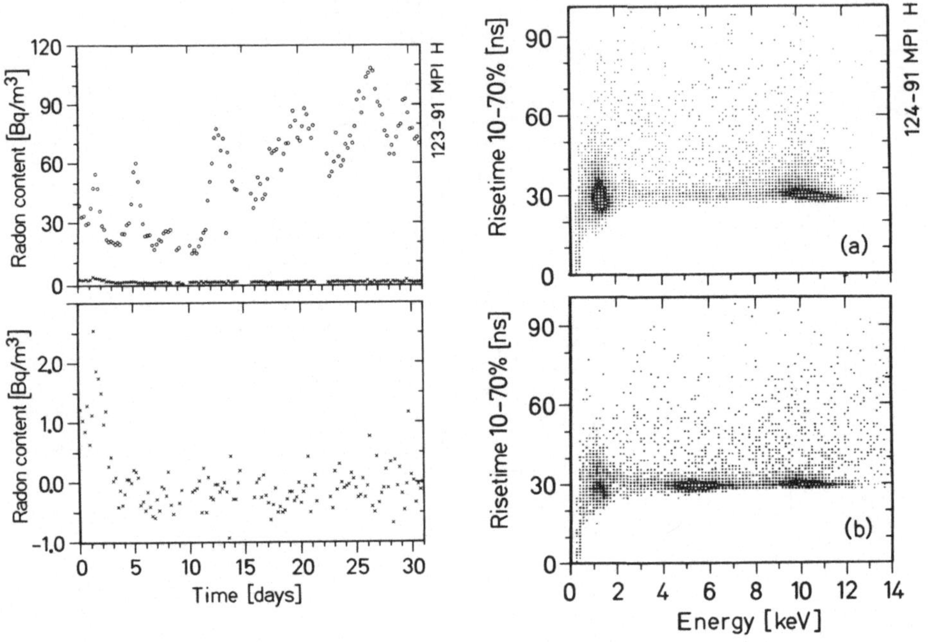

Figure 7. Radon concentration outside (upper points) and inside the shield (lower crosses, also on expanded scale) in 5 hour average as a function of time.

Figure 8. Rise time versus energy spectra of ^{71}Ge (a) and of the ^{153}Ga-Ce calibration source (b).

6. BACKGROUND CHARACTERISTICS

With continuous counter production and background selection the distribution of good counters is growing in absolute number, and the average background is slowly decreasing. More than 20 proportional counters have passed long-duration background tests, first at Heidelberg and then in the Gran Sasso Underground Laboratory. The best 9 examples measured in the passive side of the shield are listed in Table 1. The counters were filled with Ge from different reservoirs to also test their radiopurity. The enriched isotopes are alternatingly used as carrier material to check the carry-over from one run to the next. Blank denotes a blank run of the GeH_4 synthesis line at Gran Sasso. All counters were filled with the standard gas mixture Xe + 30% GeH_4. Because of the small numbers only the average is given in counts per day. The rates represent the total events (> 0.5 keV) and the events in the L- and K-peak energy region, without (all) and with rise time cut (fast). A better illustration of the situation is given by the example shown in Fig. 9 referring to the first entry in Table 1. Energy- and rise time cuts for L- and K-peak are visualized by boxes. The average background rate of Table 1 for L- plus K-fast corresponds to 0.07 cpd compared to about 0.5 cpd (66.3% efficiency) predicted by the Standard Solar model, at the beginning of a 3 weeks run.

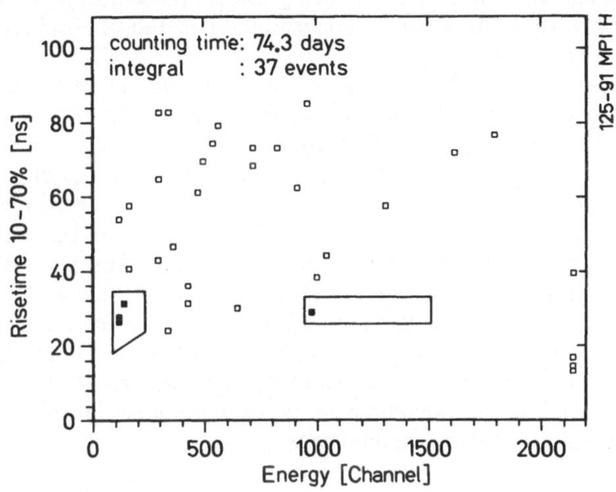

Figure 9. Risetime versus energy background spectrum of counter # 103 measured in the passive side of the shield (see also Table 1).

Despite of this good ratio which allows to measure a SSM signal to better than 8% in a four-year experiment, it is still interesting to analyze the nature of the residual background of the proportional counters.

The many years of counter development have always been accompanied by radioimpurity measurements and basic studies of background components in general [17, 21]. Table 2 gives a compilation of the most relevant background sources for the latest set of HD2-Fe-type proportional counters in the passive shield. Some source strengths can be given only with large errors, like the neutron- and γ-flux at the position of the counter, which is also due to their individual shielding thickness. Here conservative upper limits are given. The neutron flux in the LNGS has been measured [22] and the γ-flux was assumed to be similar to that determined at the MPI Low Level laboratory. The radioimpurities have been either analyzed by radiochemical methods [23] (primordial nuclides in quartz and iron) or, by gamma-ray spectroscopy [17]. For the non-secular equilibrium in the U-chain of the iron, ^{226}Ra has been additionally determined. A determination of ^{226}Ra via radon, which for a former Fe-charge turned out to be more sensitive by a factor 6, is still pending. The corresponding count rates given in the last column of Table 2 were either experimentally determined or, in the case of Fe, calculated by the Monte Carlo method [15]. Cosmogenic radionuclides produced in the material when stored at sea level have also been considered but could not be analyzed, except for short-lived ^{58}Co [$T_{1/2}$=70.9 d] in copper (about 0.3 mBq/kg). Cosmogenic ^{54}Mn in the iron of the cathodes cannot be larger than the sea level production rate corresponding to 5 mBq/kg [17]. Tritium is either outgassed in the vacuum-glow cleaning of the iron cathodes or is relatively harmless if present in the quartz, since it is shielded by the iron. In order to avoid ^{3}H in the counting gas, tritium-free water is used in the conversion of $GeCl_4$ to GeH_4 (see introduction), and in the following gas chromatography step tritium is removed. Yet, its upper limit has to be determined by an independent method. ^{85}Kr being always present in modern Xe is avoided using Xe produced before 1950 when its concentration in the atmosphere (from nuclear fuel reprocessing) was still low.

If we omit the upper limit of ^{226}Ra in Table 2 (it is clearly not stringent), then the total rate of 0.15 cpd accounted for is far below the best value observed (Table 1). Since the raw materials have been analyzed for radioimpurities and the final product is tested for its background, it has to be suggested that the difference is due to contaminations during the assembling process. This argument is also supported by the variation of the background in the counters of the same series and by the fact that there is no statistically significant difference between HD2-Fe counters and those with a silicon cathode (HD2-Si). Silicon was mainly chosen because of its even higher radiopurity (factor 10 and more for primordial radionuclides). It would be very unlikely that ^{32}Si found in silicon in other experiments, e.g. [24], would just compensate for the difference. If the argument holds, this would provide a potential for further background reduction. It should, however, be pointed out that assembling of the counters is already done under strict clean-room conditions and by taking many acid cleaning steps.

Table 1
Background measurements of some proportional counters on the passive side

Counter #	Sample	Counting time [days]	Number of events: E>0.5 keV	L peak all	L peak fast	K peak all	K peak fast
103	tank GeH_4	74.3	37	6	3	5	1
113	nat. Ge	115.8	48	10	3	5	2
106	^{72}Ge	50.7	55	11	5	8	3
110	^{74}Ge	132.6	74	15	9	8	0
114	^{76}Ge	53.3	47	10	3	3	2
99	blank	67.6	65	13	7	6	3
107	blank	67.3	53	8	1	5	1
102	blank	34.8	16	3	2	3	0
112	blank	43.0	34	6	2	0	0
	sum	639.4	429	82	35	43	12
Counts per day:			0.67	0.13	0.05	0.07	0.02

Table 2
Background sources for the GALLEX HD2-Fe-proportional counters in the passive shield

Source	Activity or Flux[a)]	Count rate > 0.5 keV [cpd]	
External Sources:			
muons	3×10^{-8} cm^{-2}sec^{-1}	0.005	
neutrons	< 10^{-6} cm^{-2}sec^{-1}		< 0.001
gamma rays	< 10^{-6} cm^{-2}sec^{-1}		< 0.02
Rn+progenies	< 0.5 Bq m^{-3}		< 0.006
K, Th, U in copper	< 2,1,1 mBq/kg		< 0.02
Internal Sources:			
K in quartz	< 0.04 mBq/kg		< 0.0001
Th in quartz	< 0.01 mBq/kg		< 0.0002
U in quartz	< 1.2 mBq/kg		< 0.03
^{60}Co in iron cathode	< 7 mBq/kg		< 0.02
K in iron cathode	~ 0.06 mBq/kg	0.001	
^{226}Ra in iron cathode	< 3 mBq/kg		< 0.2
Th in iron cathode	< 0.3 mBq/kg		< 0.017
U in iron cathode	< 0.4 mBq/kg		< 0.03
		0.006	< 0.344

a)flux at position of proportional counter

The total background of proportional counters measured on the active side is about a factor 2 higher, which applies for both the L- and K-peaks. This difference can be explained by the presence of potassium in the glass of the photomultipliers, despite their very careful selection [16]. Therefore, the passive side is preferred for the experimental runs. If, however, ^{69}Ge or ^{68}Ge have to be measured, which decay partly by β^+ and γ-emission (in case of ^{68}Ge the daughter ^{68}Ga), the NaI(Tl) is needed. With ^{69}Ge the side reaction can be monitored. In the initialization phase of the target solution ^{68}Ge was found at an unexpectedly high level, but it was later removed by heating [11]. This removal was verified also with the coincidence method. In one mode events are accepted where 511 or 1022 keV are seen in one of the NaI(Tl)-halves plus any energy in the proportional counter (PC) in another mode: 511 keV in each half plus any energy in the PC. The approximate efficiencies and backgrounds are: 39.5%, 0.22 cpd and 12%, 0.03 cpd, respectively. One may also look for a trigger event from the ^{68}Ge decay (100% EC) within the L- or K-peak and the delayed coincidence of the ^{68}Ga decay ($T_{1/2}$=68 min) by the signature

511 + 511 + any energy in the PC. Here the efficiency is about 7% with negligible background. Also for the active side, background improvements are possible since rather soon new photomultipliers will be available with at least 4 times less K and about 40 times less U/Th.

Acknowledgements

This work has been funded by the German Federal Minister for Research and Technology (BMFT) under the contract number 06HD5541. Furthermore, it has been funded by INFN (Italy), CEA (France), KfK-Karlsruhe (Germany), Krupp-Foundation (Germany), and others. The skilful work on the GALLEX spectrometer by the technical staff of the Max-Planck-Institut für Kernphysik is highly appreciated.

7. REFERENCES

1 J.N. Bahcall, Rev.Mod.Phys., 50 (1978) 881.
2 R. Davis, in: Neutrino 88, Proc. 13th Intern.Conf. on Neutrino Physics and Astrophysics, Boston 1988 (J. Schneps, T. Kafka and W.A. Mann, eds.), World Scientific Publ.Co., 1989, 518.
R. Davis, K. Lande, C.K. Lee, B.T. Cleveland and J. Ullman, in: Inside the Sun (G. Berthomieu and M. Cribier, eds.), Kluwer Acad.Publ., Dordrecht, 1989, 171.
3 J.N. Bahcall and R.K. Ulrich, Rev.Mod.Phys., 60 (1988) 297.
4 J. Turck-Chièze, S. Cohen, M. Cassé and C. Doom, Ap.J., 335 (1988) 415.
5 K.S. Hirata et al., Phys.Rev.Lett., 65 (1990) 1298.
Y. Totsuka, Nucl.Phys. B (Proc.Suppl.)19 (1991) 69.
6 J.N. Bahcall, Neutrino Astrophysics, Cambridge University Press, New York-New Rochelle-Melbourne-Sydney, 1989.
7 S. Mikheyev and A. Smirnov, Sov.J.Nucl.Phys., 42 (1986) 913.
J.N. Bahcall and W.C. Haxton, Phys.Rev., D40 (1989) 931.
8 W. Hampel, in: Neutrino 88, Proc. 13th Intern.Conf. on Neutrino Physics and Astrophysics, Boston 1988 (J. Schneps, T. Kafka and W.A. Mann, eds.), World Scientific Publ.Co., 1989, 311.
9 T. Kirsten, in: Inside the Sun (G. Berthomieu and M. Cribier, eds.), Kluwer Academic Publ., Dordrecht, 1989, 187.
10 T. Kirsten, Nuclear Physics B (Proc.Suppl.)19 (1991) 77.
11 R. Plaga, in: Proc. PASCOS 91, Boston 1991.
12 R. Wink, in: Proc. Particles and Cosmology, Baksan 1991.
13 M. Cribier, B. Pichard, J. Rich, M. Spiro, D. Vignaud, A.Besson, A. Bevilacqua, F. Caprerau, G. Dupont, P. Sire, J. Gory, W. Hampel and T. Kirsten, Nucl.Instr.Meth., A265 (1988) 574.
14 R. Davis, Jr., D.S. Harmer and K.C. Hoffman, Phys.Rev.Lett., 20 (1968) 1205.
15 R. Plaga, unpublished PhD thesis, University of Heidelberg, 1989.
16 G. Heusser, Nucl.Instr.Meth., B17 (1986) 423.
17 G. Heusser, Nucl.Instr.Meth., B58 (1991) 79.
18 M. Wójcik, Nucl.Instr.Meth., B61 (1991) 8.
19 G. Heusser and M. Wójcik, Appl.Radiat. and Isot. 43 (1992) 9.
20 A. Urban, unpublished PhD thesis, Technical Unversity Munich, 1989.
21 G. Heusser, Nucl.Instr.Meth., B17 (1986) 418.
22 P. Belli, R. Bernabei, S. d'Angelo, N. Jucci, M.P. De Pascale, L. Paoluzi, R. Santonico, N. Taborgna and G. Villoresi, Il Nuovo Cimento, 101A (1989) 959.

23 E. Pernicka, Annual Report, Max-Planck-Institut für Kernphysik Heidelberg (1981) 188; (1989) 129.
24 D.O. Caldwell, B. Magnusson, M.S. Witherell, A. De Silva, B. Sadoulet, C. Cork, F.S. Goulding, D.A. Landis, N.W. Madden, R.H. Pehl, A.R. Smith, G. Gerbier, E. Lesquoy, J. Rich, M. Spiro, C. Tao, D. Yvon and S. Zylberajch, Phys.Rev.Lett., 65 (1990) 1305.

The Sudbury Neutrino Observatory

J.R. Leslie[a] on behalf of the SNO Collaboration †

[a] Physics Department, Queen's University, Kingston, Ontario K7L 3N6, Canada.

Abstract

The Sudbury Neutrino Observatory (SNO) detector is a 1000 tonne heavy water Čerenkov detector under construction near to Sudbury in Canada. The aim of the experiment is to detect neutrinos from the sun and other astrophysical sources. The use of heavy water allows both electron neutrinos and all types of neutrinos to be observed by three complementary reactions. The detector will be sensitive to the electron neutrino flux and energy spectrum shape and to the total neutrino flux irrespective of neutrino type. These measurements will provide information which should clarify the role of neutrino oscillations, neutrino magnetic moments and solar model predictions in explanations of the observed deficit in solar neutrino flux. In the event of a supernova it will be very sensitive to muon and tau neutrinos as well as the electron neutrinos emitted in the initial burst enabling sensitive mass measurements as well as providing details of the physics of stellar collapse.

---

† The Sudbury Neutrino Collaboration: H.C. Evans, G.T. Ewan, A.L. Hallin, H.W. Lee, J.R. Leslie, J.D. MacArthur, H.-B. Mak, A.B. McDonald, W. McLatchie, B.C. Robertson, B. Sur, P. Skensved (Queen's University): C.K. Hargrove, H. Mes, W.F. Davidson, D. Sinclair (Centre for Research in Particle Physics): E.D. Earle, G.M. Milton, E. Bonvin, (Chalk River Laboratories): P. Jagam, J. Law, J.-X. Wang, J.J. Simpson (University of Guelph): E.D. Hallman, R.U. Haq (Laurentian University): A.L. Carter, D. Kessler, B.R. Hollebone (Carleton University): R. Schubank, C.E. Waltham (University of British Columbia): R.T. Kouzes, M.M. Lowry, R.M. Key (Princeton University): E.W. Beier, W. Frati, M. Newcomer, R. Van Berg (University of Pennsylvania): T.J. Bowles, B.T. Cleveland, P.J. Doe, S.R. Elliott, M.M. Fowler, R.G.H. Robertson, D.J. Vieira, D.L. Wark, J.B. Wilhelmy, J.F. Wilkerson, J.M. Wouters (Los Alamos National Laboratory): E. Norman, K. Lesko, A. Smith, R. Fulton (Lawrence Berkeley Laboratory): N.W. Tanner, N. Jelley, P. Trent, J. Barton (University of Oxford).

1 Introduction

The Sudbury Neutrino Observatory (SNO) is based on a 1000 tonne heavy water (D_2O) detector located in a very low background laboratory 2070 m underground. The 1000 tonnes of pure D_2O used in the experiment will be lent by AECL and a suitable site in the Creighton Mine near Sudbury, Ontario, has been provided by INCO Ltd. The objective of the SNO project is to measure the intensity, energy and direction of neutrinos from the sun and supernovae. A detailed description of the proposal is given in an earlier report (SNO-87-12)[1]. Improvements (some of which are described in this report) to that reference design have since been made but the same basic principles are used. The detector and laboratory are now under construction.

The use of heavy water as a detection medium and the relatively high detection efficiency enables the SNO experiment to measure:

1) the flux and energy spectrum of electron neutrinos reaching the earth
2) the total flux of all neutrino types above an energy of 2.2 MeV and
3) time variations in these with good statistical accuracy.

With these measurements, it will be possible to:

1) show clearly if neutrino oscillations are occurring and
2) independently test solar models by determining the production rate of high-energy electron neutrinos in the solar core.

These detector properties provide an opportunity to use the sun as a calibrated source of neutrinos. Using the sun as a distant source of neutrinos would increase the sensitivity of the search for neutrino oscillations by many orders of magnitude compared to present measurements at reactors and accelerators. The various detection methods and real time capability also provides a means of observing effects attributable to neutrino magnetic moment interactions. The detection of both electron neutrinos and all types of neutrinos from a supernova would give new information on the mechanism of stellar collapse.

As described in SNO-87-12, the proposed SNO detector makes use of three complementary neutrino reactions.

The charged current (CC) reaction

$$\text{(CC)} \qquad \nu_e + \mathrm{d} \rightarrow \mathrm{p} + \mathrm{p} + \mathrm{e}^- \qquad (\mathrm{Q} = -1.44\ \mathrm{MeV})$$

of the electron neutrino on the deuteron is unique to the SNO detector. It offers excellent spectral information, thereby providing a sensitivity to the MSW effect. The large cross section will result in more than ten events per day and make the detector some fifty times more sensitive than existing experiments. The reaction also offers some directional information and would identify the sun as the source of electron neutrinos. This reaction would also identify electron neutrinos from the initial burst in the collapse of a supernova.

The neutrino-electron elastic scattering (ES) reaction,

$$\text{(ES)} \qquad \nu_x + e^- \rightarrow \nu_x + e^-$$

the primary detection mechanism for light water detectors, is sensitive to all neutrino types, but is dominated by the electron neutrino. This reaction offers excellent directional information, but provides little information about the energy spectrum of electron neutrinos.

The third neutrino induced reaction occurring in the SNO detector is the neutral current (NC) disintegration of the deuteron,

$$\text{(NC)} \qquad \nu + d \rightarrow \nu + p + n \qquad (Q = -2.2 \text{ MeV})$$

observed by the detection of the gamma rays resulting from the subsequent neutron capture. This reaction is sensitive to all neutrinos equally and would be used to measure the total flux of neutrinos with a counting rate of about 10 per day above the threshold of about 2.2 MeV.

In addition to the above reactions there are the anti-neutrino induced inverse β decays of both heavy and light hydrogen which will provide valuable information in the event of an observable stellar collapse. These reactions are:

$$\overline{\nu}_e + d \rightarrow n + n + e^+ \qquad (Q = -4.03 \text{ MeV})$$

observed by the detection of a relativistic β^+ followed by the detection of two neutrons as in the NC reaction.

$$\overline{\nu}_e + p \rightarrow n + e^+ \qquad (Q = -1.80 \text{ MeV})$$

produced in the region of the light water shield between the D_2O region and the PMTs. The neutron will not yield a detectable signal as the 2.23 MeV γ ray produced by neutron capture on light water will not be above the background level.

In this paper we discuss briefly the SNO detector and its role in the detection of neutrinos from the sun and supernovae.

2 The SNO Detector

The present design of the SNO detector is shown in Figure 1. It consists of 1000 tonnes of 99.85% enriched D_2O contained in a spherical thin-walled (5 cm thick) transparent acrylic vessel which itself is immersed in 7300 tonnes of ultrapure H_2O shield. The host cavity, which is barrel-shaped and of 22 m diameter and 30 m height, is under excavation at a depth of 2070 m below surface. A waterproof plastic liner is placed inside the cavity enclosing the light water shield. Some 9600 20-cm diameter photomultiplier tubes (PMTs) are uniformly arranged in the H_2O shield on a support structure at 2.5 m distance from the acrylic vessel. Each PMT is surrounded by a reflector to increase the light collection and the effective photocathode coverage is approximately 60%. The PMT array is sensitive to Čerenkov radiation produced by relativistic electrons and other relativistic particles in the central regions of the detector.

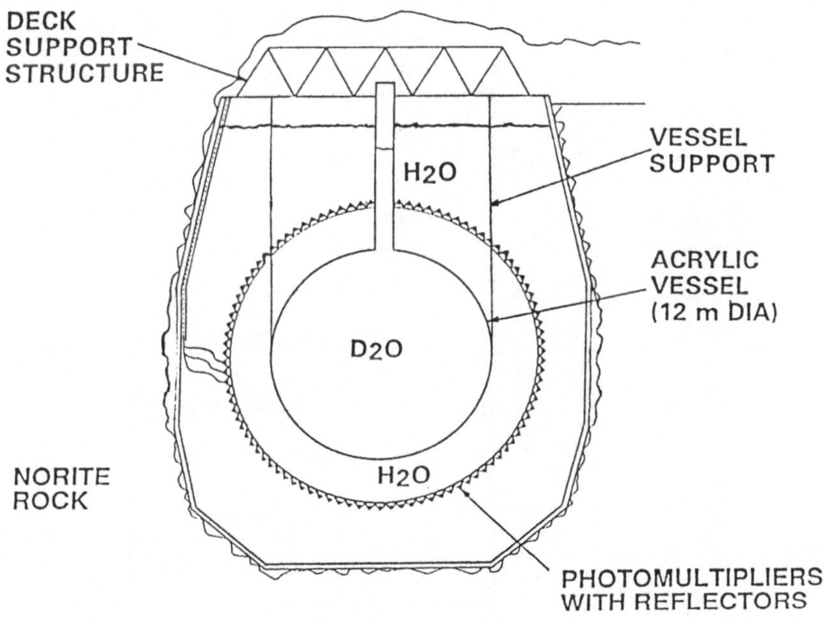

Figure 1 Line drawing of SNO detector

Neutrino interactions in the detector produce either relativistic electrons or free neutrons. The neutrons thermalize in the water and are subsequently captured generating γ-rays which produce relativistic electrons primarily through Compton scattering. The electrons from neutrino interactions or neutron captures will produce Čerenkov photons which pass through the D_2O, acrylic, and H_2O to be detected by the PMTs. The signals from the PMTs are interpreted to give the location, energy and direction of the electron based on the time of arrival of the photons at the PMTs and the locations of the particular PMTs hit. At E_e =10.0 MeV detector simulations predict a spatial reconstruction accuracy of 26 cm, an angular uncertainty 27° and an energy sensitivity 9.0 PMT hits per MeV with a 12% statistical uncertainty. The expected solar neutrino signal rates of about ten events per day require an extremely low background environment for the D_2O. There are two sources of background that must be guarded against. These sources are cosmic rays and radiation produced by naturally occurring radionuclides in the rock surrounding the detector and in the materials used in building the detector. The location of the detector is so deep that the background rate from cosmic-ray muons is negligible. Figure 2 shows the muon intensities at and depth underground of several laboratories.

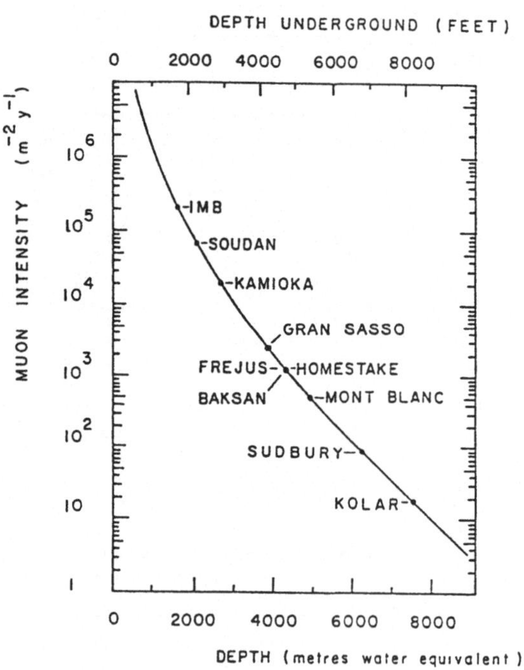

Figure 2 Cosmic Ray Muon Intensity as function of depth

The purity of materials used in the detector has to be very high. The most critical components are the D_2O, H_2O, acrylic vessel and PMTs. In the D_2O and H_2O levels of U and Th in the range of 10^{-14} gram per gram of water are required. In the acrylic vessel which has less mass the tolerable level is a few times 10^{-12} gram per gram of acrylic. The PMTs are 2.5 m from the heavy water and γ-rays are attenuated in the light water shield. To keep contributions to background to a low level the PMTs are being built using glass envelopes from a special melt of low radioactivity glass being manufactured by Schott. The U and Th content of this glass is ~5–10 times lower than normal glass.

The expected performance of the SNO detector has been calculated using an extensive series of Monte Carlo simulations of the neutrino signal processes and radioactive background processes. As an example a typical Monte Carlo calculation of the ^{8}B solar neutrino spectrum is shown in Figure 3. This shows the total spectrum and the components due to the ES, CC and NC reactions for D_2O with NaCℓ added to increase the energy of the NC signal. Below a threshold of ~5 MeV electron energy the background rises steeply and obscures the signal. The neutrinos observable from the sun are thus limited to the ^{8}B and *hep* neutrinos.

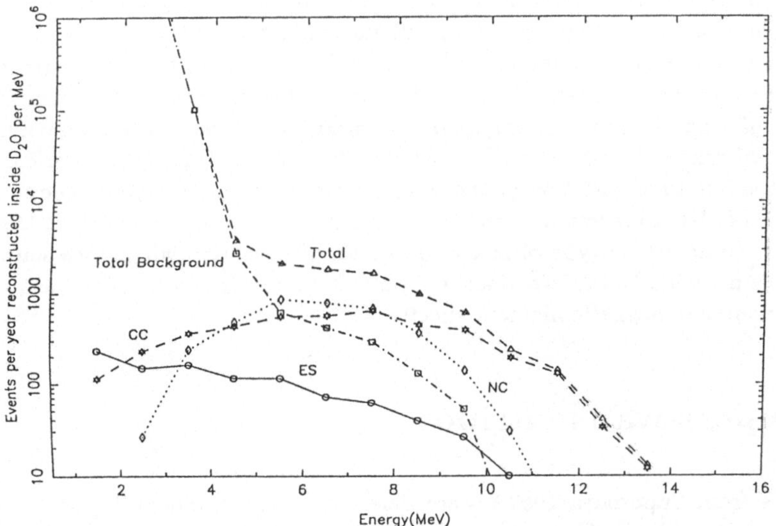

Figure 3 Monte Carlo simulation of ^{8}B spectrum in D_2O + NaCℓ

The events for the NC reaction are determined by detecting the γ-rays following the capture of the neutron released. About 2.5 tonnes of NaCℓ will be added to the heavy water. Then 83% of the neutrons will be captured by ^{35}Cℓ producing γ-

rays with energies up to 8 MeV which should be observable above the background. An additional background must be considered for this reaction, corresponding to neutrons produced by the photodisintegration of deuterium by γ-rays with energies greater than 2.2 MeV. This background will arise predominantly from radioactive decay of members of the ^{232}Th and ^{238}U decay chains contained in the D_2O. This radioactive contamination can be sampled on-line to determine the background rate accurately.

The signal for the NC reaction is the production of a free neutron and alternative schemes for the detection of this neutron, such as ^{3}He or ^{6}Li counters, are being considered. One objective is to allow real-time discrimination between events produced by neutrons and other events. Extreme purity of materials is needed for such counters as they would be located in the D_2O itself.

3 Solar Neutrinos

For some detectors the deduction of neutrino flavour oscillation parameters is dependent on the assumption that the total neutrino emission from the sun is given by the standard solar model - the unique ability of SNO to measure the total neutrino flux will make experimental interpretation independent of solar physics and of course give valuable information on the processes in the interior of the sun. The high count rate and the real time data taking of SNO will enable a study with good statistical accuracy of possible time variation of the solar neutrino flux of all types. The multiple sensitivity of SNO to reactions which are induced by CC events to detect ν_e left only, NC events to detect all type of neutrino and the ES reaction whuch has sensitivity to possible neutrino EM interactions will provide an invaluable tool for the study of possible neutrino magnetic moment effects.

4 Supernova Neutrinos

Neutrinos from Supernova 1987A were detected in the Kamioka [2] and IMB [3] experiments providing evidence that neutrinos are produced in supernovae in numbers and energy near those predicted by theory. Nearly all the events were produced by electron anti-neutrinos interacting with the H_2O.

Many calculations have been done of the production of neutrinos in a supernova and the sensitivity of several existing and proposed detectors is discussed by Burrows [4]. It is generally accepted that a short burst of ν_e will be generated by the electron capture reaction during the initial collapse. Their average energy is $\sim$15 MeV, and the burst lasts for a few tens of milliseconds. Emission of neutrinos and anti-neutrinos in the cooling phase of the protoneutron core lasts from a few seconds to a few tens of seconds.

The proposed SNO detector will make a unique contribution to the study of these neutrinos because of its sensitivity to all neutrino types, because of its low-energy threshold, (~5 MeV), and because it can distinguish the various components of the neutrino flux. This detector, existing light water Čerenkov detectors and scintillator detectors all have high sensitivity for $\overline{\nu}_e$ through the charged-current reaction $\overline{\nu}_e$ + p → n + e^+. The heavy water detector has much higher sensitivity to the ν_e flux than any other detector through reaction ν_e + d → p + p + e^-. After reconstruction, the spatial distribution of events can be used to determine the ν_e and $\overline{\nu}_e$ fluxes; the majority of the charged-current events in the D_2O are from ν_e while most of the events in the H_2O are from $\overline{\nu}_e$. It is expected that the contributions from elastic scattering reactions will be less than 10% of the total event rates. In addition, the neutral-current reaction ν_x + d → ν_x + p + n, in the heavy water can be used to measure the total neutrino intensity. Since the ν_e and $\overline{\nu}_e$ fluxes are determined independently, it is possible to deduce the total flux of other types of neutrinos. For a stellar collapse at 10 kpc, the calculated numbers of neutrino induced events in the SNO detector are listed in Table 1. The entries in the table have been calculated from ref[5].

Table 1: Predicted number of events from a stellar collapse at 10 kpc

Reaction	Target medium	Events in ν_e burst /kilotonne	Events in cooling phase /kilotonne
$\nu_e + d \rightarrow p + p + e^-$	D_2O	10	33
$\nu_x + e \rightarrow \nu_x + e$	D_2O/H_2O	1	16
$\nu_x + d \rightarrow \nu_x + p + n$	D_2O	6	760
$\overline{\nu}_e + d \rightarrow n + n + e^+$	D_2O	0	20
$\overline{\nu}_e + p \rightarrow n + e^+$	H_2O	0	120

One of the important results from SN 1987A was the upper limit on the mass of the $\overline{\nu}_e$. As is clear from Table 1 the NC reaction provides a large sensitivity to all types of neutrinos and consequently it is possible that interesting mass measurements could be obtained. The present limits are 0.25 MeV for ν_μ and 70 MeV for ν_τ. For a collapse at 10 kpc, a mass of 50 eV would give a delay of 3 seconds in the ν_τ events. Although the cooling phase may last for 10 seconds, a shift of this magnitude with respect to the ν_e and $\overline{\nu}_e$ events would be easily seen. A neutrino mass in the range 50 eV< m_ν <100 keV could be measured readily. If the mass were as high as 200 keV the events would come a year late and would be spread over half a year. This would be hard to detect against the solar neutrinos which would act as background. This sensitivity to the $\nu_{\mu,\tau}$ mass is a distinct advantage of the D_2O detector. In the event of a stellar collapse within our Galaxy it would permit the observation of finite

$\nu_{\mu,\tau}$ mass with a sensitivity far better than terrestrial experiments and, perhaps, sufficient to determine whether the total mass of neutrinos is large enough to close the Universe.

5 Project Status and Schedule

The SNO detector received funding in January 1990. The order for photo-multiplier tubes have been placed with Hamamatsu Photonics and delivery and testing of the first PMTs has commenced. A final choice of reflector material and design has been made and the PMT support structure method has been finalised. A section of the water treatment plant is in operation. Major items in the construction schedule are roughly as follows. The excavation of the cavity and the preparation of the underground laboratory will take until Nov, 1992, at which time the waterproof liner and deck will be installed. Installation of the acrylic vessel, PMTs and their support structure will take place in 1993. The initial water fill is scheduled to begin in the Nov, 1994, with a view to completing commissioning tests by April 1995.

The detector facilities at the time of initial installation will have full support for D_2O use so that the experimental procedure will be to run for a period with pure D_2O to measure the CC and ES signals and then to add neutral current detectors. The neutral current detection will be done either by adding $NaC\ell$ to the D_2O or if the attempt to produce discrete detectors of low enough radioactivity is successful to deploy these in the detector.

The SNO detector is primarily designed to measure solar neutrinos, neutrino oscillations and the MSW effect. It can participate in a Supernova Watch programme as it will be sensitive to neutrinos from supernovae except during the filling stages. It can provide unique information if such an event occurs.

During the pure D_2O fill it will be sensitive to the electron neutrinos, ν_e, produced in the initial burst as well as to $\overline{\nu}_e$. Free neutrons released by the NC reaction will be captured in deuterium and give ~6.2 MeV γ-rays which will be difficult to detect above backgrounds due to radioactivity. However if an intense burst occurs in a short time such events could be detected.

When $NaC\ell$ has been added to the D_2O the NC events will be much more clearly observed and the SNO detector will measure ν_e, $\overline{\nu}_e$, ν_μ, $\overline{\nu}_\mu$ and ν_τ, $\overline{\nu}_\tau$.

6 References

1. G.T. Ewan et al, Sudbury Neutrino Observatory Proposal, SNO-87-12 (1987).

2. K. Hirata et al, Phys. Rev. Lett. **58**, 1490 (1987).

3. R.M. Bionta et al, Phys. Rev. Lett. **58**, 1494 (1987).

4. A. Burrows proceedings of this present conference.

5. A. Burrows in Solar Neutrinos and Astronomy, AIP Conf. Proc. No. 126, 283 (1985).

Cosmic Particle Dynamics – Acceleration in Spherical Wave Fields

K. O. Thielheim
Institut für Reine und
Angewandte Kernphysik
Abteilung Mathematische Physik
University of Kiel
FRG

February 17, 1993

1 Particles From Spherical Wave Fields

Immediately after the first pulsar had been detected [1] and identified as a rotating neutron star [2], J.E. Gunn and J.P. Ostriker in their pioneer work [3, 4], investigated the dynamics of an electrically charged test particle within a spherical vacuum wave field.

They showed that rotating magnets indeed are capable to accelerate particles up to considerable values of energy. At the same time these calculations seemingly suggested that rotating magnets essentially act as *monoenergetic* sources of particles rather than as sources of particles exhibiting a *differential energy spectrum* comparable to what is observed in *cosmic particle radiation.*

Since then[1] it has become clear that particles originating from well outside the *acceleration boundary* [6] of a spherical wave field experience a large phase shift and achieve the local drift velocity in radial direction. They therefore follow the $1/r_0^2$*-law of asymptotic energy* [7].

Only particles originating from well inside the acceleration boundary, though still within the wave zone, experience a very small phase shift and attain the *Gunn–Ostriker energy*, independent of their initial position.

The spherical vacuum wave fields existing outside the transition zone of a rotating magnet will be reviewed in chapter 2.1. The equations of motion in spherical coordinates will be described in chapter 2.2. Plane wave formalism will be applied in chapter

[1]On occasion of the European Cosmic Ray Conference in Nottingham 1990 I have reported on some recent advances in the theory of particle acceleration by rotating magnetized bodies, radio pulsars for example, and the problem of the origin of high energy cosmic ray particles. The same notation is used througout the two papers. Numbers in curled brackets '{}' refer to formulae in the Nottingham paper [5].

2.3 to calculate the acceleration boundary deviding the wave zone into two regimes governed by quite different mechanisms of *particle decoupling* from the electromagnetic field. Equations of motion in spherical coordinates will be used to reproduce the Gunn–Ostriker energy [3, 4] by means of the *constant–phase approximation* in chapter 2.4.

The acceleration boundary as *all* other characteristics of particle motion further out essentially are *local phenomena* which can be described by means of a *plane–wave formalism.* The equations of motion in cartesian coordinates and their solutions for plane wave fields have been dicussed extensively in chapters 4 and 5, respectively, of the Nottingham paper [5] and need not be redicussed here. Still the $1/r_0^2$–law of asymptotic energy based on the plane wave formalism will reviewed in chapter 2.5, of this lecture.

Particles approaching the rotating magnet from outside, as from an ambient plasma, may be reflected within a certain interval of radial distance from the rotating magnet referred to as the *plasma border* [8]. The problem of *embedding a rotating magnet into a sufficiently thin, fully ionised plasma* thus is reduced to a problem of *classical scattering theory*, namely the *scattering of an electric monopole by a rotating magnetic dipole.* The deduction of the plasma border also based on plane wave formalism will be reviewed in chapter 2.6.

In the case of radio pulsars, for example, the shell corresponding to the aforementioned interval of radial distance is expected to be a *source of synchroton radiation* observable with instruments of sufficiently high spatial resolution.

Modifications, of course, are expected from effects of collective particle motion. They will be discussed elsewhere.

2 The Vacuum Case

2.1 Spherical Wave Field of a Rotating Magnet

A magnitized body rotating with its vector of magnetic dipole moment $\vec{\mu}$ inclined by an angle χ against its vector of angular velocity $\vec{\omega}$ carries electromagnetic fields which beyond the range of near field contributions, that is outside the transition zone, are pure spherical wave fields. For practical reasons it is of advantage to substitute parameters ω and μ by two other parameters, each with the dimension of a length. One is the *light radius*

$$r_L = c/\omega, \tag{1}$$

the other is the *typical radius*

$$r_T = (e\,\mu/m\,c^2)^{1/2}. \tag{2}$$

Use will be also made of the definition

$$f_L = (r_T/r_L)^2. \tag{3}$$

For the *vacuum field of a rotating magnet* the components of the electric and magnetic field vectors outside the *transition zone* are functions of the phase

$$\tilde{\Phi} = \tilde{\mathbf{x}}^0 - \tilde{\mathbf{x}}^1 \tag{4}$$

in spherical coordinates $\tilde{\mathbf{x}}^1 = r/r_L$, $\tilde{\mathbf{x}}^2 = \vartheta$, $\tilde{\mathbf{x}}^3 = \varphi$ with $\tilde{\mathbf{x}}^0 = \mathbf{x}^0 = ct/r_L$ and reduce to

$$E_r = H_r = 0 \tag{5}$$
$$E_\vartheta = H_\varphi = (\mu/r_L^2 r)\sin\chi\sin\tilde{\Phi} \tag{6}$$
$$E_\varphi = -H_\vartheta = (\mu/r_L^2 r)\sin\chi\cos\vartheta\cos\tilde{\Phi}. \tag{7}$$

Within the equatorial plane of rotation ($\vartheta = \pi/2$) the local plane wave is linearly polarized ($\alpha = 0$) with the electric vector parallel to the axis of rotation, whereas along the axis of rotation ($\vartheta = 0$) the local plane wave is circularly polarized ($\alpha = \pm\pi/4$). In between these two limiting orientations the local plane wave is elliptically polarized with the ratio of the minor to the major axis equal to the cosine of the latitudinal angle

$$\tan\alpha = \cos\vartheta\ . \tag{8}$$

The dimensionless wave amplitude is {69}

$$f_0 = r_T^2/r_L r_0.$$

2.2 Equations of Motion in Spherical Coordinates

The transition from *cartesian coordinates in configuration space* $\mathbf{x}^1 = x/r_L$, $\mathbf{x}^2 = y/r_L$, $\mathbf{x}^3 = z/r_L$ written in dimensionless form with the help of an appropriately chosen unit of length r_L which in this case is the light radius to *spherical coordinates in configuration space* $\tilde{\mathbf{x}}^1 = \mathbf{r} = r/r_L$, $\tilde{\mathbf{x}}^2 = \vartheta$, $\tilde{\mathbf{x}}^3 = \varphi$ with $x = r\sin\vartheta\cos\varphi$, $y = r\sin\vartheta\sin\varphi$, and $z = r\cos\vartheta$, where the time coordinate $\tilde{\mathbf{x}}^0 = \mathbf{x}^0 = c\tilde{t}/r_L = ct/r_L$ is left unchanged, corresponds to the following *coordinate transformation in configuration space*

$$d\mathbf{x}^1 = \sin\vartheta\cos\varphi\, d\tilde{\mathbf{x}}^1 + \mathbf{r}\cos\vartheta\cos\varphi\, d\tilde{\mathbf{x}}^2 - \mathbf{r}\sin\vartheta\sin\varphi\, d\tilde{\mathbf{x}}^3 \tag{9}$$
$$d\mathbf{x}^2 = \sin\vartheta\sin\varphi\, d\tilde{\mathbf{x}}^1 + \mathbf{r}\cos\vartheta\sin\varphi\, d\tilde{\mathbf{x}}^2 + \mathbf{r}\sin\vartheta\cos\varphi\, d\tilde{\mathbf{x}}^3 \tag{10}$$
$$d\mathbf{x}^3 = \cos\vartheta\, d\tilde{\mathbf{x}}^1 - \mathbf{r}\sin\vartheta\, d\tilde{\mathbf{x}}^2 \tag{11}$$

and

$$d\tilde{\mathbf{x}}^1 = \sin\vartheta\cos\varphi\, d\mathbf{x}^1 + \sin\vartheta\sin\varphi\, d\mathbf{x}^2 + \cos\vartheta\, d\mathbf{x}^3 \tag{12}$$
$$d\tilde{\mathbf{x}}^2 = \mathbf{r}^{-1}\cos\vartheta\cos\varphi\, d\mathbf{x}^1 + \mathbf{r}^{-1}\cos\vartheta\sin\varphi\, d\mathbf{x}^2 - \mathbf{r}^{-1}\sin\vartheta\, d\mathbf{x}^3 \tag{13}$$
$$d\tilde{\mathbf{x}}^3 = -(\mathbf{r}\sin\vartheta)^{-1}\sin\varphi\, d\mathbf{x}^1 + (\mathbf{r}\sin\vartheta)^{-1}\cos\varphi\, d\mathbf{x}^2 \tag{14}$$

respectively.

The square of the norm of the differential of coordinates in Minkowsky space also written in dimensionless form then is $(ds)^2 = \tilde{g}_{jk} d\tilde{\mathbf{x}}^j d\tilde{\mathbf{x}}^k$ with the metric tensor $\tilde{g}_{00} = 1$, $\tilde{g}_{11} = -1$ $\tilde{g}_{22} = -\mathbf{r}^2$, $\tilde{g}_{33} = -(\mathbf{r}\sin\vartheta)^2$ and $\tilde{g}_{jk} = 0$ for all other values of indices and the inverse metric tensor $\tilde{g}^{00} = 1$, $\tilde{g}^{11} = -1$, $\tilde{g}^{22} = -\mathbf{r}^{-2}$, $\tilde{g}^{33} = -(1/\mathbf{r}\sin\vartheta)^2$ and $\tilde{g}^{jk} = 0$ for all other values of indices.

Since the coordinate transformation under consideration is nothing but a transition from one to another set of orthogonal coordinates in three–dimensional configuration space the components of the velocity vector with respect to spherical coordinates $\tilde{\mathbf{u}}^j = d\tilde{\mathbf{x}}^j/ds$, i.e. $\tilde{\mathbf{u}}^0 = \mathbf{u}^0 = \gamma$ and $\vec{\tilde{\mathbf{u}}} = (\tilde{\mathbf{u}}^1, \tilde{\mathbf{u}}^2, \tilde{\mathbf{u}}^3)^T := (\mathbf{u}_r, \mathbf{r}^{-1}\mathbf{u}_\vartheta, (\mathbf{r}\sin\vartheta)^{-1}\mathbf{u}_\varphi)^T$, so that

$\|\mathbf{u}\|^2 = \gamma^2 - \mathbf{u}_r^2 - \mathbf{u}_\vartheta^2 - \mathbf{u}_\varphi^2 = 1$, likewise transform according to (9) to (11) and (12) to (14), respectively.

Since use is made of curvilinear coordinates *Christoffel Symbols*

$$\tilde{\mathbf{\Gamma}}^j_{k\,l} = \left(\frac{\tilde{g}^{jm}}{2}\right)\left(\frac{\partial \tilde{g}_{m\,k}}{\partial \tilde{\mathbf{x}}^l} + \frac{\partial \tilde{g}_{m\,l}}{\partial \tilde{\mathbf{x}}^k} - \frac{\partial \tilde{g}_{k\,l}}{\partial \tilde{\mathbf{x}}^m}\right), \tag{15}$$

are introduced the non–vanishing components of which are $\tilde{\mathbf{\Gamma}}^1_{2\,2} = -\mathbf{r}$, $\tilde{\mathbf{\Gamma}}^1_{3\,3} = -\mathbf{r}\sin^2\vartheta$, $\tilde{\mathbf{\Gamma}}^2_{3\,3} = -\sin\vartheta\cos\vartheta$, $\tilde{\mathbf{\Gamma}}^2_{1\,2} = \tilde{\mathbf{\Gamma}}^2_{2\,1} = \tilde{\mathbf{\Gamma}}^3_{1\,3} = \tilde{\mathbf{\Gamma}}^3_{3\,1} = 1/\mathbf{r}$ and $\tilde{\mathbf{\Gamma}}^3_{2\,3} = \tilde{\mathbf{\Gamma}}^3_{3\,2} = \cot\vartheta$. The covariant derivation of the velocity vector with respect to dimensionless eigentime, i.e. the covariant vector of acceleration

$$D\tilde{\mathbf{u}}^j/ds = d\tilde{\mathbf{u}}^j/ds + \tilde{\mathbf{\Gamma}}^j_{k\,l}\tilde{\mathbf{u}}^k\tilde{\mathbf{u}}^l \tag{16}$$

written in dimensionless form owns the components

$$\begin{aligned}
D\tilde{\mathbf{u}}^0/ds &= d\gamma/ds && (17)\\
D\tilde{\mathbf{u}}^1/ds &= d\mathbf{u}_r/ds - (\mathbf{u}_\vartheta^2 + \mathbf{u}_\varphi^2)/\mathbf{r} && (18)\\
D\tilde{\mathbf{u}}^2/ds &= d\mathbf{u}_\vartheta/ds - (\mathbf{u}_\varphi^2\cot\vartheta - \mathbf{u}_r\mathbf{u}_\vartheta)/\mathbf{r} && (19)\\
D\tilde{\mathbf{u}}^3/ds &= d\mathbf{u}_\varphi/ds + (\mathbf{u}_\vartheta\mathbf{u}_\varphi\cot\vartheta + \mathbf{u}_r\mathbf{u}_\varphi)/\mathbf{r}. && (20)
\end{aligned}$$

Analogously, the electromagnetic field may be described by field vectors with components defined with respect to spherical coordinates,

$$\begin{aligned}
\vec{\tilde{E}} &= (\tilde{E}^1, \tilde{E}^2, \tilde{E}^3)^T && (21)\\
&:= (E_r, \mathbf{r}^{-1}E_\vartheta, (\mathbf{r}\sin\vartheta)^{-1}E_\varphi)^T && (22)
\end{aligned}$$

and

$$\begin{aligned}
\vec{\tilde{H}} &= (\tilde{H}^1, \tilde{H}^2, \tilde{H}^3)^T && (23)\\
&:= (H_r, \mathbf{r}^{-1}H_\vartheta, (\mathbf{r}\sin\vartheta)^{-1}H_\varphi)^T, && (24)
\end{aligned}$$

respectively, which then are related to the corresponding components defined with respect to cartesian coordinates again through (9) to (11) and (12) to (14) respectively,

$$\begin{aligned}
\tilde{E}^1 &= \sin\vartheta\cos\varphi\, E^1 + \sin\vartheta\sin\varphi\, E^2 + \cos\vartheta\, E^3 && (25)\\
\tilde{E}^2 &= \mathbf{r}^{-1}\cos\vartheta\cos\varphi\, E^1 + \mathbf{r}^{-1}\cos\vartheta\sin\varphi\, E^2 - \mathbf{r}^{-1}\sin\vartheta\, E^3 && (26)\\
\tilde{E}^3 &= -(\mathbf{r}\sin\vartheta)^{-1}\sin\varphi\, E^1 + (\mathbf{r}\sin\vartheta)^{-1}\cos\varphi\, E^2. && (27)
\end{aligned}$$

The components of the mixed representation of the field tensor defined with respect to spherical coordinates in configuration space, $\tilde{F}^j{}_k = (\partial\tilde{\mathbf{x}}^j/\partial\mathbf{x}^l)F^l{}_m(\partial\mathbf{x}^m/\partial\tilde{\mathbf{x}}^k)$ are,

$$(\tilde{F}^j{}_k) = \begin{pmatrix} 0 & E_r & \mathbf{r}E_\vartheta & \mathbf{r}\sin\vartheta E_\varphi \\ E_r & 0 & \mathbf{r}H_\varphi & -\mathbf{r}\sin\vartheta H_\vartheta \\ \mathbf{r}^{-1}E_\vartheta & -\mathbf{r}^{-1}H_\varphi & 0 & \sin\vartheta H_r \\ (\mathbf{r}\sin\vartheta)^{-1}E_\varphi & (\mathbf{r}\sin\vartheta)^{-1}H_\vartheta & -(\sin\vartheta)^{-1}H_r & 0 \end{pmatrix} \tag{28}$$

Also, dimensionless forms of the electric and magnetic field vectors are introduced through $\vec{\tilde{f}} = (er_L/mc^2)\vec{\tilde{E}}$ and $\vec{\tilde{h}} = (\epsilon r_L/mc^2)\vec{\tilde{H}}$ corresponding to $\tilde{\mathbf{F}}^j{}_k = (er_L/mc^2)\tilde{F}^j{}_k$.

These components enter into the equation of motion[2]

$$D\tilde{\mathbf{u}}^j/ds = \tilde{\mathbf{F}}^j{}_k\mathbf{u}^k \tag{29}$$

and consequentley

$$d\tilde{\mathbf{u}}^j/ds = -\tilde{\mathbf{\Gamma}}^j_{kl}\mathbf{u}^k\mathbf{u}^l + \tilde{\mathbf{F}}^j{}_k\mathbf{u}^k. \tag{30}$$

Each component of this Lorentz equation of motion on its right side incorporates a *geometry term*, which is independent from the given electromagnetic field, and a *Lorentz force term*. The geometry terms can be referred to as *effective force terms*.

The components of the Lorentz equation of motion are

$$\frac{d\gamma}{ds} = f_L\{\mathbf{u}_\vartheta \sin\tilde{\Phi} - \mathbf{u}_\varphi \cos\vartheta \cos\tilde{\Phi}\}/\mathbf{r} \tag{31}$$

$$\frac{d\mathbf{u}_r}{ds} = (\mathbf{u}_\vartheta{}^2 + \mathbf{u}_\varphi{}^2)/\mathbf{r} + f_L\{\mathbf{u}_\vartheta \sin\tilde{\Phi} - \mathbf{u}_\varphi \cos\vartheta \cos\tilde{\Phi}\}/\mathbf{r} \tag{32}$$

$$\frac{d\mathbf{u}_\vartheta}{ds} = (\mathbf{u}_\varphi{}^2 \cot\vartheta - \mathbf{u}_r\,\mathbf{u}_\vartheta)/\mathbf{r} + f_L\,(v/\mathbf{r})\sin\tilde{\Phi} \tag{33}$$

$$\frac{d\mathbf{u}_\varphi}{ds} = -(\mathbf{u}_\vartheta\,\mathbf{u}_\varphi \cot\vartheta + \mathbf{u}_r\,\mathbf{u}_\varphi)/\mathbf{r} - f_L\,(v/\mathbf{r})\cos\vartheta\cos\tilde{\Phi}. \tag{34}$$

Under given premises

$$v = \hat{\tilde{\Phi}} = \frac{d\tilde{\Phi}}{ds} = \gamma - \mathbf{u}_r \tag{35}$$

and consequentley

$$\mathbf{u}_r = (1 + \mathbf{u}_\vartheta{}^2 + \mathbf{u}_\varphi{}^2 - v^2)/2\,v. \tag{36}$$

2.3 The Acceleration Boundary

The notion of an *acceleration boundary* has been introduced to describe the merging of two ranges of the radial coordinate which distinguish by the character of particle decoupling from a spherical wave field [6]. The premises are that the particle is initially at rest ($\vec{u}(\Phi_0) = 0$) where the electric field strength has its maximum ($\Phi_0 = \pi/2$).

In a spherical wave field the dynamics of an electrically charged particle is subject to a competition between two quantities: One is the decay length of the spherical wave amplitude in radial direction, the other is the interval by which, in a half period[3], the particle would be shifted in direction of wave propagation through drift motion within the corresponding plane wave field.

[2] *Attractive gravitational forces*, as well as *repelling radiation reaction forces*, are found to be negligible in comparison with the forces due to the spherical wave field under the premises of the standard set of parameters in regions outside the transition zone as one may expect from the orders of magnitude involved [8, 9].

[3] In consistency with the original definition [6] it is convenient to refer to one half period instead of to one full period as in {72}.

By definition, the acceleration boundary is located where the two ranges merge, which is the case where these two quantities become equal,

$$r_B = (\pi/4)^{1/3} r_L f_L^{2/3} \cdot [(1 + 3\cos^2\vartheta)/(1 + \cos^2\vartheta)]^{1/3}, \tag{37}$$

that is

$$r_B = 0.923 \cdot r_L (r_T/r_L)^{4/3} \cdot [(1 + 3\cos^2\vartheta)/(1 + \cos^2\vartheta)]^{1/3}. \tag{38}$$

Using the standard set of parameters the acceleration boundary is $r_B = 2.8 \cdot 10^{12}$ cm for protons with $\vartheta_0 = \pi/2$.

Remarkably, the deduction of the acceleration boundary (37) or (38) is based on the plane wave formalism rather than on the spherical wave formalism.

Within the inner range of the radial coordinate, where the wave amplitude is high while the decay length of the wave amplitude is small compared with particle translation by drift, the particle, before decoupling from the electromagnetic wave field, will experience a phase shift which is small compared with π and consequently exhibit a monotonous development of energy.

Within the outer range of the radial coordinate, where the wave amplitude is low while the decay length of the wave amplitude is large compared with particle translation by drift, the particle, before decoupling from the electromagnetic wave field, will experience a phase shift which is large compared with π. Consequently, the particle will exhibit an oscillatory development of energy, attaining the local drift velocity [7].

Results of numerical integrations of the equations of motion in spherical coordinates nicely confirm these predictions.

Fig. (1) is for particles originating from within the equatorial plane of rotation $\vartheta_0 = \pi/2$ corresponding to linear polarization. The initial phase is $\Phi_0 = \pi/4$. Different curves are for different initial values r_0 of the radial coordinate.

As is obvious from this diagram test particles originating from positions sufficiently near to the rotating magnet immediately decouple from the spherical wave field. This phenomenon manifests itself in a monotonous development of energy as a function of radial coordinate.

Alternatively, test particles originating from postions sufficiently far from the rotating magnet exhibit an oscillatory development of energy before approaching their respective asymptotic energy values.

2.4 The Gunn–Ostriker Energy

If the inititial position of the particle under consideration, though within the wave zone, still is comparatively near to the rotating magnet the amplitude of the spherical wave field may be very high and the particle may be accelerated very efficiently to almost the velocity of light.

The particle then is expected to experience only a small shift in phase compared to π before leaving the range of influence of the rotating magnet.

It is then justified to make use of the *constant phase approximation*, e. g.

$$\tilde{\Phi} \cong \tilde{\Phi}_0 = \frac{\pi}{2}. \tag{39}$$

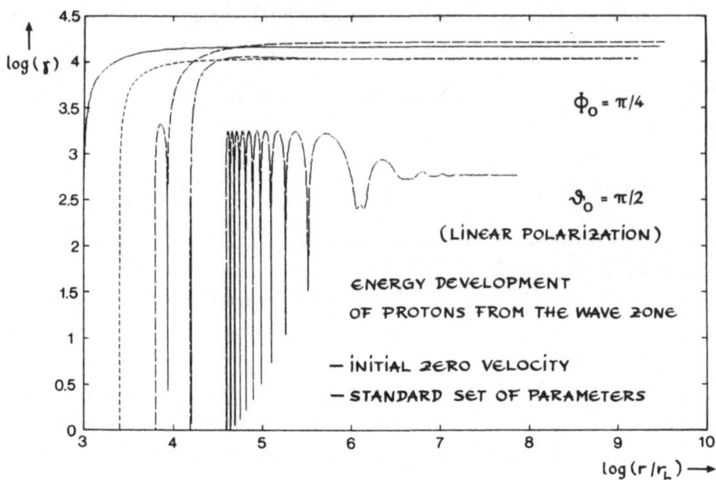

Figure 1: Energy development in spherical vacuum wave, numerical results.

Also, for the purpose in mind, it will be sufficient to consider *orbits confined to the neighbourhood of the equatorial plane of rotation*, e. g.

$$\vartheta \cong \vartheta_0 = \frac{\pi}{2} \tag{40}$$

with

$$\mathbf{u}_\vartheta = \hat{\vartheta} = \frac{d\vartheta}{ds} \tag{41}$$

restricted by

$$|\mathbf{u}_\vartheta| \ll f_L. \tag{42}$$

The equations of motion (32) to (35) then reduce to

$$\frac{d\gamma}{ds} = f_L \frac{\mathbf{u}_\vartheta}{\mathbf{r}} \tag{43}$$

$$\frac{d\mathbf{u}_r}{ds} = \frac{\left(\mathbf{u}_\vartheta^2 + \mathbf{u}_\varphi^2\right)}{\mathbf{r}} + f_L \frac{\mathbf{u}_\vartheta}{\mathbf{r}} \tag{44}$$

$$\frac{d\mathbf{u}_\vartheta}{ds} = -\frac{\mathbf{u}_r \mathbf{u}_\vartheta}{\mathbf{r}} + f_L \frac{v}{\mathbf{r}} \tag{45}$$

$$\frac{d\mathbf{u}_\varphi}{ds} = -\frac{\mathbf{u}_r \mathbf{u}_\vartheta}{\mathbf{r}}. \tag{46}$$

(47)

The equation of motion (45) delivers

$$\frac{d(\mathbf{r}\,\mathbf{u}_\vartheta)}{d\tilde{\Phi}} = f_L \tag{48}$$

which can be integrated to

$$\mathbf{u}_\vartheta = f_L \frac{\tilde{\Phi} - \tilde{\Phi}_0}{\mathbf{r}}, \tag{49}$$

so that

$$\mathbf{u}_\vartheta \to 0 \text{ for } \mathbf{r} \to \infty. \tag{50}$$

The equation of motion (46) delivers

$$\frac{d(\mathbf{r}\,\mathbf{u}_\varphi)}{ds} = 0 \tag{51}$$

and therefore

$$\mathbf{u}_\varphi = \frac{\mathbf{r}_0\,\mathbf{u}_\varphi(\tilde{\Phi}_0)}{\mathbf{r}}, \tag{52}$$

so that with the initial condition

$$\vec{\mathbf{u}}(\tilde{\Phi}_0) = 0 \tag{53}$$

the particle orbit is found to be contained in a meridian plane since

$$\mathbf{u}_\varphi = 0. \tag{54}$$

The equation of motion (44) then reduces to

$$\frac{d\mathbf{u}_r}{ds} \cong \frac{\mathbf{u}_\vartheta^2}{\mathbf{r}} + f_L \frac{\mathbf{u}_\vartheta}{\mathbf{r}} \tag{55}$$

and with (42) even further to

$$\frac{d\mathbf{u}_r}{ds} \cong f_L \frac{\mathbf{u}_\vartheta}{\mathbf{r}}. \tag{56}$$

Dividing the equation of motion (45) by (56) delivers

$$\frac{d\mathbf{u}_\vartheta}{d\mathbf{u}_r} \cong \frac{v - \mathbf{u}_r \mathbf{u}_\vartheta / f_L}{\mathbf{u}_\vartheta}. \tag{57}$$

Substraction of the equation of motion (55) from (43) with (56) delivers

$$\frac{dv}{d\mathbf{u}_r} \cong -\frac{\mathbf{u}_\vartheta}{f_L} \tag{58}$$

from which by introduction into (57)

$$\mathbf{u}_\vartheta \frac{d\mathbf{u}_\vartheta}{d\mathbf{u}_r} \cong \frac{d(v\,\mathbf{u}_r)}{d\mathbf{u}_r} \tag{59}$$

and consequently

$$\mathbf{u}_\vartheta^2 \cong 2v\mathbf{u}_r, \tag{60}$$

so that with (50)

$$v \to 0 \text{ for } \mathbf{r} \to \infty. \tag{61}$$

From (58) with (60)

$$\frac{dv}{\sqrt{v}} = -\frac{\sqrt{2}}{f_L}\sqrt{\mathbf{u}_r}\, d\mathbf{u}_r \tag{62}$$

and consequently

$$\sqrt{v} = 1 - \frac{2^{1/2}}{3f_L}\mathbf{u}_r^{3/2}, \tag{63}$$

and with (61) for $r \to \infty$

$$\mathbf{u}_{GO} = 3^{2/3} \cdot 2^{-1/3} f_L^{2/3} \tag{64}$$

that is

$$\mathbf{u}_{GO} \cong 1.65 \cdot (r_T/r_L)^{4/3}. \tag{65}$$

2.5 The $1/r_0^2$–Law of Asymptotic Energy.

As I have suggested earlier particles originating from positions well outside the acceleration boundary, $r_0 \gg r_B$, can be expected to attain the local drift velocity {54} with {69} in the direction of wave propagation before decoupling from the spherical wave field and consequently follow an $1/r_0^2$–law of asymptotic energy [7]

$$\mathbf{u}_r(\infty) = \overline{\mathbf{u}^1} = (1/4)(r_L/r_0)^2(r_T/r_L)^4. \tag{66}$$

Again, this prediction is nicely confirmed by results of numerical integration of the equations of motion shown in Fig. (2). The diagram in this case is for $\vartheta_0 = \pi/4$ and for initial values of phase $\Phi_0 = 0, \pi/6, \pi/3$ and $\pi/2$.

The $1/r_0^2$ – law of asymptotic energy is also confirmed by results of averaging procedures applied to the equations of motion (32) to (35) to describe the mean development of particle energy [10].

The $1/r_0^2$ – law of asymptotic energy is of particular interest with respect to differential energy spectra generated by appropiate source functions describing the birth of electrically charged particles through decay or ionisation of neutral particles which may have invaded the electromagnetic wave field from outside. A constant source function, for example, will generate a differential energy spectrum of the form $S(E)\, dE \sim E^{-2.5}\, dE$.

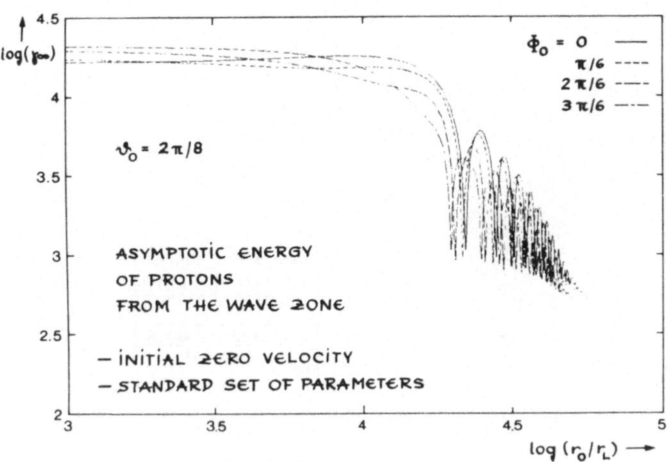

Figure 2: Asymptotic particle energy in spherical vacuum wave, numerical results.

2.6 The Plasma Border.

For sufficiently high values of the frequency of rotation ω and of the magnetic dipole moment μ, a magnet rotating within a fully ionized, ambient plasma is expected to be able to sweep away all charged particles from its immediate vicinity [8] if the ambient plasma is of sufficiently low density. Under such conditions, a rotating magnetised body can evacuate a region of space up to a certain radial distance $r = r_P$, called the *plasma border* {71}.

$$r_P = (2^{-1/2})(\beta\,\gamma)^{-1} r_L (r_T/r_L)^2.$$

Using the standard set of parameters the plasma border is $r_p = 4.7 \cdot 10^{14}$ cm , i.e. $r_p = 1.5 \cdot 10^{-4}$ pc for protons with $\beta = 2^{-1/2}$ corresponding to $\beta\gamma = 1$.

Features of particle dynamics discussed so far are illustrated in Fig. (3).

References

[1] A. Hewish et al., Observation of a Rapidly Pulsating Radio Source., Nature 217 (1968) 709–713.

[2] T. Gold, Rotating Neutron Stars as the Origin of the Pulsating Radio Sources, Nature 218 (1968) 731–732.

[3] J.E. Gunn and J.P. Ostriker, Acceleration of High–Energy Cosmic Rays by Pulsars, Phys. Rev. Lett. 22 (1969) 728–773.

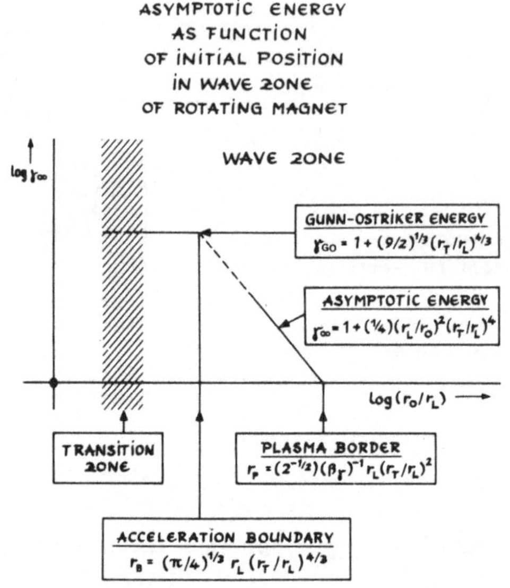

Figure 3: Schematic diagram illustrating asymptotic particle energy as a function of initial position.

[4] J.E. Gunn and J.P. Ostriker, On the Motion and Radiation of Charged Particles in Strong Electromagnetic Waves. 1. Motion in Plane and Spherical Waves., APJ 165 (1971) 523–541.

[5] K.O. Thielheim, Particle Acceleration by Rotating Magnetized Bodies and the Problem of the Origin of High Energy Cosmic Particles, Nucl. Phys. B (Proc. Suppl.) 22 B (1991) 60–79.

[6] K.O. Thielheim, Acceleration Boundary for an Electrically Charged Particle within the Field of a Rotating Magnetic Dipole, J. Phys. 20 (1987) L203–L206.

[7] K.O. Thielheim, Particle Acceleration by Rotating Magnets – Pulsars as Cosmic Accelerators, Fundamentals of Cosmic Physics 13 (1989) 357–399.

[8] K.O. Thielheim, Pulsars as particle accelerators – Some features of the machine become visible
In: Shapiro, M. M. and Wefel, J. P. (eds.): Cosmic Gamma Rays, Neutrinos and

Related Astrophysics, Proc. Nato Advanced Study Institute, Erice, Italy, April 20–30 , 1988, 523-536, Kluwer Academic Publishers

[9] K.O. Thielheim, Pulsars as Cosmic Ray Accelerators, the Critical Frequency — Proc. Europ. Particle Accelerator Conference, Rome, Italy, June 7–11, 1988, 532–534

[10] K.O. Thielheim, Particle Acceleration in Spherical Wave Fields, Pro. IAU Coll. 128, Lagow, Poland, June 17 - 23, 1989, 109 - 111.

Neutrinos from Active Galactic Nuclei

Victor J. Stenger

Department of Physics and Astronomy, University of Hawaii
2505 Correa Rd., Honolulu, Hawaii 96822, USA

Abstract

Active galactic nuclei (AGNs) are predicted to be intense sources of neutrinos of energies as high as 1000 TeV. The combined neutrinos from all AGNs are estimated to produce a flux of muons in the earth that exceeds the cosmic ray background for muon energies above 300 GeV. The DUMAND II array, currently under construction, would see thousands of these events per year. Such observations would not only provide a glimpse of the interior of active galaxies, but allow the study of neutrino interactions at energies two orders of magnitude higher than available at accelerators. At these energies, attenuation in the earth is appreciable and the core of the earth could be seen.

1. INTRODUCTION

In a recent calculation, Stecker, Done, Salamon and Sommers predict that the flux of high-energy neutrinos emitted by active galactic nuclei (AGNs) will be so high as to be detectable in underground and underwater experiments currently either operating or under construction.[1]

In this model, protons are accelerated in the vicinity of a massive black hole at the center of the galaxy – possibly by accretion shock, although the precise acceleration mechanism is not critical. While the model itself is not new, recent UV and X-ray data have made it possible to make more reliable quantitative predictions. The accelerated protons interact with the photon field surrounding the black hole. These photons are thought to be responsible for the observed "40 eV bump" characteristic to many AGN spectra. At an energy of about 10^7 GeV, protons interacting with 40 eV photons will produce the resonance reactions:

$$
\begin{array}{l}
p + \gamma \rightarrow \Delta^{+} \rightarrow \pi^{+}\ n \\
\qquad\qquad\qquad \downarrow\!\!\rightarrow \mu^{+}\ \nu_{e}\ \bar{\nu}_{\mu} \\
\qquad\qquad \rightarrow \pi^{o}\ p \\
\qquad\qquad\quad \downarrow\!\!\rightarrow 2\gamma
\end{array} \tag{1}
$$

which will be expected to dominate. The γ-rays do not escape the photon field, but interact and eventually degrade in energy to produce the observed X-rays. Thus existing very high energy γ-ray telescopes would not be expected to see a signal, with only the neutrinos escaping from the central regions of the AGN.

The observation of neutrinos from active galactic nuclei would not only provide unique information on the nature of the most powerful energy sources in the universe, but comprise a sample of neutrino interactions with matter at energies at least two orders of magnitude higher than is available with current particle accelerators. Furthermore, neutrinos of such great energy will experience significant attenuation in the earth, making possible a measurement of the earth's density profile and the direct observation of the earth's core.

Here I present a calculation of the flux of muons that would exist underground from AGN neutrinos at the predicted flux levels, and the energy spectrum and zenith angle distribution expected for these muons. Some remarks are also made about their detectability in the Deep Underwater Muon and Neutrino Detector (DUMAND II) currently under construction.

2. CALCULATION OF THE MUON FLUX

Suppose a muon detector is located at a depth d below the surface of the earth. Consider a ν_{μ} of energy E_{ν} directed toward the detector that travels a distance x through the earth and then interacts. Let $Z = \int \rho dx$ be the column density of matter traversed by the neutrino, where ρ is the matter density. If σ_T is the total $\nu_{\mu}N$ cross section, including neutral currents, then the neutrino flux hitting the earth will be attenuated prior to the interaction point by a factor $\exp(-N_A \sigma_T Z)$, where N_A is Avogadro's number.

If the interaction is charged current, a muon of energy $(1-y)E_{\nu}$ will be produced, where y is the conventional Bjorken variable. Let E_{μ} be the muon energy at the detector, and R_{μ} be the mean range of a muon that starts with energy $(1-y)E_{\nu}$ and ends with E_{μ}. The average flux of muons of energy E_{μ} per unit energy passing through the detector at and angle θ with respect to the zenith can then be obtained from the numerical integration,

$$F_\mu(E_\mu,\theta) = N_A \int_{E_\mu}^{\infty} F_\nu(E_\nu)\, dE_\nu \int_0^{1-\frac{E_\mu}{E_\nu}} R_\mu[E_\mu,(1-y)E_\nu]\; e^{-N_A \sigma_T Z(\theta)} \frac{d\sigma_\nu}{dy}\, dy \qquad (2)$$

where $F_\nu(E_\nu)$ is the differential flux of neutrinos hitting the earth and $d\sigma_\nu/dy$ is the differential cross section for $\nu_\mu N$ charged current interactions.

Note that the column density Z is a function of θ, which leads to a zenith angle dependence of the flux even when the neutrino source distribution is isotropic. The exponential technically remains inside the integral over Bjorken y since Z depends on the muon range R_μ, which depends on y.

Most underground and underwater muon detectors look for neutrinos below the horizon, where cosmic ray muons are negligible and the only known source of a signal is neutrinos. Deeper detectors, such as DUMAND, can also look somewhat above the horizon, so the zenith angle range $-1 < \cos\theta < 0.2$ has been considered. In this calculation, water density has been used above the horizon. Below the horizon, the earth density profile as a function of depth in the earth has been taken from Stacey.[2]

As already noted, the energy of neutrinos from AGNs predicted by the model are orders of magnitude higher than available from the highest energy particle accelerators on earth. Thus any estimate of νN interaction rates must necessarily be based on theory. While the Standard Model enables a fairly reliable extrapolation to higher energies, some uncertainties result from the unknown details of quark and gluon structure functions. The cross section formulas used here are those calculated by Reno and Quigg.[3] The actual observation of neutrino events from AGNs may thus provide information on very high energy neutrino interactions.

3. MUON SPECTRA AND ANGULAR DISTRIBUTIONS

The integral spectrum of muons implied by the flux given in Ref. 1 for the sum of all AGNs, integrated over the zenith angle θ, is shown in Fig. 1. This is compared with the muons expected from neutrinos produced by primary cosmic rays hitting the atmosphere. We see that the spectral shape for AGN muons is flat almost out to 10^4 GeV. Half of the muons hitting the detector have energies above 3–4x10^4 GeV. Atmospheric neutrinos, on the other hand, produce a sharply falling muon energy spectrum.

The muon energy spectra are essentially independent of zenith angle (not shown). The zenith angle dependence of the muon flux is given in Fig. 2 for muon thresholds of 1 GeV and 10 TeV, again for both AGNs and atmospheric neutrinos. The distributions are clearly not sensitive to the energy spectrum, and the extreme flatness of the AGN spectrum compared to atmospheric neutrinos is again in evidence. The non-isotropic nature of AGN muons is a direct result of the attenuation in the earth of very high energy neutrinos.

Fig. 2 is the zenith angle distribution expected from the sum of all AGNs in the Stecker et al. calculation. The distributions expected from two point sources, the Seyfert galaxy NGC4151 and the quasar 3C273 are presented in Fig. 3, again using the neutrino spectra predicted in Ref. 1. The harder spectrum of the quasar results in even greater attenuation at large zenith angles, where the neutrinos must pass through more matter to get near the detector.

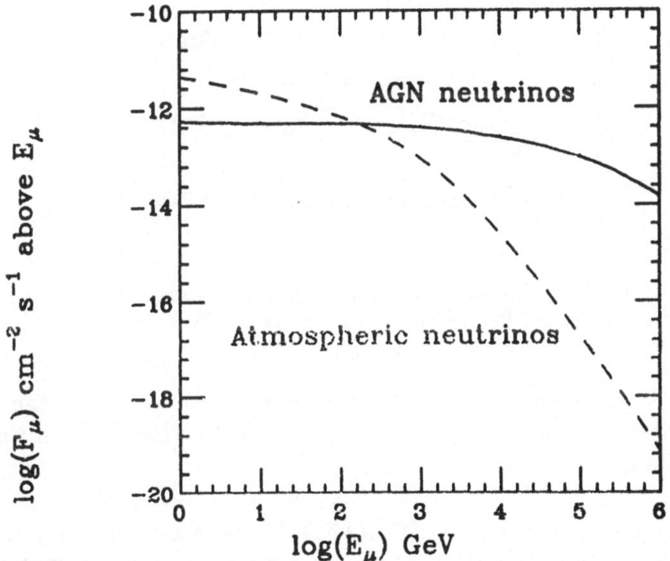

Fig. 1. Integral muon energy spectrum for a detector at a depth of 4.5 km, integrated over all zenith angles, for neutrinos from Active galactic nuclei compared with atmospheric neutrinos.

4. CAPABILITY OF DUMAND

The second phase DUMAND array now under construction for deployment in the ocean off the Big Island of Hawaii,[4] is ideally suited for the detection of neutrinos from active galactic nuclei. The effective detection area of DUMAND II as a function of muon energy is shown in Fig. 4. Folding in the expected E^{-2} neutrino energy spectrum from binary pulsar point source candidates, such as Cygnus X-3, gives an effective area of 20,000 m^2. The predicted AGN spectrum is even flatter, giving an effective area of 28,000 m^2.

The mean energy of muons in the first case is a few TeV, compared to about 15 TeV for AGNs. This also leads to a significant improvement in muon pointing accuracy, from a median error of about 1° for binary pulsar sources to 0.5^o for AGNs.

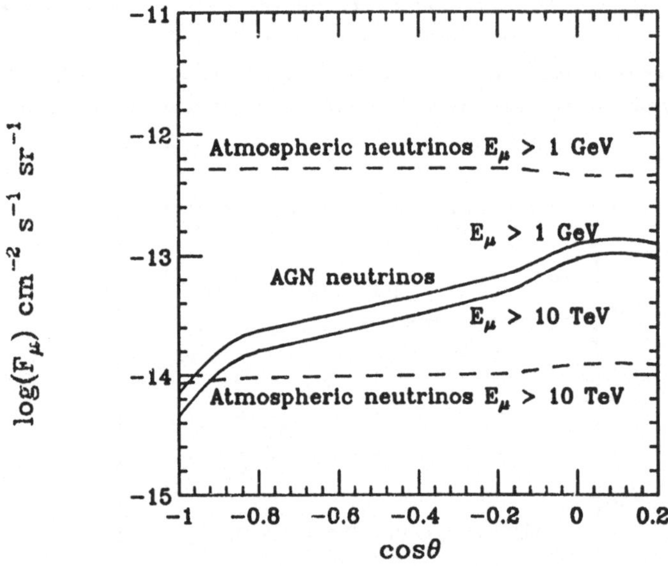

Fig. 2. Distributions in $\cos\theta$, where θ is the zenith angle of the muon, for AGN and atmospheric neutrino events, for two muon energy thresholds.

The event rates predicted for DUMAND II are 4,400 per year for the sum of all AGNs, 2 per year for NGC4151, and 0.2 per year for 3C273. It seems that the detection of point source AGNs will have to await DUMAND III. However, the event rate from the sum of all AGNs is substantial and easily detectable.

DUMAND II will be capable of detecting point sources of high energy neutrinos that produce 5–10 events per year. This is possible because of the low background of the detector, less than one event per year in each 1° pixel from atmospheric neutrinos or various backgrounds that can fake neutrino events. Diffuse sources, such as the sum of AGNs, must be distinguished from the atmospheric background. However, this can be done by measuring the dE/dx of muons traversing the array. When a muon has several TeV or greater, nuclear scattering and radiative processes produce Cherenkov light that is proportional to muon energy. Monte Carlo calculations have shown that an AGN signal can be distinguished from atmospheric neutrinos in DUMAND II at a signal–to–noise ratio of about 10% or greater.[5]

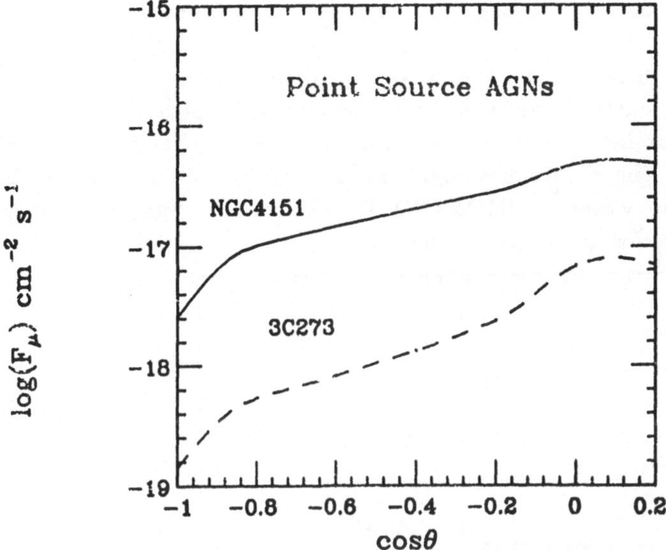

Fig. 3. Distributions in $\cos\theta$, where θ is the zenith angle of the muon, for neutrinos from two AGN point sources.

5. CONCLUSIONS

If the Stecker et al. predictions of the very high energy neutrino flux from active galaxies are correct within an order-of-magnitude, they should be detectable with DUMAND II. However only the net flux from all AGNs is significant. Point sources appear to be currently out of reach.

At the predicted rate, current experiments, in particular IMB and Kamiokande, might have been expected to see the first hints. Since neither positive nor negative reports from these experiments have appeared at this writing, I assume the question is still open. However, note that the flux does not exceed the atmospheric flux until one goes above about 300 GeV in muon energy. Thus experiments seeking this effect must make a high energy cut. DUMAND II will be able to do this, since it observes muons over a path length of 100 m or more and can estimate energy, at least crudely, from dE/dx.

The observation of AGN neutrinos would be remarkable in several respects. Not only would they provide a peek into the centers of the most energetic objects in the universe, but they would represent neutrino interactions at energies two orders of magnitude higher than those from earthbound accelerators. At these energies, the attenuation in the earth is substantial and this can be used to both study the very high energy νN cross section and do neutrino tomography of the earth.

This paper has focussed on the muons produced by ν_μ charged current interactions in the earth. As has been previously reported, the Glashow resonance can be searched for in the reaction $\bar{\nu}_e e^- \rightarrow W^- \rightarrow$hadronic cascade at 6.4 PeV, either via the Cherenkov light produced by the cascade or by an acoustic signal in he water.[6] DUMAND II will contain hydrophones in the array that will listen for messages of this nature.

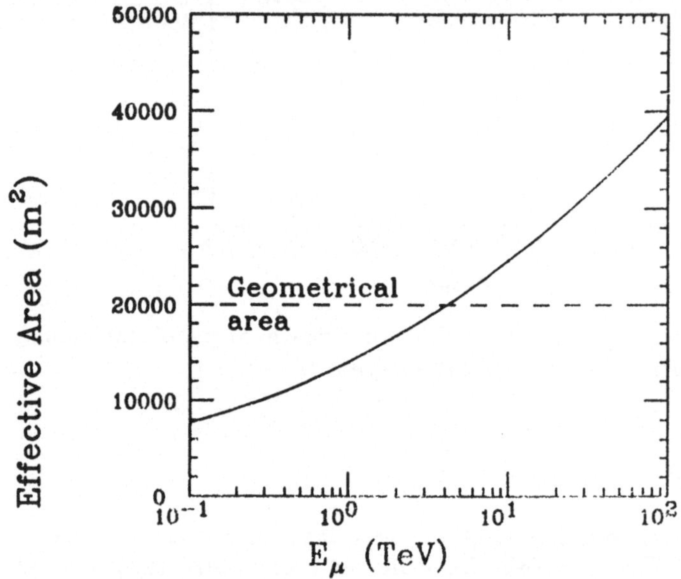

Fig. 4. The effective detection area of DUMAND as a function of muon energy. The geometrical area is 20,000 m^2.

6. REFERENCES

1 F.W. Stecker, C. Done. M.H. Salamon, P. Sommers, Phys. Rev. Lett. **66** (1991) 2697.

2 Frank P. Stacey, "Physics of the Earth, John Wiley, New York, 1976.

3 C. Quigg, M.H. Reno, T.P. Walker, Phys. Rev. Lett. **57** (1986) 774; M.H. Reno and C. Quigg, Phys. Rev. **D 37** (1988) 657.

4 D. Samm, these Proceedings.

5 A. Okada, private communication.

6. J.G. Learned and T. Stanev, Proceedings of the 3rd Venice Meeting on Neutrino telescopes, February 1991.

Characteristics of Cosmological Evolution and Large Scale Structures in the Cosmos [1, 2]

Peter Minkowski
Institute for Theoretical Physics
University of Bern
Sidlerstr. 5, CH-3012 Bern, Switzerland

March 1992

Abstract

Recent findings of large scale structures in the cosmos (way beyond 10 Mpc) are analyzed in the light of the characteristic densities, temperatures, mass and time scales governing the assumed (standard) cosmological evelution.

---

[1]Work supported in part by Schweizerischer Nationalfonds

[2]Contribution presented at the Int. Conference "Trends in Astroparticle Physics" 10-12 October 1991, Aachen, Germany

1 Structures and Cosmological Embedding

Large scale structures [1] [2], have been revealed to extend in a seemingly selfsimilar pattern beyond the distance of 100 Mpc. This is illustrated in figures 1 and 2.

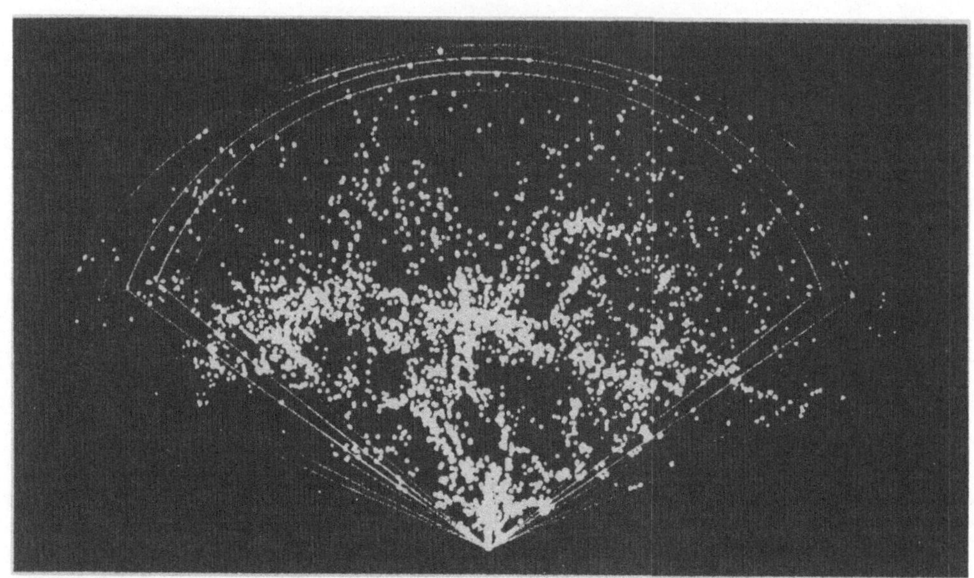

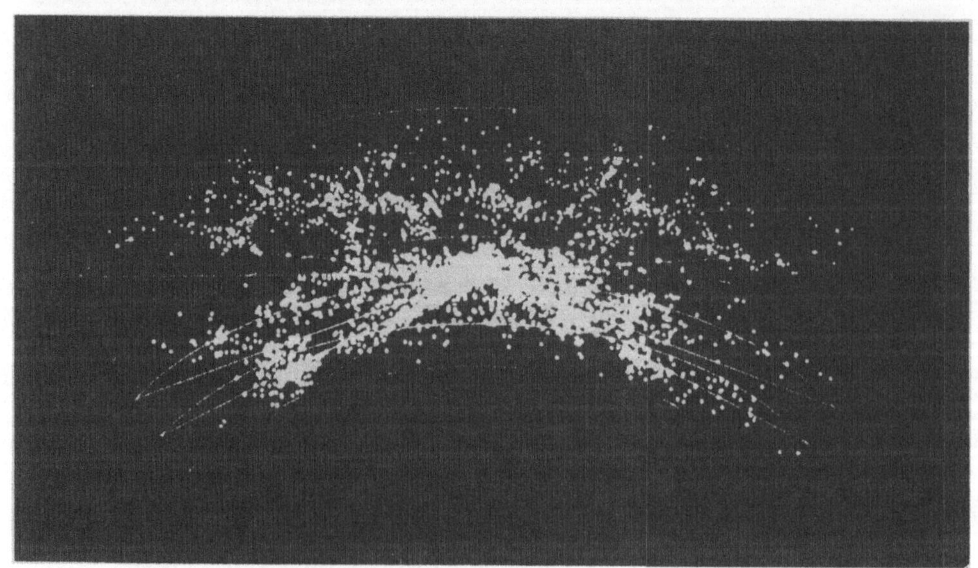

Figure 1: The four slices of the Center for Astrophysics redshift survey show how galaxies cluster on large scales. The separate slice with yellow dots (uppermost slice in the bottom view) covers northern declinations from $8^{0}.5$ to $14^{0}.5$. In this slice the 'Great Wall' structure is revealed by the clustering of (yellow) dots. The adjacent slices include declinations from $26^{0}.5$ to $44^{0}.5$. The top view shows the four slices superimposed, where the 'Great Wall' structure extends more or less horizontally all accross. Reproduced from [3] .

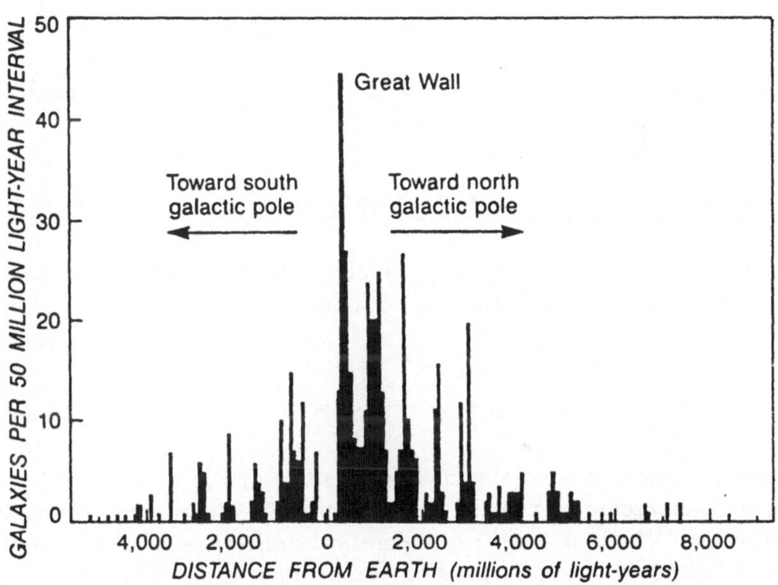

Figure 2: Data from the pencil beam serveys of ref. [2] , using the Anglo-Aystralian-Telescope for the south galactic pole pencil and the Kitt Peak Telescope for the north galactic pole pencil. The plot shows density of galaxies (number of galaxies per 5.10^{7} ly) versus distance. The average spacing of the prominent spikes is of the order of 130 Mpc $\approx 4.24\ 10^{8}$ ly . Reproduced from [4] .

On the other hand observation of the spectral density of the background microwave radiation by the Cosmic Background Explorer (COBE) satellite [5] and by antennas operated during a rocket flight [6], have revealed a remarkably homogeneous black body radiation with the temperature

$$T = 2.735 \pm 0.060\ ^{o} K$$

The dipolar distortion effect due to the proper velocity of the Milky Way relative to the homogeneous background amounts to a temperature variation of $\approx \pm 3mK$ and is easily accounted for [7] , [8] .
The discussion of promordial abundances of light elements has been exstended to include ^{7}Li : (H , D , ^{3} He , ^{4} He , ^{7} Li) [9] - [11] . Concordance with all these abundances is achieved for the ratio of (all) nucleons to photons today
$\eta = \eta_{[10]} .10^{-10}$ in the range

$$2.5 \lesssim \eta_{[10]} \lesssim 6 \;;\; \eta_{[10]} = 4 \;\leftrightarrow\; \Omega_{0b} h^2_{50} = 0.06$$
$$\Omega_{0b} = \varrho_b / \varrho_{cr}\ (today) \tag{1}$$

In eq. 1 $\varrho_{b,cr}$ denote baryon - and critical energy density respectively. In the following ϱ denotes the total energy density.

2 Cosmological Evolution

We thus redraw an outline of the cosmological evolution based on the following set of characteristic input quantities valid today :

$$H_0 = h_{50} (50\ km / sec) / Mpc \;;\; h_{50} = 1$$
$$\Omega_0 = 0.2 (= 0.2 \pm 0.1) [12] \;;\; \Omega_0 = \varrho / \varrho_{cr}\ (today) \tag{2}$$

We do not conclude from the energy densities in eqs. (1) and (2) that a significant nonbaryonic part of the energy density is established, although such a part is suggested. We do not see any theoretical preference for Euclidean time sections $\Omega = 1$, although a mass excess on scales much larger than 20 Mpc could lead to this value.
Todays distance scale is set by the Hubble distance :

$$R^H_0 (50) = c / H_0 \;;\; R^H_0 (50) = \begin{cases} 1.85 & .10^{28} & cm \\ 1.96 & .10^{10} & ly \\ 0.938 & .10^{33} & \hbar c / (1eV) \\ 1.15 & .10^{61} & l_{Pl} \end{cases} \tag{3}$$

Planck units are given in the appendix.

We shall only consider the hyperbolic case in the following. We introduce function characterizing the energy density relative to the critical one :

$$\delta = \Omega^{-1} - 1 = \frac{\varrho_{cr} - \varrho}{\varrho} \quad ; \quad \delta_0 = 4$$

$$\varrho_{cr} = \frac{3}{8 \pi G_N} H^2 = \frac{3}{8 \pi} \frac{m_{Pl}}{l_{Pl} (R^H)^2} \tag{4}$$

Todays critical density corresponding to the initial data in eqs. (2 , 3) is :

$$\varrho_{0\,cr} = \begin{cases} 4.69 \quad .10^{-30} & g \,/\, cm^3 \\ 2.63 \quad .10^{3} & eV \,/\, (c^2 \, cm^3) \\ 2.02 \quad .10^{-11} & eV^4 \,/\, (\hbar^3 \, c^5) \\ 1.33 \quad .10^{8} & \hbar \, c \,/\, cm^4 \end{cases} \tag{5}$$

The curvature radius is thus given by :

$$R = (1 - \Omega)^{-1/2} R^H \quad ; \quad R_0 = \begin{cases} 2.07 \quad .10^{28} & cm \\ 2.19 \quad .10^{10} & ly \end{cases} \tag{6}$$

<u>Todays relic photons and neutrinos</u>

The initial data in eqs. (2) and (3) are extended to the temperature of the background microwave radiation :

$$T_{0\gamma} = \begin{cases} 2.75 & {}^{o}K \\ 2.37 \quad .10^{-4} & eV \end{cases}$$

$$T_{0\nu} = \left(\frac{4}{11} \right)^{1/3} T_{0\gamma} = \begin{cases} 1.96 & {}^{o}K \\ 1.69 \quad .10^{-4} & eV \end{cases} \tag{7}$$

Based on the data in eq. (7) the photon energy density today is :

$$\varrho_\gamma = \frac{\pi^2}{15} T_\gamma^4 \quad ; \quad \varrho_{0\gamma} = \begin{cases} 4.81 \quad .10^{-34} & g \,/\, cm^3 \\ 0.27 & eV \,/\, (c^2 \, cm^3) \\ 2.08 \quad .10^{-15} & eV^4 \,/\, (\hbar^3 \, c^5) \end{cases} \tag{8}$$

$$\Omega_\gamma = \varrho_\gamma / \varrho_{cr} \quad ; \quad \Omega_{0\gamma} = 1.03 \,.10^{-4}$$

The number density of photons becomes :

$$\varrho_{n\gamma} = \frac{2\zeta(3)}{\pi^2} T_\gamma^3 \quad ; \quad \varrho_{n0\gamma} = \begin{cases} 422 & / \, cm^3 \\ 3.24 \; .10^{-12} & eV^3 / (\hbar c)^3 \end{cases} \tag{9}$$

The corresponding quantities for a massless neutrino flavor are :

$$\begin{aligned} \varrho_\nu &= \frac{7}{8} \frac{\pi^2}{15} T_\nu^4 \quad ; \quad \varrho_{0\nu} = 2.33 \; .10^{-5} \, \varrho_{0\,cr} \\ \varrho_{n\nu} &= \frac{2\eta(3)}{\pi^2} T_\nu^3 \quad ; \quad \varrho_{n0\nu} = \frac{3}{11} \varrho_{n0\gamma} = 115 \, / \, cm^3 \end{aligned} \tag{10}$$

Irrespective of the masses of three light neutrino flavors the inequality

$$\varrho_{0\gamma} + 3\,\varrho_{0\nu} \geq 1.73 \; .10^{-4} \, \varrho_{0\,cr} \tag{11}$$

holds.

<u>Matter dominated era</u>

The evolution equations are :

$$\begin{aligned} \dot{R}^2 &= 1 + \frac{R_0^*}{R} \quad ; \quad R_0^* = \frac{1}{\delta_0} R_0 = \frac{1}{\delta_0} (1 - \Omega_0)^{-1/2} R_0^H \\ R(t) &= R_0^* \, r(t) \quad ; \quad t = R_0^* \, \tau / c \\ r(t) &= \sinh^2(\chi/2) = \delta(t) \quad ; \quad \tau = \frac{1}{2} [\sinh(\chi) - \chi] \\ R_0^* &= 0.517 \; .10^{28} \, cm \quad ; \quad t_0 = 1.66 \; .10^{10} \, y \end{aligned} \tag{12}$$

In the matter dominated era, the quantity δ in eq. (4) is proportional to the scale parameter R :

$$\frac{\delta(R_2)}{\delta(R_1)} = \frac{R_2}{R_1} \tag{13}$$

Transition from radiation to matter dominated era

We assume here that neutrinos are massive and nonrelativistic during most of the matter dominated. Let t_{-1} denote the time of transition between radiation and and matter dominated eras. It follows :

$$1 + z_{-1} = \frac{R_0}{R_{-1}} = \frac{\Omega_0}{\Omega_{0\gamma}} \approx 2000 \tag{14}$$

We note that the decoupling of photons and hydrogen recombination at $1 + z \approx 1000$ occurs in the present scenario well within the matter dominated era. The characteristic quantities at t_{-1} are :

$$\chi_{-1} = 0.0894 \quad ; \quad \tau_{-1} = 0.60 \,.10^{-4} \quad ; \quad t_{-1} = 3.26 \,.10^{5}\, y$$

$$R_{-1} = \begin{cases} 1.04 \,.10^{25} \; cm \\ 1.10 \,.10^{7} \; ly \end{cases} \quad ; \quad \delta_{-1} = 1 / 500$$

$$T_{-1\gamma} = \begin{cases} 5500 \; {}^{\circ}K \\ 0.474 \; eV \end{cases} \quad ; \quad \varrho_{-1} = \begin{cases} 7.50 \,.10^{-21} \; g / cm^3 \\ 3.23 \,.10^{-2} \; eV^4 / (\hbar^3 c^5) \end{cases} \tag{15}$$

t_{-1} in eq. (15) does not have any absolute meaning. It only determines the time elapsed between the transition time from radiation to matter dominated eras :

$$\Delta t_{[0-1]} = t_0 - t_{-1} = 1.66 \,.10^{10}\, y \tag{16}$$

Radiation dominated era

$$\dot{R}^2 = 1 + \left(\frac{R^*_{-1}}{R} \right)^2 \quad ; \quad R^*_{-1} = \frac{1}{\sqrt{\delta_{-1}}} R_{-1}$$

$$R(t) = R^*_{-1} \tilde{r}(t) \quad ; \quad t = R^*_{-1} \tilde{\tau} / c$$

$$\tilde{r}(t) = \sinh(\tilde{\chi}) = (\delta)^{1/2} \quad ; \quad \tilde{\tau} = \cosh(\tilde{\chi}) - 1$$

$$R^*_{-1} = \begin{cases} 2.31 \,.10^{26} \; cm \\ 2.45 \,.10^{8} \; ly \end{cases} \quad ; \quad \tilde{t}_{-1} = 2.45 \,.10^{5}\, y \quad ; \quad \tilde{\tau}_{-1} = 1.00 \,.10^{-3} \tag{17}$$

Going further backward in time in the radiation dominated era we are dealing with increasingly small values of $\tilde{r}$, $\tilde{\chi}$, $\tilde{\tau}$, giving rise to the approximate relation :

$$\tilde{\tau} \approx \tilde{r}^{2} / 2 \quad ; \tag{18}$$

In the radiation dominated era, the quantity δ in eq. (4) is proportional to the square of the scale parameter R :

$$\frac{\delta (R_2)}{\delta (R_1)} = \left(\frac{R_2}{R_1} \right)^2 \approx \frac{t_2}{t_1} \tag{19}$$

The final phase of nucleosynthesis corresponds to

$$T_{-2\,\gamma} \approx 0.1\ MeV \downarrow T_{-2\,\nu}$$

At time t_{-2} the characteristic quantities are :

$$R_{-2} = \begin{cases} 4.91 & .10^{19} \quad cm \\ 51.9 & \quad ly \end{cases} \quad ; \quad t_{-2} = 174\ sec$$
$$\delta_{-2} = 4.49\ .10^{-14} \tag{20}$$

With $e^{\pm}$, $\overset{(-)}{\nu}_{e\,,\,\mu\,,\,\tau}$, γ (and nucleons) in equilibrium we have at $T_{-3\,\gamma} = T_{-3\,\nu} \approx 1\ MeV$:

$$R_{-3} = \begin{cases} 3.50 & .10^{18} \quad cm \\ 3.70 & \quad ly \end{cases} \quad ; \quad t_{-3} = 0.88\ sec$$
$$T_{-3} = 1\ MeV \quad ; \quad \delta_{-3} = 2.29\ .10^{-16} \tag{21}$$

The onset of nucleosynthesis corresponds to the point : $T_{-4} \approx 10\ MeV$ where the quantities in eqs. (20) , (21) become :

$$R_{-4} = \begin{cases} 3.50 & .10^{17} \quad cm \\ 1.17 & .10^{7} \quad lsec \\ 1.77 & .10^{28} \quad \hbar\, c \,/\, MeV \end{cases} \quad ; \quad t_{-4} = 0.88\ .10^{-2}\ sec$$
$$T_{-4} = 10\ MeV \quad ; \quad R_{-4}\, T_{-4} = 1.77\ .10^{29} \quad ; \quad \delta_{-4} = 2.29\ .10^{-18} \tag{22}$$

The small values of δ simply reflect the large ratios of scale factors in eqs. (19) and (13) .

The product $R\,T$ of temperature and scale factor stays very roughly constant during the earlier evolution in the radiation dominated era, even though phase transitions (quark gluon plasma, weak phase transition, ...) and nonequilibrium conditions do generate some changes.
The large value of $R\,T$ in eq. (22) is suggestive of a preceding inflationary phase.

Transition from inflationary to radiation dominated era

We assume the transition temperature $T_{-5} \approx 10^{16}\,GeV$ between inflationary and radiation dominated eras. This choice shall serve here just to fix ideas :

$$R_{-5} = \begin{cases} 0.350 & cm \\ 1.17\ .10^{-11} & lsec \\ 1.77\ .10^{13} & \hbar\, c \,/\, GeV \end{cases} \quad ; \quad t_{-5} = 0.88\ .10^{-38}\ sec$$

$$T_{-5} = 10^{16}\,GeV \quad ; \quad R_{-5}\,T_{-5} \approx 1.77\ .10^{29} \quad ; \quad \delta_{-5} = 2.29\ .10^{-54} \tag{23}$$

In the radiation dominated era pressure and energy density were in the relation $p = \varrho \,/\, 3$. Inflation sets in whence the pressure becomes (sufficiently) negative. We assume that at t_{-5} the pressure changes discontinuously and becomes

$$p = -\,\varrho \quad \longrightarrow \quad \dot{\varrho} = 0 \tag{24}$$

t_{-5} in eq. (23) again does not have any absolute meaning. It only determines the duration of the radiation dominated era :

$$\Delta\, t_{[-1-5]} = \tilde{t}_{-1} - t_{-5} = 2.45\ .10^{5}\ y \tag{25}$$

Inflationary era

Eqs. (12) and (17) take the form

$$\dot{R}^{2} = 1 + \left(\frac{R}{R^{*}_{-5}} \right)^{2} \quad ; \quad R^{*}_{-5} = \sqrt{\delta_{-5}}\, R_{-5}$$

$$R(t) = R^{*}_{-5}\, \hat{r}(t) \quad ; \quad t = R^{*}_{-5}\, \hat{\tau} / c$$

$$\hat{r}(t) = \sinh(\hat{\tau}) = (\delta)^{-1/2} \tag{26}$$

$$R^{*}_{-5} = \begin{cases} 5.30 \; .10^{-28} \; cm \\ 1.77 \; .10^{-38} \; lsec \\ 2.68 \; .10^{-14} \; \hbar c / GeV \end{cases} \quad ; \quad \hat{t}_{-5} = 1.11 \; .10^{-36} \; sec$$

$$\hat{\tau}_{-5} = 62.45$$

Eqs. (13) and (19) become

$$\frac{\delta(R_2)}{\delta(R_1)} = \left(\frac{R_1}{R_2} \right)^{2} \tag{27}$$

It is instructive to determine the time t_{-6} early in the inflationary era, such that

$$\frac{R_{-5}}{R_{-6}} = R_{-4}\, T_{-4} = 1.77 \; .10^{29} \quad \rightarrow$$

$$\sinh(\hat{\tau}_{-6}) = \sinh(\hat{\tau}_{-5}) / (R_{-4}\, T_{-4}) = 3.73 \; .10^{-3}$$

$$\hat{\tau}_{-6} \approx \sinh(\hat{\tau}_{-6}) = 3.73 \; .10^{-3} \tag{28}$$

$$R_{-6} = 10^{-16} \; \hbar c / GeV \quad ; \quad t_{-6} = \begin{cases} 0.664 \; .10^{-40} \; sec \\ 1.00 \; .10^{-16} \; \hbar / GeV \end{cases}$$

It is rewarding to see, that indeed the time and extension scales become of comparable order of magnitude.

Since at t_{-6} , $\hat{\tau}$ is much smaller than 1, for earlier times the approximate relation

$$R(t) \approx c t \tag{29}$$

holds. At such microscopic time and extension scales the averaged semiclassical Einstein equations necessarily fail. Nevertheless we note that for $t \rightarrow 0$ within the inflationary era - for the hyperbolic case - the gravitational interaction becomes negligible and thus a qualitatively correct limiting behaviour cannot be ruled out. This limit generates an infinite event horizon, which eliminates - with the seservations just stated - the paradox of acausal homogeneity of the evolution at some time in the late stages.

Appendix

Planck units

$$\begin{aligned}
m_{Pl} &= \left(\frac{\hbar c}{G_N} \right)^{1/2} &&= \begin{cases} 1.2211.10^{19}\ GeV / c^2 \\ 2.1768.10^{-5}\ g \end{cases} \\
l_{Pl} &= \frac{\hbar}{m_{Pl}\, c} &&= 1.61610.10^{-33}\ cm \\
\varrho_{n,Pl} &= \frac{1}{l_{Pl}^3} &&= 2.37\ .10^{98} / cm^3 \\
\varrho_{Pl} &= \frac{m_{Pl}}{l_{Pl}^3} &&= 5.16\ .10^{93}\ g / cm^3
\end{aligned} \tag{30}$$

References

[1] M. Geller and J. Huchra, *Science* **246** (1989) 897

[2] R. Broadhurst, R. S. Ellis, D. C. Koo and A. S. Szalay, *Nature* **343** (1990) 726

[3] M. J. Geller and J. P. Huchra , *Sky and Telescope* **82** (1991) 134

[4] D. N. Schramm, *Sky and Telescope* **82** (1991) 140

[5] J. C. Mather et al, *Ap. J.* **354** (1990) L37

[6] H. Gush, M. Halpern and E. Wishnow, *Phys. Rev. Lett.* **65** (1990) 537

[7] A. C. S. Readhead et al, *Ap. J.* **346** (1989) 566

[8] R. B. Partridge, *Rep. Prog. Phys.* **51** (1988) 647

[9] T. Walker, G. Steigman, D. N. Schramm, K. Olive and H. Kang, *Ap. J.* **376** (1991) 51

[10] D. N. Schramm, Fermilab preprint FERMILAB-Conf-91/266-A, 1991

[11] E. W. Kolb and M. S. Turner, The Early Universe and The Early Universe: Reprints, *Addison-Wesley Inc.* , New York , 1988

[12] A. Sandage and G. A. Tammann, The Dynamical Parameters of the Expanding Universe as they Constrain Homogeneous World Models with and without Λ , in [11] : Reprints, p. 15.

The Galaxy Distribution as a Voronoi Foam

Vincent Icke

Sterrewacht Leiden, Postbus 9513, 2300 RA Leiden, The Netherlands

Abstract

Underdense regions in the cosmic mass distribution expand with respect to the background and become more and more spherical. This physical mechanism produces a geometrical model, *Voronoi foam*, for the asymptotic distribution ("skeleton") of the cosmic mass on 10-200 Mpc scales. Voronoi foam is a packing of polyhedral cells uniquely determined by the sites of the initial underdense regions. The walls are identified as the pancakes, the edges as the filaments, and the vertices as the clusters in the galaxy distribution.

The Voronoi tessellation is a good model for the galaxy distribution on megaparsec scales. It does not describe that distribution on small scales, but it forms a useful complement to N-body techniques, which have a low dynamic range in wavenumber space, and which suffer from sampling noise and boundary effects on very large scales. The Voronoi model enables one to study, in arbitrary detail, what the consequences of a cellular galaxy distribution are.

The correlation function of the vertices in the Voronoi tessellation is a power law with a slope and amplitude totally in accordance with the observed cluster-cluster correlation, indicating that this function could be due to cellular geometry of the galaxy distribution on megaparsec scales, rather than to the dynamics of these structures.

The Voronoi model has one failing in common with other theories of cosmic structure formation: the implied fluctuations of the microwave background are too large, unless the usual fudging is applied. For a complete review of this subject, see Icke & Van de Weygaert (1991).

1. INTRODUCTION

The observations of megaparsec structure in the Universe (usually called by the nondescript name "large scale" structure in the literature) date back to the work of Shapley & Ames (1932), whose catalogue of bright stellar systems showed the dramatic excess of galaxies in some regions of the sky, notably the Virgo Cluster (see Oort 1983). Even in those early days, it was evident that the distribution of luminous matter on megaparsec scales is distinctly non-Poissonian. Interpretation of this distribution was initiated by Rubin (1954) and Limber (1953, 1954), who introduced the two-point correlation function, and by Neymann & Scott (1955), who used the angular autocorrelation function of the Lick survey counts, but it was not until Totsuji & Kihara (1969) and Peebles (1980, and references therein) that the two-point correlation function of galaxies was reliably estimated.

Oort (1970) emphasised the importance of larger structures because these, due to their long evolutionary timescales, can be expected to give information about the earliest times of structure formation. In keeping with this, Icke (1972, 1973) hypothesised that such large aggregates (at least as large as 15 Mpc) form before galaxies (a scenario which has been pioneered by Zel'dovich and his co-workers in the context of the "pancake" picture of galaxy formation), and he tried to find evidence for coherent objects on a scale of 15–20 Mpc by studying the distribution of galaxies in position-velocity space, a technique that is now known as "slicing" (Fig.1).

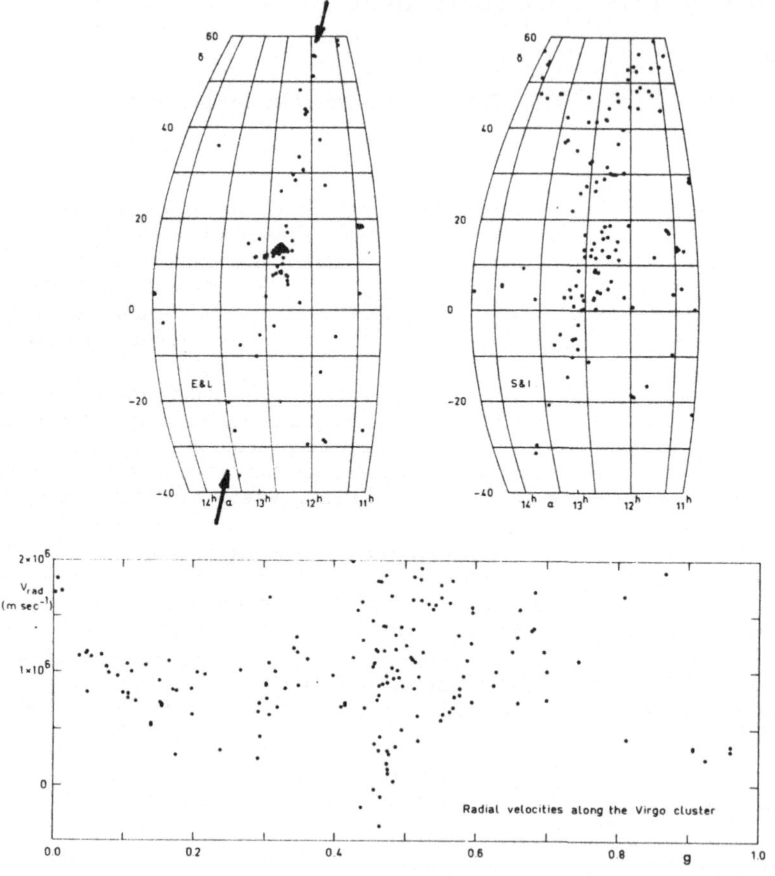

Figure 1. *A slice of the Universe in the constellation Virgo. Left: elliptical and lenticular galaxies. Right: spiral and irregular galaxies. Bottom: radial velocities of these galaxies, as a function of their position along the major axis of the cluster (indicated by the arrows). Today's larger scale surveys usually show such data in the form of pie-wedge diagrams (as in Fig.2). After Icke (1972).*

Icke's efforts to corroborate his hypothesis were largely unsuccessful, due to a conspicuous lack of good observations. Appreciable improvements of the data by Humason, Mayall, & Sandage (1956) were not published until the beginning of the 1980's, when several extensive galaxy redshift surveys became available (Sandage & Tammann 1981; Fisher & Tully 1981; Huchra *et al.* 1983). These redshift surveys provided evidence for the early suggestion by Einasto (1978; see also Einasto, Joêveer, & Saar 1980) of cell-like structures in the distribution of galaxies. The mapping of e.g. the Local Supercluster (Tully 1982), the Coma Supercluster (Gregory & Thompson 1978), and the enormous 140 Mpc long chain of galaxies forming the Perseus Supercluster (Giovanelli, Haynes, & Chincarini 1986) showed that galaxies cluster in pancake-like and filamentary structures, while the redshift surveys also revealed the existence of voids in the distribution of galaxies (Davis *et al.* 1982; Kirshner *et al.* 1981,1987).

These voids are enormous regions, tens of megaparsecs in extent, wherein few or no galaxies are found. In particular the discovery of the Boötes void (Kirshner *et al.* 1981) made clear how large these voids can be. Since 1986 the publication of new redshift slices from the second CfA redshift survey (De Lapparent, Geller, & Huchra 1986; see also Da Costa *et al.* 1988 for the southern sky) confirmed the existence of a cellular or sponge-like arrangement of galaxies. The recent discovery by the CfA group (Geller & Huchra 1989) of a $60 \times 170 \times 5h^{-1}$Mpc "Great Wall" of galaxies, and the discovery by Broadhurst *et al.* (1990) of a regular "spiky" redshift distribution of galaxies in a narrow pencil beam survey along the South and North galactic poles, are suggestive of cellular structures on scales as large as 100 Mpc.

The general impression from all these observations is a universe in which the galaxies are situated in walls (pancakes), denser filaments, and very dense nodes, forming a network which surrounds huge voids (Fig.2). Further details about the observed properties of the large scale galaxy distribution can be found in the reviews by Oort (1983), Geller (1988), Bahcall (1988) and Rood (1988).

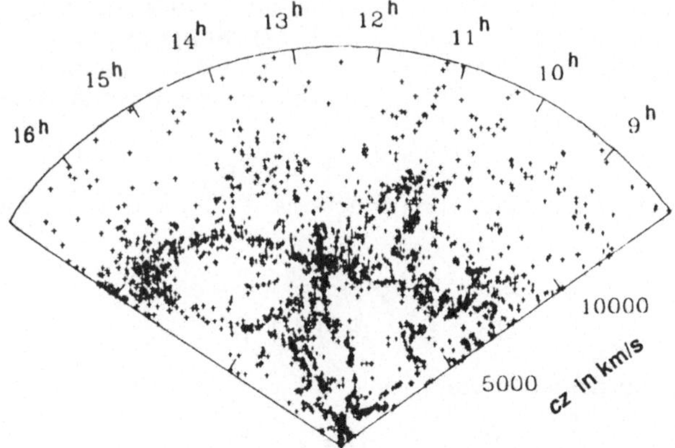

Figure 2. *A combination of CfA redshift survey slices; reproduced from Geller & Huchra (1989), with the kind permission of the American Association for the Advancement of Science (AAAS). See also De Lapparent, Geller, & Huchra (1986).*

2. THEORY

Most theories of the formation of structure on megaparsec scales are based on the gravitational instability scenario (e.g. Peebles 1980). Comprehensive reviews treating these theories were presented by Jones (1976), Efstathiou & Silk (1983), Szalay (1988), Shandarin & Zel'dovich (1989), Bond (1990), and Efstathiou (1990).

On scales comparable with the particle horizon, the formation of structure must be due to perturbations of a FRW-type universe (Lifschitz 1946), which is analytically difficult and not securely linked to observations. On small scales, galaxy formation is dominated by dissipative processes and is theoretically still quite intractable. But on intermediate scales, Newtonian gravitational instability should suffice for describing the formation of structure. Hydrodynamical details such as pressure effects are unlikely to be important in the progenitors of structures in the 10–500 Mpc regime, so we can restrict ourselves to a "dust" equation of state. The potential Φ near any point (x, y, z) of a selfgravitating medium can be written as

$$\Phi = \sum_{ijk} a_{ijk} x^i y^j z^k \ . \tag{2.1}$$

Near a density maximum, the leading terms are the quadratic ones, which, by a suitable orientation of Cartesian coordinates, can be written as

$$\Phi = Ax^2 + By^2 + Cz^2 + \cdots \ . \tag{2.2}$$

Neglecting terms of higher than second order, this is the potential of a homogeneous ellipsoid. That should be no surprise: the smallest closed contours in any topographical map are ellipses.

The collapse of high-density regions can thus be approximated by considering the homologous motion of ellipsoids. Consider, then, the Newtonian collapse of a homogeneous spheroid (Lin, Mestel, & Shu 1965; Icke 1972). Suppose that a particle of such a mass distribution were initially located at (a, b, c), and that at some later time t it had moved to the point $(aX(t), bY(t), cZ(t))$, then the density ρ would evolve according to

$$\rho(t) = \rho_0 / XYZ \ . \tag{2.3}$$

The equations of motion for the scaling functions X, Y, and Z are found as follows. The potential Φ obeys

$$\begin{aligned}\Phi &= k(\alpha x^2 + \beta y^2 + \gamma z^2) = \\ &= k(\alpha a^2 X^2 + \beta b^2 Y^2 + \gamma c^2 Z^2) \ ,\end{aligned} \tag{2.4}$$

and Poisson's Equation demands that

$$k(\alpha + \beta + \gamma) = 2\pi G \rho \ . \tag{2.5}$$

The components of the gravitational force are $-\partial\Phi/\partial x = -\frac{1}{X}\partial\Phi/\partial a$ *et cycl.*, so that the equations of motion become

$$-\frac{1}{X}\frac{d^2X}{dt^2} = 2\pi G \rho \alpha \ ; \quad -\frac{1}{Y}\frac{d^2Y}{dt^2} = 2\pi G \rho \beta \ ; \quad -\frac{1}{Z}\frac{d^2Z}{dt^2} = 2\pi G \rho \gamma \ , \tag{2.6}$$

where the functions α, β, and γ are defined as

$$\alpha = abc \int_0^\infty \frac{ds}{(a^2 + s)\Delta} \quad \text{et cycl.}, \tag{2.7}$$

$$\Delta^2 \equiv (a^2 + s)(b^2 + s)(c^2 + s) \ , \tag{2.8}$$

(*cf.* Chandrasekhar 1969, Ch. 3). Here a, b, and c are identified with the axes of the ellipsoid.

Now comes a crucial observation, first made by Lynden-Bell (1964): without loss of generality, one can order the axes according to $a > b > c$, in which case $\alpha < \beta < \gamma$, so that Eqs.(2.6) give

$$-\frac{1}{X}\frac{d^2X}{dt^2} < -\frac{1}{Y}\frac{d^2Y}{dt^2} < -\frac{1}{Z}\frac{d^2Z}{dt^2} \,. \tag{2.9}$$

Consequently, the axial ratios $a : b : c$ always increase with time, and *slight initial asphericities are amplified during the collapse.* This secular increase of aspherical perturbations provides an explanation for the pancake-like, and later filamentary, appearance of megaparsec structures (cf. Zel'dovich 1970, 1978). Note also that, for the contraction described, the velocities inside the ellipsoid are linear functions of position: *the collapse produces a Hubble-type velocity field.*

In order to avoid nonlinearities and other complications in high-density regions, we may view the development of structure in a selfgravitating pressure-free medium by considering the evolution of the *low*-density regions. These are the progenitors of the observed voids. The arguments presented above can still be applied, except that the sense of the final effect is reversed: because a void is effectively a region of negative density in a uniform background, the voids expand as the overdense regions collapse, while *slight asphericities decrease as the voids become larger* ("Bubble Theorem", Icke 1984). The proof holds strictly only on a non-expanding background, though this should be no objection for structures which are much smaller than the particle horizon. Because $|\delta\rho/\rho|$ does not exceed unity in a void, the approximation will remain good for a longer period, except, of course, near the outer parts of the voids, where the matter gets swept up.

According to the above, taking voids as the dominant dynamical component of the Universe, one may think of the megaparsec structure as a close packing of spheres of different sizes, out of which matter flows in a slightly super-Hubble expansion towards the interstices of the spheres. Thus, *the importance of the Bubble Theorem is that it provides a specific physical mechanism for producing the non-Poissonian matter distribution in the large scale Universe.*

3. VORONOI FOAM

Continuing the above argument, I can construct the "skeleton" of the mass distribution by considering the locus of points towards which the matter streams out of the voids. Suppose that some cosmic process produces a collection of regions where the density is slightly less than average (the origin and statistical properties of the requisite fluctuations is a very important unsolved problem). As we have seen, these regions are the seeds of the voids, because underdense patches become expansion centres, from which matter flows away until it encounters similar material flowing out of an adjacent void. Making the approximation that the excess Hubble parameter is the same in all voids, the matter must collect on planes that perpendicularly bisect the axes connecting the expansion centres.

For any given set of expansion centres, or *nuclei*, the arrangement of these planes define a unique process for the partitioning of space, a *Voronoi tessellation* (Voronoi 1908; Stoyan, Mecke, & Kendall 1987). A particular realisation of this process (i.e. a specific subdivision of N-space according to the Voronoi tessellation) may be called a *Voronoi foam* (Icke & Van de Weygaert 1987). In three dimensions a Voronoi foam consists of a packing of Voronoi cells, each cell being a convex polyhedron enclosed by the bisecting planes between the nuclei and their neighbours. A Voronoi foam consists of four geometrically distinct elements: the polyhedral cells (*voids*), their walls (*pancakes*), edges (*filaments*) where three walls intersect, and nodes (*clusters*) where four filaments come together.

In the cosmological context, each Voronoi cell is a void. The planes are identified with the "walls" in the galaxy distribution (see e.g. Geller & Huchra 1989), the filaments are identified with the elongated "super"clusters (Icke 1972; Oort 1983), and the vertices correspond to the virialised Abell clusters (Abell 1958).

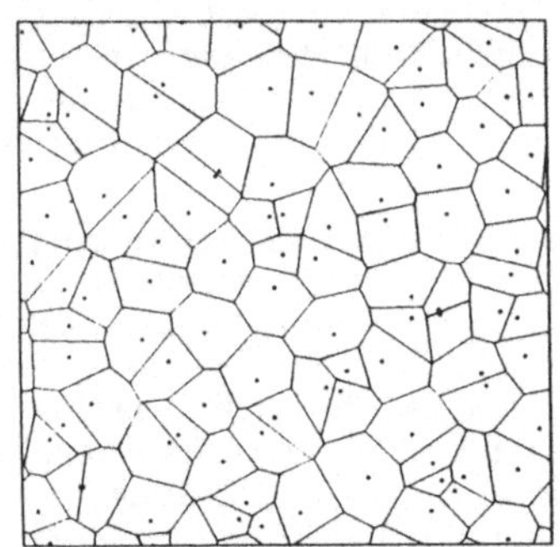
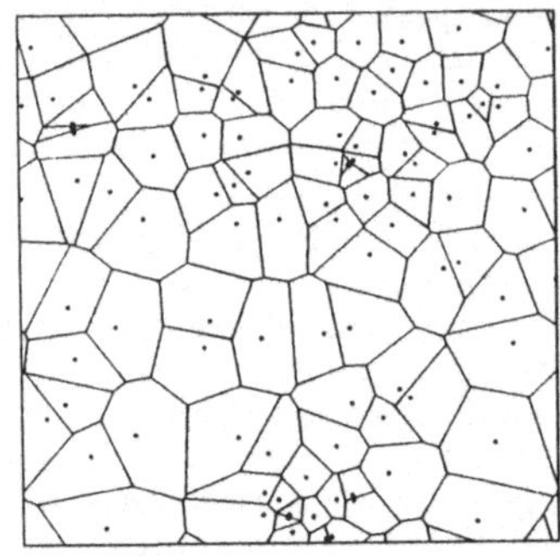

Figure 3. *Two examples of two-dimensional Voronoi foams, for different values of the correlation between the nuclei, which are indicated by dots. After Icke & Van de Weygaert (1987).*

We have constructed two- and three-dimensional Voronoi foams geometrically (Icke & Van de Weygaert 1987; Van de Weygaert & Icke 1989). The example in Fig.3 shows the characteristic appearance of two-dimensional Voronoi foams; the similarity with numerical simulations of gravitational clustering of collisionless particles is indeed quite striking (Melott 1983; Matsuda & Shima 1984). In three dimensions, the similarity is even more remarkable, in projection as well as in a slice (compare Fig.2 and Fig.6; see also Centrella & Melott 1983). A stereogram of three geometrically constructed Voronoi cells is given in Fig.4, showing the disposition of the walls, filaments, and nodes.

The advantage of using these geometrically constructed models is that one is not restricted by the resolution or number of particles. A cellular structure can be generated over a part of space beyond the reach of any N-body experiment. This makes the Voronoi model particularly suited for studying the properties of galaxy clustering in cellular structures on very large scales, for example in very deep pencil beam surveys, and for studying the clustering of clusters in these models.

If one relaxes the constraint of equal excess Hubble expansion in every void, one obtains a *Johnson-Mehl tessellation* (Johnson & Mehl 1939), which would be more realistic. This structure closely resembles the cells produced in the "adhesion" model by Kofman & Shandarin (1988). In Johnson-Mehl tessellations the walls are not bisecting planes but hyperboloids with their axes along the lines connecting the nuclei. It is very difficult to construct such a tessellation geometrically, but one might do it approximately by using a kinematical approach* such as the one described in the next section.

* This was done by Icke (unpublished, 1989); it is found that the difference is indeed small, due to the fact that the deviation between a hyperboloidal and a plane wall is of *second order* in the "excess Hubble constant" of the voids.

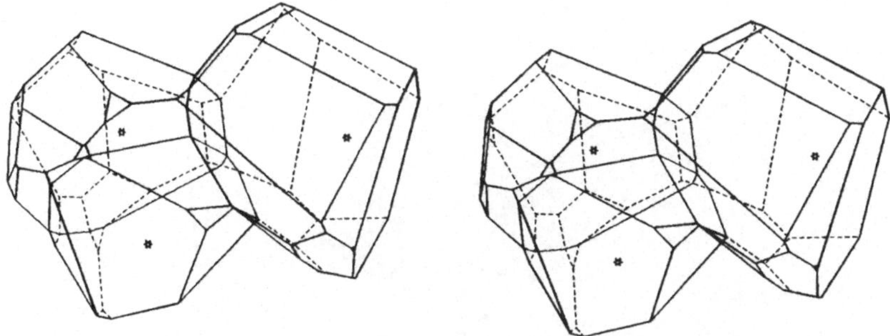

Figure 4. *Stereo pair of three Voronoi cells and their nuclei (stars). The cell edges ("filaments") are drawn in thick lines, the edges of the three walls where the cells touch are thinly drawn, and hidden lines are dashed. The picture is meant for a stereo viewer; if one wants to view these with the "crossed-eyes" method, the left and right images must be reversed. Courtesy R. van de Weygaert.*

The Voronoi foams are expected to give a good asymptotic description of the matter distribution at late times in several physical models. Models to which the Voronoi foams apply can be either gravitational instability models in which the structure formation is dominated by the negative density fluctuations, as in models with fluctuation spectra having a high-frequency cutoff or considerable power at low wavelengths; or models in which the driving force is due to explosions. The Voronoi foams outline the "skeleton" of the mass distribution, around which matter assembles during the evolution of the Universe.

4. KINEMATICS OF VORONOI CELLS

The Voronoi tessellation can be considered as the skeleton of the mass distribution. The galaxy distribution itself depends on the initial fluctuation spectrum, and on the details of the small-scale interactions of the gravitating matter. N-body simulations are preferred for solving that problem, but even those may not suffice. It seems probable that the small scale structure in the baryonic matter is considerably influenced (maybe even determined) by dissipational processes. But it can be very useful to distribute galaxies on a Voronoi skeleton according to some plausible description, in order to study the clustering properties in a cellular model of megaparsec structure.

The kinematical model is based on the notion that when matter streams out of the voids towards the Voronoi skeleton, cell walls form when material from one void encounters that from an adjacent one. If the matter is collisionless (as in some species of dark matter), only self-gravity will tend to hold it together. Moreover, if galaxies form during the expansion of the voids, the formation of the walls may be amplified by dynamical friction between the galaxies (Toomre and Toomre 1972; Binney and Tremaine 1987, Ch.7).

Accordingly, the structure formation scenario of the kinematical model is as follows. Within a void, the mean distance between galaxies increases uniformly in the course of time; this amounts to an excess Hubble expansion about the cell nucleus. When a galaxy tries to enter an adjacent cell, the gravity of the wall, aided and abetted by

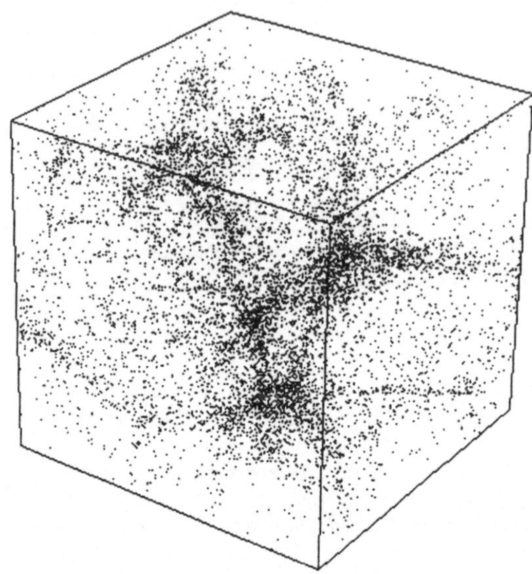

Figure 5. *Perspective cube showing the distribution of 10,000 galaxies expanding away from 15 Voronoi nuclei, at dimensionless time* $\tau = 1.5$*; the amplitude of the galaxy-galaxy two-point correlation function at this time corresponds approximately to what is observed today.*

dissipational processes, will slow its motion down. On the average, this amounts to the disappearance of its velocity component perpendicular to the cell wall. Thereafter, the galaxy continues to move within the wall, until it tries to enter the next cell; it then loses its velocity component towards that cell, so that the galaxy continues along a filament. Finally, it comes to rest in a node, as soon as it tries to enter a fourth neighbouring void. In a Voronoi foam, there are exactly four cells adjoining each node, and the above process is unique. An immediate consequence of this kinematic behaviour is, that the density in the walls quickly becomes smaller than in the filaments, which, in turn, remain less dense than the nodes, where all matter eventually congregates. This is the main reason why I identify the nodes with the rich Abell clusters.

Voronoi cell formation with periodic boundary conditions, as described above, is shown in Fig.5. When taking slices, as in Fig.6, one sees a striking resemblance with the observed redshift distributions of galaxies (e.g. De Lapparent, Geller, & Huchra 1986; see Fig.2, reproduced from Geller & Huchra 1989). Of course the picture is expected to differ considerably in the nonlinear areas. In particular, the velocity fields near the clusters and filaments are expected to be dominated by their gravity, and not by the expansion of the void.

5. COMPARISON WITH OBSERVATIONS

To quantify the comparison between the Voronoi tessellation and observations one has to derive the statistical properties of the clustering. Quantification of the complicated three-dimensional pattern in the megaparsec galaxy distribution, however, has proved to be a difficult task. A complete statistical description of any point process

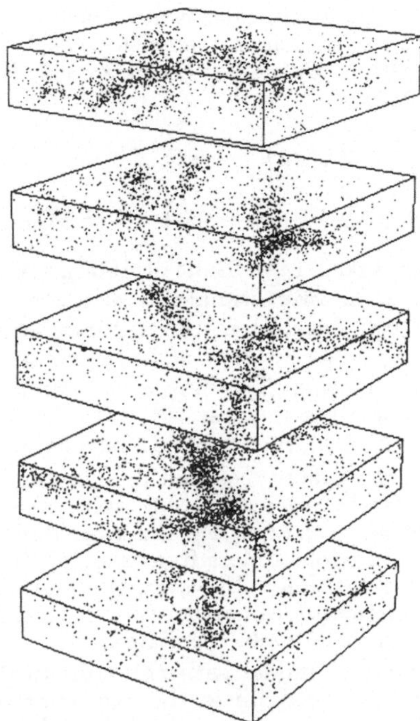

Figure 6. *The cube of Fig.5, sliced into five equal slabs.*

demands higher order N-point correlation functions (Peebles 1980), yet it is practically impossible to measure $N \geq 4$ correlation functions from galaxy catalogues. This has led to the use of several techniques highlighting different geometrical properties.

The most widely used statistic is the two-point correlation function, $\xi(r)$, and its sky-projected equivalent, the angular two-point correlation function $w(\theta)$. If δN is the number of objects found in a spherical shell with volume δV and radius r, centered on a randomly chosen object, then

$$\delta N = n\,(1+\xi(r))\,\delta V \;, \tag{5.10}$$

where n is the average number density of objects. In the case of galaxies, n is unknown, in part because the Universe has been insufficiently sampled (the deeper the redshift surveys go, the larger structures they reveal: *cf.* Geller & Huchra 1989). For the CfA1 survey Davis & Peebles (1983) found

$$\xi_{gg}(r) = (r_0/r)^{\gamma} \;, \tag{5.11}$$

$$\gamma = 1.77 \pm 0.04; \qquad r_0 = 5.4 \pm 0.3\ h^{-1}\ \text{Mpc}\;. \tag{5.12}$$

so that it seems fairly certain that ξ can be adequately described by a power law over a decade and a half in r. In particular, there is no good evidence for the existence of a preferred length scale in the small scale galaxy clustering, though there are some indications from the CfA2 survey that ξ_{gg} has a shoulder at a few megaparsec.

Evidence for structure on megaparsec scales is also provided by the two-point correlation function of Abell clusters, which has the same slope as, but a far larger amplitude than, the galaxy-galaxy correlation (Bahcall & Soneira 1983; Klypin & Kopylov 1983; Postman, Geller, & Huchra 1986):

$$\xi_{cc}(r) = (r_0/r)^{\gamma} , \tag{5.13}$$

$$\gamma = 1.8 \pm 0.2; \qquad r_0 = 26 \pm 4\ h^{-1}\ \mathrm{Mpc} , \tag{5.14}$$

up to a scale of $100h^{-1}$ Mpc (taken from Bahcall 1988). Although there are some indications that r_0 in Eq.(5.5) is an overestimate, due to contamination effects in the Abell catalogue (Sutherland 1988; Dekel *et al.* 1989; Olivier *et al.* 1990), the difference in amplitude between ξ_{cc} and ξ_{gg} suggests that at least one of them does not properly sample the underlying mass distribution. The difference in amplitude can, of course, be attributed to a dependence of the mass-to-luminosity ratio on the local mass density (a common, but so far utterly unphysical, fudge called "biasing"; see e.g. Kaiser 1984).

Comparing observations of the galaxy distribution and the theory of the evolution of the gravitational field requires knowledge of how light traces the gravitational potential; in other words, knowledge of the biasing. Since we do not know what the dark matter is, nor how its kinetic behaviour differs from that of the baryons, I assume for simplicity that light traces mass faithfully. However, the pair correlation behaviour of Abell clusters and of Voronoi vertices does not depend on this, provided that the dark mass and the luminous matter are positively correlated.

In order to compare the Voronoi model with the statistical properties of the observed galaxy distribution, we have determined the two-point correlation functions of the kinematic simulations and of the Voronoi vertices (Van de Weygaert & Icke 1989).

Because we identify the vertices with Abell clusters, the comparison could be done without further assumptions. When doing this, I can use the fact that *the Voronoi node distribution is a topological invariant* in co-moving coordinates, and does not depend on the way in which the walls, filaments, and nodes are populated with galaxies. Thus the statistics of the nodes should provide a robust measure of the Voronoi properties.

For a random distribution of 1000 expansion centres, generating a Voronoi tessellation with 6733 vertices, we got the remarkable result shown in Fig.7. In order to make a proper comparison between theory and observations, we must specify a single scaling parameter, with the dimension of length: the only free parameter of the Voronoi tessellation can always be specified in this way. We might, for example, choose the mean distance between nuclei. Observationally, however, this would be a very awkward choice. Instead, we applied our scaling by expressing r in units of $d/2$, where d is the mean separation between vertices (in the model) or clusters (in the observations). The vertex-vertex correlation function thus obtained is a power law from small scales up to the scale of one Voronoi cell, beyond which it vanishes:

$$\xi_{vv}(r) = (r_0/r)^{\gamma} , \tag{5.15}$$

$$\gamma = 2.0; \qquad r_0 = 1.13 , \tag{5.16}$$

(in dimensionless units; taking d equal to 55 h^{-1} Mpc for Abell clusters of richness class $\mathrm{R} \geq 1$ we get $r_0 = 31\ h^{-1}$ Mpc in physical units). The uncertainties shown in Fig.7 are purely Monte Carlo scatter.

Thus, both the slope (≈ -2) and the amplitude of the vertex-vertex correlation function are in good accordance with the cluster-cluster correlation. Although several other models of megaparsec structure formation, including the popular cold dark matter

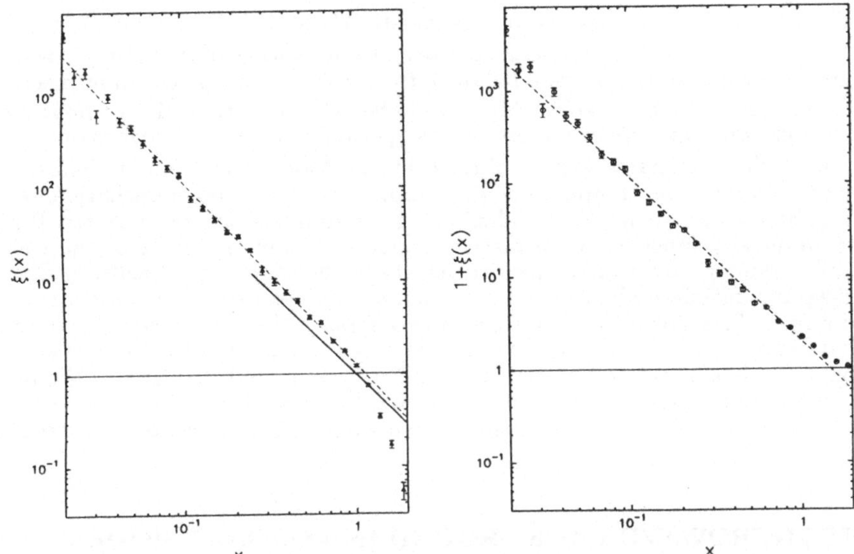

Figure 7. *The spatial vertex-vertex correlation function of a Voronoi foam with 1000 Poissonian nuclei. The left panel shows* ξ*, the right one* $1+\xi$*. Both have been plotted as functions of the dimensionless quantity* $x \equiv 2r/d$*, in which d is the mean separation between vertices. The heavy solid line is an approximate fit to the observational data. After Van de Weygaert & Icke (1989).*

scenario (e.g. White *et al.* 1987), have succeeded in obtaining the right *slope*, they did not succeed in reproducing the *amplitude*. It should be noted that our fit is not obtained arbitrarily, but comes about spontaneously by choosing the natural normalization in which the mean density of Voronoi vertices equals the observed density of Abell clusters.

This normalization, which gives $r_0 = 31\ h^{-1}$ Mpc, also determines the average size of the Voronoi cells. When this size is used in a calculation of the distribution of redshifts of galaxies populating a Voronoi foam in a deep pencil beam survey, one obtains results which are in excellent agreement with the observations (Broadhurst *et al.* 1990; Coles 1990; Van de Weygaert 1991).

The broad conclusion I draw from this is that *the cluster-cluster correlation function is determined by the geometrical properties of the galaxy distribution*, i.e. their cell-like or sponge-like arrangement. Because this conclusion is purely based on the geometry of the structures which have formed, instead of their dynamics, I consider this as the key result in the study of the Voronoi tessellations and their relation to megaparsec structures.

Recently Broadhurst *et al.* (1990) found a striking, quasi-periodic, galaxy redshift distribution in a pencil beam survey. The mean spacing between the peaks in the redshift distribution was found to be $128h^{-1}$ Mpc. Another remarkable discovery was the evidence for "pleated walls" spanning hundreds of megaparsec (Geller & Huchra 1989; Fig.2). Both of these phenomena arise quite naturally within a Voronoi model. In the Voronoi picture the "Great Wall" occurs because adjacent Voronoi walls form large pleated sheets. Both Coles (1990) and Van de Weygaert (1990) interpreted the Broadhurst *et al.* results on the basis of the Voronoi model. Coles (1990) used some new analytical results on Voronoi tessellations derived by Møller (1989). He showed that in a Voronoi foam resulting from Poisson nuclei with a mean density of $n \simeq (100h^{-1}\mathrm{Mpc})^{-3}$ (i.e. using the normalization discussed above), the mean distance between intersections of different cells is $< \lambda_1 > \simeq 137h^{-1}$ Mpc, while he estimated that along about 5% of the lines of sight one can expect a more or less "periodic" redshift distribution.

These analytical estimates were confirmed by the Monte Carlo simulations of Van de Weygaert (1991). He computed a three-dimensional Voronoi tessellation resulting from 2500 Poissonian nuclei. Galaxies, whose luminosities are selected from a Schechter luminosity function, were placed randomly within the walls of the cellular model, while taking account of the proper density in each topological feature. Simulations of deep magnitude-limited pencil beam surveys (cones) through these structures were performed. The redshift distribution in two opposite beams, showing periodicity, is shown in Fig.8, together with the corresponding pair counts. On the basis of some 30 pairs of opposing beams out to a depth of $z = 0.5$ it is estimated that 15% of the beams show the observed regular pattern, with a spacing between the peaks on the order of $105h^{-1} - 150h^{-1}$ Mpc (normalizing on the mean distance between clusters, as described above).

6. THE MICROWAVE BACKGROUND IN VORONOI MODELS

The Voronoi model has successfully met several statistical tests, but a grave objection was raised by Coles & Barrow (1990) concerning the smoothness of the cosmic microwave background radiation (CMBR). They estimated the redshift at which void shells coalesce in a universe filled with homogeneous undecelerated expanding shells around Poissonian nuclei (assuming that the self-similar non-linear expansion phase is not yet reached), whilst requiring that the voids did not percolate at recombination; that the voids grow large enough to coalesce (i.e. cellular megaparsec structure has formed); and that the temperature fluctuations of the CMBR produced by the primordial underdensities are less than the observed limits.

Coles & Barrow found that the Sachs-Wolfe effect leads to fluctuations which are only marginally in agreement with observations if a period of re-ionization is invoked. In that case a redshift of coalescence of $z_c \approx 5.16$ is obtained, while Saarinen, Dekel & Carr (1987) showed that $z_c \sim 5 - 10$ is needed to give the shells time enough to cool and fragment. Moreover, for adiabatic density fluctuations, or fluctuations with non-zero velocities on the last scattering surface, the model is completely excluded by small-angle limits, unless there is extensive re-ionization.

The Voronoi model for the mass distribution in the Universe is in accordance with all statistical data known at present. However, it remains to be seen if the model can pass one of the most stringent cosmological tests, namely confrontation with the observed smoothness of the CMBR. A detailed study of the CMBR in an evolving kinematic Voronoi distribution is under way (Icke & Poelman 1992); preliminary results show CMBR peak-to-peak fluctuations in the range between $\Delta T/T \approx 10^{-4}$ and $\Delta T/T \approx 10^{-2}$, depending on when the Voronoi evolution is assumed to begin. Thus it appears that the objections by Coles & Barrow will be sustained. The usual recourse to dark matter and biasing would help, but at the expense of simplicity.

As with other models, the key problem is: given an initial fluctuation small enough to keep the CMBR smoother than 10^{-5}, does the amplitude of the consequent structure become large enough by today?

In the Voronoi model, which focuses on the key role of the evolution of the voids, it is easy to calculate the relative growth of the other features (walls, filaments, and nodes).

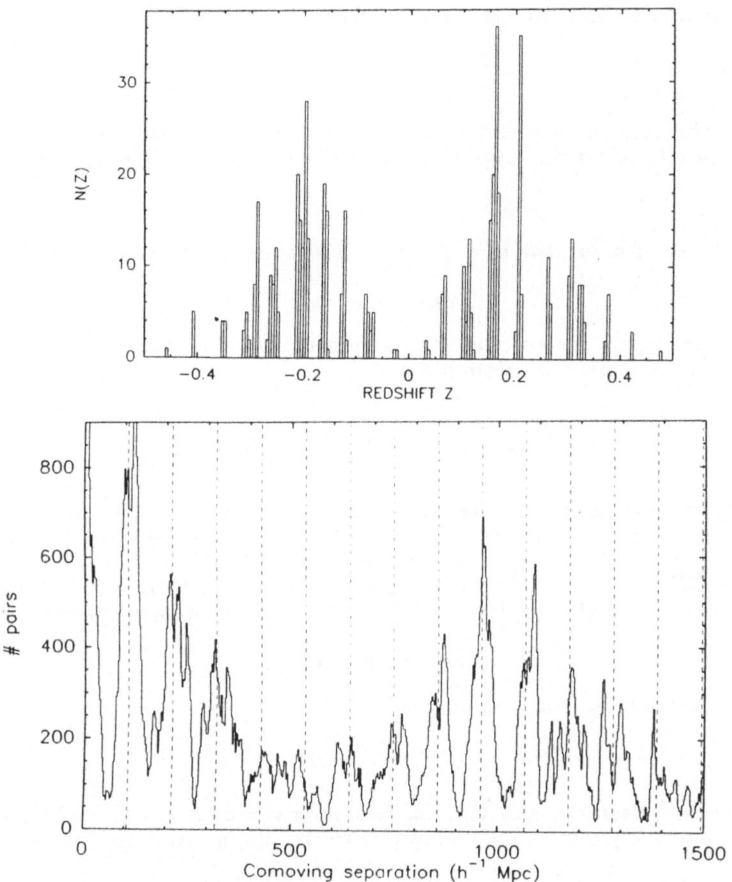

Figure 8. *A deep pencil-beam redshift survey through a Voronoi foam. Top: The redshift distribution for pencil beam surveys taken in opposite directions through a 3-D Voronoi tessellation. Positive redshift means the galaxy is in the cone pointing in one direction, while negative redshift means the opposite direction. Bottom: The corresponding pair-count correlation diagram. The dashed line indicates scales as multiples of* $107h^{-1}$ *Mpc. After Van de Weygaert (1991).*

Let m_v, m_w, m_f, m_n be the mass in voids, walls, filaments, and nodes, respectively. The first three features lose mass in a Voronoi "cascade": void→wall→filament→node. Each feature gains mass from the one immediately above it in the hierarchy. The mass loss term in N dimensions has the form $-NmH^*\,dt$, where H^* is the excess Hubble expansion and dt is the time increment.

In the quasi-linear case, $H^* \propto t^{-1/3}$, because the excess velocity increases as $H^*Rx = v \propto t^{1/3}$, and in the Einstein-De Sitter case $R \propto t^{2/3}$. The power law dependence means that one may absorb H^* into the time by defining $d\tau = H^* dt$. Consequently,

$$\tau = \tau_1 (t/t_1)^{2/3} , \tag{6.17}$$

where t_1 is the time at which the Universe becomes transparent to radiation. The constant τ_1 is related to the amplitude δ_1 of the voids at decoupling; one readily finds that

$$3\tau_1 = \delta_1 . \tag{6.18}$$

The amount dm of mass lost in a dimensionless time interval $d\tau$, in N dimensions, is then

$$dm = -Nm \, d\tau . \tag{6.19}$$

These equations allow one to relate the dimensionless time parameter τ to the cosmic time and to the redshift z at decoupling:

$$\tau = \frac{1}{3}\delta_1 (t_0/t_1)^{2/3} (t/t_0)^{2/3} = \frac{1+z}{3}\delta_1 (t/t_0)^{2/3} . \tag{6.20}$$

The effective excess Hubble parameter H^* is *the same* in all topological features. This is one of the many pretty properties of the Voronoi model. Let N be a node, let P be the point where a Delaunay line intersects a wall, let W be a point in the wall, and let the angle WNP be called θ. If I indicate the distance NP by a, the distance PW by b, and NW by c, then the velocity at W is cH^*. The component perpendicular to the wall is

$$cH^* \cos\theta = aH^* , \tag{6.21}$$

whereas the component along the wall is

$$cH^* \sin\theta = bH^* . \tag{6.22}$$

Thus, *the excess velocity in any Voronoi feature is simply found by multiplying H^* with the length along the feature.* This allows us to use the above formula for $N = 3, 2, 1$, and 0.

The mass gain is found simply by reversing the sign of the loss term of the feature higher in the hierarchy. This gives the following equations:

$$dm_v/d\tau = \quad - 3m_v , \tag{6.23}$$

$$dm_w/d\tau = 3m_v - 2m_w , \tag{6.24}$$

$$dm_f/d\tau = 2m_w - m_f , \tag{6.25}$$

$$dm_n/d\tau = m_f . \tag{6.26}$$

These equations are solved by noting that the N-dimensional mass loss equation has a solution of the type $\psi \exp(-N\tau)$:

$$m_v = e^{-3\tau} , \tag{6.27}$$

$$m_w = 3e^{-2\tau}(1 - e^{-\tau}) , \tag{6.28}$$

$$m_f = 3e^{-\tau}(1 - e^{-\tau})^2 , \tag{6.29}$$

$$m_n = (1 - e^{-\tau})^3 . \tag{6.30}$$

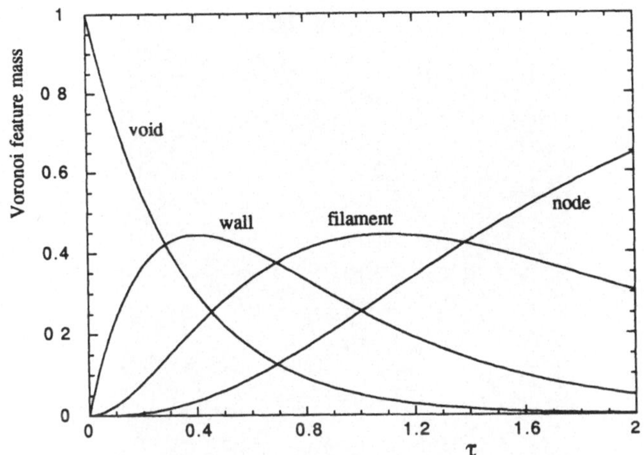

Figure 9. *Evolution of the mass encompassed by the four Voronoi features: voids, walls, filaments, and nodes. Ordinate: mass fraction, abscissa: dimensionless time parameter* $\tau = \tau_1 (t/t_1)^{2/3}$*, where* t_1 *is the cosmic time at decoupling, and* $\tau_1 = \delta_1/3$*, where* δ_1 *is the initial density amplitude of the voids.*

If I call $m_h = m_w + m_f + m_n$ the total mass of the high density regions, while $m_l = m_v + m_w + m_f$ is the mass in low density features, I find that

$$m_h/m_v = \exp(3t^{2/3}) - 1 \ , \tag{6.31}$$

$$m_l/m_n = \left(1 - \exp(-t^{2/3})\right)^{-3} - 1 \ . \tag{6.32}$$

The scaling of the time is given by the initial amplitude. Both mass ratios can be compared with published galaxy surveys. This gives us a way to determine what time it is in the simulations and in the Universe. When one determines the best fitting values (see Fig.9), one finds $\tau \approx 1$ (note that this is in reasonable agreement with age determinations from the two-point correlation of galaxies in the kinematic Voronoi model; cf. Fig.5). That means, via Eq.(6.20), that the initial amplitude must be on the order of 2×10^{-3}. The Sachs-Wolfe effect (i.e. the gravitational redshift incurred by photons which must climb out of a potential well in which they were formed) is then about 70 times larger than that which is allowed by CMBR observations.

Numerical calculations of the fate of CMBR photons on their way through an evolving Voronoi universe show, that the Rees-Sciama effect (i.e. the growth of $\delta T/T$ due to the fact that photons falling into an evolving cosmic potential well see a deeper well when they climb out) is on the order of a few times 10^{-6}, which is allowed by the CMBR observations.

Apparently, there is not really enough time to make the contrast between voids and the rest, or between nodes (=Abell clusters) and the rest, as large as seems necessary, unless some special pleading is invoked; e.g. observational selection effects, or a particular dependence of the M/L ratio on the local mass density (="biasing"). When none of these alternatives provides a satisfying solution, the interpretation of megaparsec structure formation dominated by the expansion of voids is wrong. This would not only

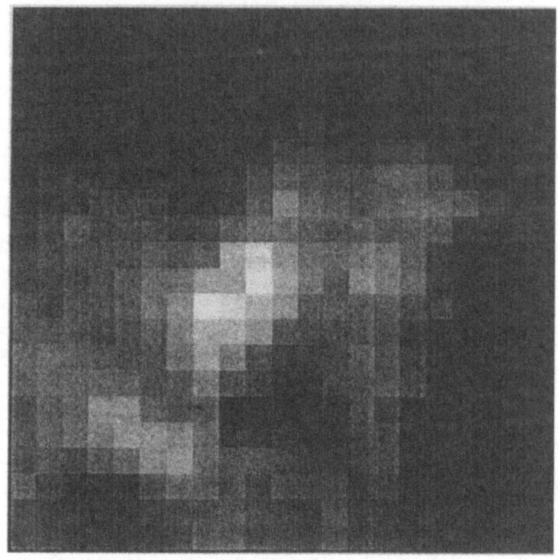

Figure 10. *Corrugations in a photon sheet, calculated numerically by tracing CMBR photons on their way through an evolving Voronoi universe. The inhomogeneities tend to an asymptotic pattern ("freeze out"), imprinted by the initial potential wells and their early evolution. Black/white amplitude is about* 10^{-3}*; calculations made by Ylva Poelman.*

exclude the Voronoi model, but also similar models in which the voids play an important role (like pancake models, or neutrino-dominated models).

7. CONCLUSIONS

The structure of our Universe on a scale of about 10–500 Mpc can be plausibly modeled by means of pressure-free Newtonian gravitational collapse. This produces a three-dimensional galaxy distribution which is non-Poissonian, because the regions which initially had a slight overdensity collapse and become more irregular, while the low-density regions expand and become more regular in the course of time. Thus, selfgravity produces a generic structure of closely packed superHubble bubbles, out of which matter flows. This matter first collects on the perpendicular bisecting planes between the spheres; from there, it flows to the lines where three walls intersect; and finally it collects in the nodes where four filaments come together. Asymptotically, this distribution is described by a statistical process known as Voronoi tessellation. Visual comparison between Voronoi foams and the observed distribution of galaxies shows a promising similarity. Detailed statistical study shows that the two-point correlation of Voronoi foam is indistinguishable from that which is obtained for the actual galaxy distribution: *the Voronoi distribution is statistically indistinguishable from the real thing.* This holds in particular for the slope and the amplitude of the two-point correlation function of Abell clusters and of three-dimensional Voronoi vertices. Likewise, the cluster multiplicity function is properly reproduced. The most serious problem (which the Voronoi model has in common with other theories of structure formation) is that the CMBR fluctuations are too large.

Acknowledgements. It is a pleasure to thank J.H. Oort for his constant interest in this work, B. Jones for his unflagging encouragement and enthusiasm, and Y. Poelman for her help with the work on the microwave background. I am especially indebted to R. van de Weygaert for his brilliant contributions to the Voronoi foam model.

References

Abell, G.O., 1958. *Astrophys. J. Suppl.* **3**, 211.

Bahcall, N.A., 1988. *Ann. Rev. Astron. Astrophys.* **26**, 631.

Bahcall, N.A., & Soneira, R., 1983. *Astrophys. J.* **270**, 20.

Binney, J., & Tremaine, S., 1987. *Galactic Dynamics*, Princeton Univ. Press, Princeton.

Bond, J.R., 1990. In *Frontiers in Physics - From Colliders to Cosmology*, Proceedings of the Lake Louise Winter Institute, eds. Campbell, B., & Khanna, F. (Singapore: World Scientific), in press.

Broadhurst, T.J., Ellis, R.S., Koo, D.C., & Szalay, A.S., 1990. *Nature* **343**, 726.

Centrella, J.M., & Melott, A.L., 1983. *Nature* **305**, 196.

Chandrasekhar, S., 1969. *Ellipsoidal Figures of Equilibrium*, Yale Univ. Press, New Haven.

Coles, P., 1990. *Nature* **346**, 446.

Coles, P., & Barrow, J.D., 1990. *Monthly Notices Roy. Astron. Soc.* . **244**, 557.

Davis, M., Huchra, J., Latham, D.W., & Tonry, J., 1982. *Astrophys. J.* **253**, 423.

Davis, M., & Peebles, P.J.E., 1983. *Astrophys. J.* **267**, 465.

Dekel, A., Blumenthal, G.R., Primack, J.R., & Olivier, S., 1989. *Astrophys. J. (Letters)* **338**, L5.

De Lapparent, V., Geller, M.J., & Huchra, J.P., 1986. *Astrophys. J. (Letters)* **302**, L1.

Efstathiou, G., 1990. in *Physics of the Early Universe*, Proceedings of the Thirty-Sixth Scottish Universities Summer School in Physics 1989, eds. Peacock, J.A., Heavens, A.F., & Davies, A.T., (SUSSP publications, Edinburgh), p. 361.

Efstathiou, G., & Silk, J., 1983. *Fundam. Cosm. Phys.* **9**, 1.

Einasto, J., 1978. In IAU Symposium No. 79, *The Large Scale Structure of the Universe*, eds. Longair, M.S., & Einasto, J., (Dordrecht: Reidel), p. 52.

Einasto, J., Joêveer, M., & Saar, E., 1980. *Monthly Notices Roy. Astron. Soc.* **193**, 353.

Fisher, J.R., & Tully, R.B., 1981. *Astrophys. J. Suppl.* **47**, 139.

Geller, M., 1988. In *Large scale structures in the Universe*, 17th Saas-Fee course, eds. Martinet, L., & Mayor, M., Geneva Observatory, p. 69.

Geller, M., & Huchra, J., 1989. *Science* **246**, 897.

Giovanelli, R., Haynes, M.P., & Chincarini, G.L., 1986. *Astrophys. J.* **300**, 77.

Gregory, S.A., & Thompson, L.A., 1978. *Astrophys. J.* **222**, 784.

Huchra, J.P., Davis, M., Latham, D., & Tonry, L.J., 1983. *Astrophys. J. Suppl.* **52**, 89.

Humason, M.L., Mayall, N.U., & Sandage, A.R., 1956. *Astron. J.* **61**, 97.

Icke, V., 1972. *Formation of Galaxies Inside Clusters*, Ph. D. Thesis, Leiden.

Icke, V., 1973. *Astron. Astrophys.* **27**, 1.

Icke, V., 1984. *Monthly Notices Roy. Astron. Soc.* **206**, 1P.

Icke, V., & Poelman, Y., 1992. In preparation.

Icke, V., & Van de Weygaert, R., 1987. *Astron. Astrophys.* **184**, 16.

Icke, V., & Van de Weygaert, R., 1991. *Quart. J. Roy. astron. Soc.* **32**, 85.

Johnson, W.A., & Mehl, R.F., 1939. *Trans. Am. Inst. Min. Metal Eng.* **135**, 416.

Kaiser, N., 1984. *Astrophys. J. (Letters)* **284**, L9.

Kirshner, R.P., Oemler, A., Schechter, P.L., & Shectman, S.A., 1981. *Astrophys. J. (Letters)* **248**, L57.

Kirshner, R.P., Oemler, A., Schechter, P.L., & Shectman, S.A., 1987. *Astrophys. J.* **314**, 493.

Klypin, A.A., & Kopylov, A.I., 1983. *Soviet Astr. Lett.* **91**, 267.

Kofman, L.A., & Shandarin, S.F., 1988. *Nature* **334** 129.

Lifschitz, E.M., 1946. *J. Phys. U.S.S.R.* **10**, 116.

Limber, D.N., 1953. *Astrophys. J.* **117**, 134.

Limber, D.N., 1954. *Astrophys. J.* **119**, 655.

Lin, C.C., Mestel, L., & Shu, F.H., 1965. *Astrophys. J.* **142**, 1431.

Lynden-Bell, D., 1964. *Astrophys. J.* **139**, 1195.

Matsuda, T., & Shima, E., 1984. *Prog. Theor. Phys.* **71**, 855.

Melott, A.L., 1983. *Monthly Notices Roy. Astron. Soc.* **205**, 637.

Møller, J., 1989. *Adv. Appl. Appl. Prob.* **21**, 37.

Neymann, J., & Scott, E.L., 1955. *Astron. J.* **60**, 33.

Olivier, S., Blumenthal, G.R., Dekel, A., Primack, J.R., & Stanhill, D., 1990. *Astrophys. J.* **356**, 1.

Oort, J.H., 1970. *Astron. Astrophys.* **7**, 381.

Oort, J.H., 1983. *Ann. Rev. Astron. Astrophys.* **21**, 373.

Peebles, P.J.E., 1980. *The Large-Scale Structure of the Universe*, Princeton Univ. Press.

Postman, M., Geller, M.J., & Huchra, J.P., 1986. *Astr. J.* **91**, 267.

Rood, H.J., 1988. *Ann. Rev. Astron. Astrophys.* **26**, 245.

Rubin, V.C., 1954. *Proc. Nat. Acad. Sciences* **40**, 541.

Saarinen, S., Dekel, A., & Carr, B., 1987. *Nature* **314**, 598.

Sandage, A., & Tammann, G.A., 1981. *A Revised Shapley-Ames Catalog of Bright Galaxies, Carnegie Inst. Washington, Publ. 633*

Shandarin, S.F., & Zel'dovich, Ya.B., 1989. *Revs. Mod. Phys.* **61**, 185.

Shapley, H., & Ames, A., 1932. *Harvard Obs. Ann.* **88**, 43.

Stoyan, D., Kendall, W.S., & Mecke, J., 1987. *Stochastic Geometry and Its Applications*, Akademie-Verlag, Berlin.

Sutherland, W.J., 1988. *Monthly Notices Roy. Astron. Soc.* **234**, 159.

Szalay, A., 1988. In *Large scale structures in the Universe*, 17th Saas-Fee course, eds. Martinet, L., & Mayor, M., Geneva Observatory, p. 173.

Toomre, A., & Toomre, J., 1972. *Astrophys. J.* **178**, 623.

Totsuji, H., & Kihara, T., 1969. *Publ. Astron. Soc. Japan* **21**, 221.

Tully, R.B., 1982. *Astrophys. J.* **257**, 389.

Van de Weygaert, R., 1991. *Monthly Notices Roy. Astron. Soc.* **249**, 159

Van de Weygaert, R., & Icke, V., 1989. *Astron. Astrophys.* **213**, 1.

Voronoi, G., 1908. *J. reine angew. Math.* **134**, 198.

White, S.D.M., Frenk, C.S., Davis, M., & Efstathiou, G., 1987. *Astrophys. J.* **313**, 505.

Zel'dovich, Ya.B., 1970. *Astron. Astrophys.* **5**, 84.

Zel'dovich, Ya.B., 1978. In IAU Symposium No. 79, *The Large Scale Structure of the Universe*, eds. Longair, M.S., & Einasto, J. (Dordrecht: Reidel), p. 409.

A Novel Possibility for Neutrino Masses and Decay

Saul Barshay

III. Physikalisches Institut, Technische Hochschule Aachen, D-5100 Aachen, Germany

Abstract

I consider the possibility that a sigma model is relevant for calculating the masses and mixings of Dirac neutrinos. New processes involving pairs of Goldstone bosons are predicted. Heavy neutrinos decay into lighter ones and a pair of Goldstone bosons, $\nu_2 \rightarrow \nu_1 + b + \bar{b}$. For a dark-matter mass of ν_2 of $\sim$ 15 eV, a lifetime $\lesssim 10^{21}$ sec is possible; for a mass of ν_2 of $\sim$ 17 KeV, a lifetime $\lesssim 10^6$ sec is possible.

After hearing this morning's talks, and anticipating those from tomorrow, it seems fair to say that basic physics issues underlying these wonderful experimental adventures are: (1) the possible "disappearance" of some of the Sun's (electron) neutrinos due to some mechanism involving finite neutrino mass; and (2) the possible discovery of distant, point-sources of very high-energy neutrinos, and gamma rays - or perhaps even exotic objects - via detection of interactions producing charged leptons, muons in particular.

I would like to make some remarks upon these matters, that is on neutrino masses and decay, and upon the role of hypothetical Goldstone bosons[1]. If neutrinos have mass, the heavier ones presumable decay. The question is: how? The answer presumably has something to do with the way neutrinos acquire mass.

Consider a neutral Dirac fermion, a "neutron", n. It acquires its mass m_n from the mechanism shown in Fig. 1.

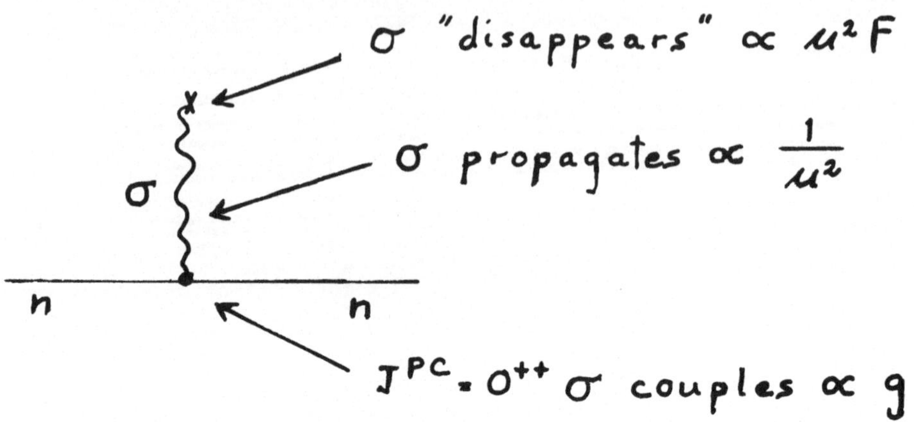

Fig. 1 Mechanism for the mass m_n

The result is clearly,

$$m_n = g\left(1/\mu^2\right)\left(\mu^2 F\right) = gF \quad \left(\Rightarrow g = m_n/F\right) \tag{1}$$

Consider a second, neutral Dirac fermion, a lighter "neutron", n'. For simplicity, take its mass as "initially" zero, <u>but</u> couple n' a little to n. Then n' acquires a mass $m_{n'}$, from the mechanism shown in Fig. 2.

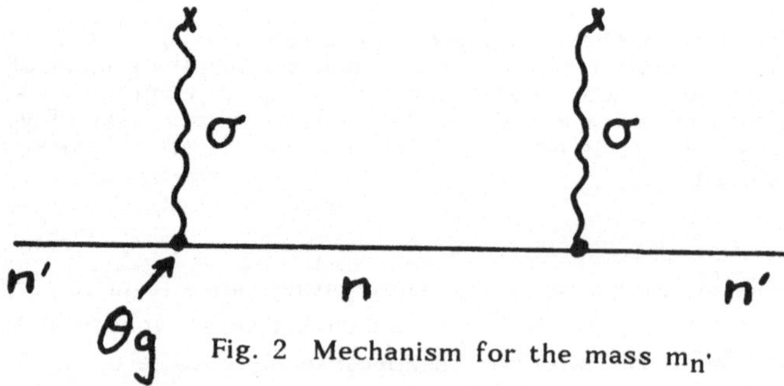

Fig. 2 Mechanism for the mass $m_{n'}$

The result is

$$\begin{aligned} m_{n'} &= \left(\Theta g\right)^2 \left(1/m_n\right) \left(\left(1/\mu^2\right) \cdot \mu^2 F\right)^2 \\ &= \left(\Theta m_n/F\right)^2 \left(1/m_n\right) \left(F\right)^2 \\ &= \Theta^2 m_n \end{aligned} \tag{2}$$

Now n decays, because of the process shown in Fig. 3.

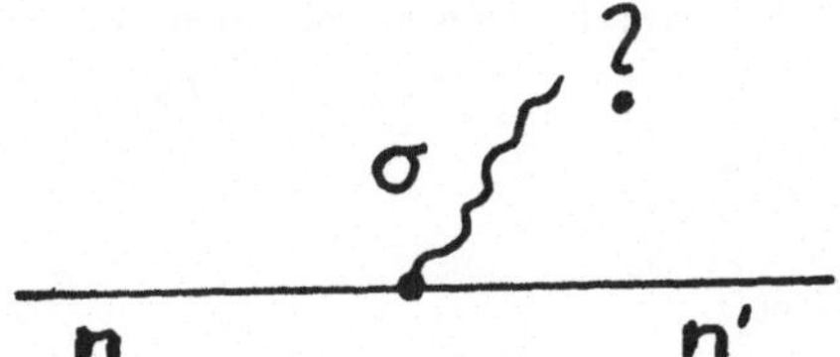

Fig. 3 Transition via a virtual σ

What does it decay into, n' and a σ? No, not generally, because the σ boson is heavy and not real. But, it is always present virtually. So, what does σ decay into? Answer: into two Goldstone bosons, b (assume b is approximately massless, i.e. $m_b << (m_n - m_{n'})$). Therefore, we have the decay[1],

$$n \rightarrow n' + b + \bar{b} \tag{3}$$

This is so, if the mass generation of m_n arises because of a spontaneously broken, global symmetry[2,3]. The decay mechanism is shown in Fig. 4.

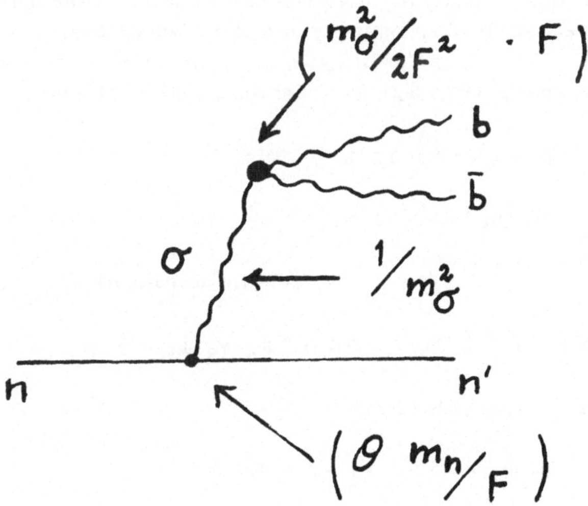

Fig. 4: The decay $n \rightarrow n' + b + \bar{b}$, via σ

This leads to the lifetime[F1],

$$\tau(n \rightarrow n' + b + b) = \left\{ \left(\Theta m_n / F \right)^2 \left(m_n / F \right)^2 \left(m_n / 6144 \pi^3 \right) \right\}^{-1}$$

To get a feeling for the consequences of these considerations, I put in some numbers. (1) For an assumed mass m_n = 15 eV, which would be relevant for dark-matter consideration[4], and using $\Theta^2 \gtrsim 10^{-3}$, which is an empirical upper limit[5] for $\nu_\tau - \nu_\mu$ mixing, one has

$$m_{n'} = \Theta^2 m_n \gtrsim 10^{-2} \text{ eV} \tag{4a}$$

$$\tau_n \sim 8 \times 10^{20} \text{ sec} \tag{4b}$$

This is for[F2] $F \cong 300$ MeV. So, $m_{n'}$ is a mass in the range relevant[6] to resolution of the solar neutrino anomaly. The lifetime means that decays are occuring in the present universe. (2) Assuming[7,8,9,10] a mass m_n = 17 KeV, and $|\Theta| \gtrsim 0.03$, one has

$$m_{n'} = \Theta^2 m_n \sim 17 \text{ eV} \qquad \text{(dark matter ?)} \tag{5a}$$

$$\tau_n \sim 8 \times 10^5 \text{ sec} \quad \text{(a factor}^{11} \text{ in the development of very large scale structure}^{12,13} \text{ in the universe?)} \tag{5b}$$

Again, this is for $F \cong 300$ MeV. Why this F? I don't know. (Do you know why $F_{el-wk} \cong 300$ GeV?) But, what has been sketched here is ~ the σ-model of pion-nucleon physics[2] ($F_\pi \cong 100$ MeV). The symbol ~, here means without the second "neutron", n'; and with only a neutral π^0 as the pseudoscalar, Goldstone boson required[3] by a spontaneously broken, chiral symmetry[2], which breaking allows the nucleon to acquire mass. That is, a "primary" mass term ($\propto m_n^0 \neq 0$) for n is forbidden in the Lagrangian $\mathcal{L}(\psi_n, \bar{\psi}_n)$, by invariance under the chiral transformation[2],

$$\begin{aligned} \psi_n &\Rightarrow \psi_n + \delta\psi_n = \psi_n - \frac{i\beta}{2}\gamma_5\psi_n \\ \bar{\psi}_n &= \psi_n^\dagger\gamma_0 \Rightarrow \bar{\psi}_n + \delta\bar{\psi}_n = \bar{\psi}_n - \frac{i\beta}{2}\bar{\psi}_n\gamma_5 \end{aligned} \tag{6a}$$

(β is infinitesimal)

which gives, $$\delta\left\{-m_n^o\bar{\psi}_n\psi_n\right\} = i\,\beta\, m_n^o\bar{\psi}_n\gamma_5\psi_n \neq 0 \tag{6b}$$

Rather, postulate[2] in $\mathcal{L}$, interaction terms

$$-\frac{g_n}{2}\bar{\psi}_n(\sigma + i\gamma_5 b)\psi_n - \frac{g_n}{2}\bar{\psi}_n(\sigma^\dagger + i\gamma_5 b^\dagger)\psi_n \tag{7}$$

Under the chiral transformation, now including

$$\begin{aligned} \sigma &\Rightarrow \sigma + \delta\sigma = \sigma - \beta b \\ b &\Rightarrow b + \delta b = b + \beta\sigma \end{aligned} \tag{8a}$$

one has, $$\delta\left\{ \overline{\psi}_n\,(\sigma + i\,\gamma_5\, b)\,\psi_n \right\} = 0 \tag{8b}$$

Now, with $\sigma \to F + \tilde{\sigma}$, the vacuum expectation value $\langle\sigma\rangle = F \neq 0$ ($\langle\tilde{\sigma}\rangle = 0$); therefore $\mathcal{L}$ contains an n mass term,

$$g_n\, F\, \overline{\psi}_n \psi_n\,, \qquad g_n = m_n / F \tag{9}$$

The specifically new element here[1] is a term in $\mathcal{L}$ which explicitly breaks the chiral symmetry, linking n' to n via σ; giving mass to n'.

$$\tilde{\mathcal{L}} = \tilde{\mathcal{L}}^{\dagger} = -i\,(\Theta g_n)\,\overline{\psi}_n\,\gamma_5 \psi_{n'}\sigma - i\,(\Theta g_n)\,\overline{\psi}_{n'}\,\gamma_5\,\psi_n\,\sigma^{\dagger} \tag{10}$$

The γ_5 in the interaction results[F3] in a positive mass shift from the mechanism shown in Fig. 3, i.e. $m_{n'} = 0 \Rightarrow m_{n'} > 0$.

The masses and interactions in the bosonic sector[F4] follow from part of $\mathcal{L}$,

$$\mathcal{L}_\lambda = -\frac{\mu^2}{2}(\sigma^{\dagger}\sigma + b^{\dagger}b) - \frac{\lambda^2}{4}(\sigma^{\dagger}\sigma + b^{\dagger}b)^2$$

$$\underset{\sigma \to F + \tilde{\sigma}}{\Longrightarrow} \quad -\frac{m_\sigma^2}{2}\,\tilde{\sigma}^{\dagger}\tilde{\sigma} - \frac{m_b^2}{2}\, b^{\dagger} b - \frac{\lambda^2}{2}\, F\,(\tilde{\sigma} + \tilde{\sigma}^{\dagger})\, b^{\dagger} b$$

$$- \frac{m_b^2}{2}\, F\,(\tilde{\sigma} + \tilde{\sigma}^{\dagger}) + \ldots$$

with $$m_\sigma^2 = \mu^2 + 2\lambda^2 F^2\,, \quad m_b^2 = \mu^2 + \lambda^2 F^2 \tag{11}$$

For $F \neq 0$, but $\langle\tilde{\sigma}\rangle = 0$, we must have $m_b^2 = 0$

$$(\mu^2 < 0) \tag{12}$$

The Goldstone boson is b, massless but for explicit symmetry-breaking terms in $\mathcal{L}$. Also,

$$m_\sigma^2 = \lambda^2 F^2 \Longrightarrow \lambda^2 = m_\sigma^2 / F^2 \tag{13}$$

So the effective decay coupling in Fig. 5 is $\frac{\lambda^2}{2} F = m_\sigma^2 / 2F$.

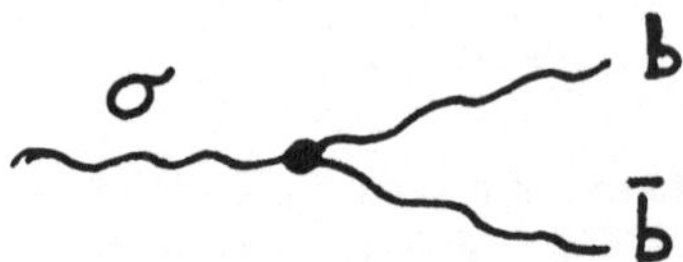

Fig. 5 The decay $\sigma \to b + \bar{b}$

Limiting[F5] $|\lambda| \lesssim 10$, gives $m_\sigma \lesssim 3$ GeV for $F \cong 300$ MeV. Then,

$$\Gamma_\sigma = \left(\frac{m_\sigma}{F}\right)^2 \frac{m_\sigma}{64\pi} \sim 0.5\, m_\sigma \sim 1.5 \text{ GeV} \tag{14}$$

In conclusion, it is worthwhile to keep some general issues before ones' eyes. What are the physical uses of neutrino masses?

(1) (a) $m_{\nu_\tau} \sim 15$ eV, dark-matter contributor[4]?

(b) $\Theta^2 \lesssim 10^{-3} \Rightarrow m_{\nu_\mu} \lesssim 10^{-2}$ eV , solar $\nu_e \to \nu_\mu$ transition[6] ?

(c) $\tau_{\nu_\tau}(\nu_\tau \to \nu_\mu + b + \bar{b}) < 10^{21}$ sec

($t_{universe} \sim 5 \times 10^{17}$ sec)

$\Rightarrow$ decays occuring now ?

(2) (a) $m_{\nu_\tau} \sim 17$ KeV, more very large-scale structure[12,13] in the universe, because[11, F7]

$\tau_{\nu_\tau}(\nu_\tau \to \nu_\mu + b + \bar{b}) < 10^6$ sec.

(b) $\Theta^2 \lesssim 10^{-3} \Rightarrow m_{\nu_\mu} \gtrsim 15$ eV, a dark-matter mass follows "naturally".

What are the physical uses of (~ massless) Goldstone bosons b, coupled to neutrinos (via γ, largely unobservable directly)?

(1) Possibly sporadic, distant "point-sources" of Goldstone bosons in the sky, which bosons induce anomalous high-energy interactions[14] (probably rich in $\mu(\tau)$-pairs) here?

(2) A contribution to dark matter i.e. m_b not exactly zero?

It is 30 years since Goldstone's paper[3]. We have yet to observe a "real" Goldstone boson[F6]. Or a Higgs-like σ. Do they exist in Nature?

Footnotes

F1. Proportionality to m_n for a two-body phase space goes over to proportionality to $m_n^2 \cdot m_n$ for a three-body phase space.

F2. Note $g_n = \frac{m_n}{F} \cong 0.5 \times 10^{-7}$. For m_n = 17 KeV, $g_n \sim 0.5 \times 10^{-4}$.

F3. Note that an admixture of 1 here, brings in parity violation at the level of mass generation (not removable by a choice of relative phase between ψ_n and $\psi_{n'}$). For γ_5 only, a third n" interacting with n and with n', brings in parity violation not removable by a choice of phases.

F4. The fields γ and b are not Hermitian.

F5. The relevant strength measure is $\lambda^2/8\pi^2$.

F6. The pion in Ref. 2 is the best-known "pseudo" Goldstone boson. However, the pion is also expected as a quark-antiquark bound state. As such, it does have a rather small mass.

F7. As of this date (April, 1992), the experimental case for a 17 KeV neutrino is weakened. See the brief review in "Pursuing the 17 KeV Neutrino", A. Hime, Los Alamos preprint, LA-UR-92-946. The argument (Ref. 6) for a neutrino mass $\lesssim 10^{-2}$ eV remains. Thus, it would be useful to consider possible consequences of a neutrino decaying, with emission of two Goldstone bosons, in the present universe.

References

1. S. Barshay, "A Sigma Model for Neutrino Masses and Mixings and Decay", Modern Physics Letters A, 6 (1991) 3583.

2. M. Gell-Mann and M. Lèvy, Nuovo Cimento 16 (1960) 705.

3. J. Goldstone, Nuvo Cimento 19 (1961) 155.

4. R. Cowsik and J. McClelland, Phys. Rev. Lett. 29 (1972) 669.

5. Review of Particle Properties, Particle Data Group, Phys. Lett. B239 (1990) 1.

6. J.N. Bahcall and H.A. Bethe, Phys. Rev. Lett. 65 (1990) 2233, and references therein.

7. J.J. Simpson, Phys. Rev. Lett. 54 (1985) 1891.

8. J.J. Simpson and A. Hime. Phys. Rev. D39 (1989) 1825; D39 (1989) 1837.

9. A. Hime and N.A. Jelley, Phys. Lett. B257 (1991) 441.

10. D.R.O. Morrison, "Review of 17 KeV Neutrino Experiments", CERN-PPE/91-140; D.H. Perkins, Conference summary, these proceedings.

11. J.R. Bond and G. Efstathiou, Phys. Lett. B265 (1991) 245.

12. W. Saunders et al., Nature 349 (1991) 32.

13. D. Lindley, Nature 349 (1991) 14.

14. S. Barshay, "Goldstone Bosons from Sources in the Sky?", in preparation.

Added note (June, 1992). The specific hypothesis for the solar neutrino issue put forth in reference 6, has $m_{\nu_\mu} \gtrsim 10^{-3}$ eV, $\Theta^2 \gtrsim 10^{-2}$; therefore $m_{\nu_\tau} \sim m_{\nu_\mu}/\Theta^2 \lesssim 0.1$ eV is hardly relevant for dark matter. Here, I have argued that $m_{\nu_\mu} \sim 10^{-2}$ eV, $\Theta^2 \sim 10^{-3}$, such that $m_{\nu_\tau} \sim m_{\nu_\mu}/\Theta^2 \sim 10$ eV is relevant for dark matter. (Assumed, $\Theta^2_{\mu\tau} \gtrsim \Theta^2_{\mu e} = \Theta^2$.) Then few, (if any) of the main-cycle neutrinos from the Sun are effected (note Fig. 1 in H.A. Bethe and J.H. Bahcall, Phys. Rev. D44 (1991) 2964). This situation appears relevant today ($\sin^2\Theta_{\mu e} \sim 1/2 >>> \Theta^2_{\mu\tau}$ is also possible).

A look at stellar collapse by neutral currents

Carlo Bemporad

INFN and Dipartimento di Fisica, Università di Pisa, Italy

Abstract

The study of stellar gravitational collapse by neutral current neutrino interactions is discussed as a development of present Gran Sasso experiments.

1. INTRODUCTION

The study of neutral current interactions by solar or stellar gravitational collapse ν's and the comparison with charged current interactions is an important and fascinating field of research; recent solar-ν data from HOMESTAKE and KAMIOKANDE exps. [1-2] and their possible suggestive interpretation [3-4], make these investigations even more valuable and worth pursuing.

At the Gran Sasso national laboratory MACRO and LVD are sensitive to stellar gravitational collapse mainly via charged currents. It is interesting to discuss a possible additional experiment, more sensitive to neutral current interactions. For this purpose, heavy water (HW) is the obvious ν-target. A HW quantity of ≈ 200 *tons* would be sufficient for adding potentialities to MACRO, while it would be too small for a new experiment on solar-ν_e detection; moreover solar-ν_e detection needs a painstaking work and a special experiment design in order to reduce all types of natural radioactivity background. MACRO, a large surface experiment, will never reach the low noise level needed for solar-ν_e studies.

Another possibility for detecting stellar collapse NC ν-interactions is to rely on the ν-excitation of the $(1^+, T = 1)$ 15.1 MeV ^{12}C level and on its γ de-excitation.

We briefly list the physics items, relative to stellar gravitational collapse, one might investigate by a new "NC-experiment":

1) Detection of ν neutral current interactions. 2) Study of energy equipartition among all ν types. 3) Study of possible MSW effects in stellar matter during collapse. 4) Limit on $\nu_{\mu,\tau}$ masses from the time-profile of the stellar collapse $\nu_{\mu,\tau}$ burst.

2. STELLAR COLLAPSE ν-INTERACTIONS IN WATER, HW AND ^{12}C

2.1. Neutrino fluxes and spectra

When presenting predictions relative to the experimental sensitivity to a stellar gravitational collapse, it is important to state which kind of model one is using in the calculations. We adopt a constant temperature thermal ν-spectrum of the Fermi-Dirac type (fig.1):

$$\Phi(E_\nu) = \frac{A_0 E_\nu^2}{1 + \exp[E_\nu/T]} \,.$$

The values of the parameters were taken from the statistical analysis on SN1987 data by Bludman and Schindler [5]. It is assumed that the ν_e and $\overline{\nu}_e$ neutrinospheres have a temperature $T_{\nu_e} = T_{\overline{\nu}_e} = 3.3\ MeV$, while all other neutrinospheres have a temperature roughly twice that value: $T_{\nu_{\mu,\tau}} = T_{\overline{\nu}_{\mu,\tau}} = 6.6\ MeV$. As a consequence the constants A_0 for the different ν-types are related by: $A_0(\nu_{\mu,\tau}) = A_0(\nu_e)/16$; $A_0(\nu_e) = 5.21\ 10^{55}\ MeV^{-3}$. The total energy radiated through neutrinos of any type is: $E = 5.862\ A_0\ T^4 \approx 3.4\ 10^{53}\ ergs$. Results for any other choice of parameters can easily be obtained.

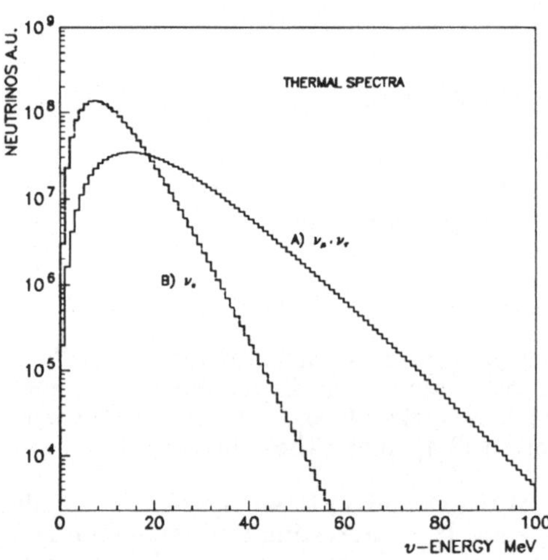

fig.1. Thermal ν-spectra for all ν-flavours.

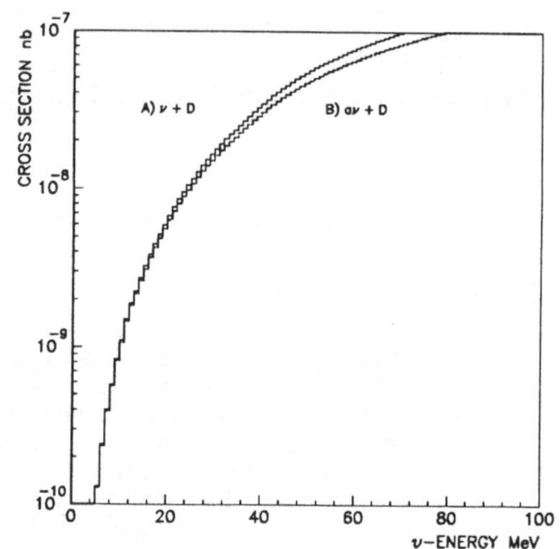

fig.2. The ^2D ν-disintegration cross sections.

2.2. Neutrino cross sections

CC ν-cross sections on ^2D are the ones of ref.[6-7]; the ^2D ν-disintegration cross section we use is the one by Bahcall [8] and is presented in fig.2; (a more recent evaluation by S.Ying et al. predicts a factor of two larger values [9]). ν-cross sections on electrons and free protons are the ones of ref.[6]. The NC ν-cross section on ^{12}C for the superallowed (iso-vector, axial vector) transition to the $(1^+, T = 1)$ 15.1 MeV level is the one by Fukugita et al. [10].

2.3. Number of events generated by ν's in heavy water

A stellar gravitational collapse at the galactic center induces a rather large number of events in HW; the events are concentrated in a time interval of $\approx 10\ s$; the detection of a collapse by a HW-implemented MACRO is possible and the natural radioactivity background is a manageable problem. The number of events generated in HW by neutral and charged current ν-interactions is presented in Table I (no detection efficiency factor is applied on reactions other than ES). All numbers refer to 100 *tons* of D_2O HW.

2.4. Number of events generated by ν's in Carbon

For sake of comparison, Table II presents the total number of CC $\overline{\nu}_e$-events in a scintillator experiment of 1000 *tons* sensitive mass (like MACRO) and the total number of NC events, obtainable in a scintillator experiment of similar mass, to which $\approx$ 3000 *tons* of carbon are added (no detection efficiency factor is applied).

Table I

Stellar Collapse Events in 100 tons of Heavy Water			
Current Type	Reaction Type	Neutronization Burst	Cooling Stage
CC	$\nu_e + d \rightarrow p + p + e^-$	0.5	2
CC	$\overline{\nu}_e + d \rightarrow n + n + e^+$	0	1
NC	$\nu_x + d \rightarrow \nu_x + p + n$	0	42
ES	$\nu_x + e \rightarrow \nu_x + e$	0	1

Table II

Stellar Collapse Events in a liquid scintillator exp. with added graphite			
Current Type	Reaction Type	Neutronization Burst	Cooling Stage
CC	$\overline{\nu}_e + p \rightarrow n + e^+$	0	220*
NC	$\nu_x +^{12} C \rightarrow \nu_x +^{12} C^*(15)$	0	47 **

* *(reference CC-reaction in* 1000 *tons of liquid scintillator)*

** *(NC-reaction in* 1000 *tons of liquid scintillator plus* $\approx$ 3000 *tons of carbon)*

A look at Table I shows that the only relatively abundant events, generated in 100 *tons* of HW by stellar gravitational collapse ν's, are the 42 NC ^2D ν-disintegrations. Most of the events ($\approx$ 85%) are induced by $\nu_{\mu,\tau}$'s and $\overline{\nu}_{\mu,\tau}$'s.

2.5. Comparison between NC experiments based on ^{12}C and Heavy Water

A NC experiment needs 1000 *tons* of liquid scintillator and $\approx$ 3000 *tons* of carbon for collecting $\approx$ 45 NC events; the same number of NC events is obtainable by only 100 *tons* of HW and $\approx$ 300 *tons* of liquid scintillator. The ^{12}C reaction has a 15.1 MeV threshold; the ^2D reaction has a 2.2 MeV threshold and can therefore also be induced by ν_e's and $\overline{\nu}_e$'s (ν_e's have softer spectra than $\nu_{\mu,\tau}$'s). ^2D is easily broken by γ's from natural radioactivity background; a HW experiment must be carefully shielded. Low tritium content ($< 0.05\ \mu Ci\ Kg^{-1}$), high purity ($H_2O < 0.15\%$) HW is an extremely expensive material; leasing is advisable. Total costs for HW use depend on the HW correct handling. It is important to mantain the "virgin"-HW high purity and to avoid the addition of substances which might later be almost impossible to remove.

3. THE DETECTION OF ν NC-INTERACTIONS IN HW

3.1. "External" NC-event detection

The only relatively abundant events are the ones from NC ^2D ν-disintegration; one faces the problem of an efficient n-detection. Table III presents parameters relative to n-slowing-down and n-diffusion in water, HW and carbon (scintillator parameters are roughly similar to the ones of water). R_s and R_d are r.m.s. radii for "slowing-down" and "diffusion"; τ_s and τ_d are the "slowing-down time" and the "capture time". The HW parameters are for 100% pure HW and the carbon ones are for graphite ($\rho = 1.62\ g\ cm^{-3}$).

R_d and τ_d are reduced to 215 cm and 43900 μs for HW 99.85% pure.

Table III

Parameters for Neutron Slowing-Down and Diffusion in Water, Heavy Water and Carbon			
Parameter	H_2O	D_2O	C
R_s (cm)	14	27	46
τ_s (μs)	10	46	150
R_d (cm)	6.6	250	116
τ_d (μs)	235	60900	11300

Table III shows that, if the HW is sufficiently pure and if the HW container is small (for instance: a 30 cm thickness) the neutron has a high probability of escaping from the HW container. Gadolinium doped liquid scintillator (0.05% $\leftrightarrow$ 0.1% by weight), surrounding the HW containers, can efficiently capture neutrons. Several γ-rays are emitted for each n-capture; $E_{tot} = \sum_i E_{\gamma_i} \approx 8\ MeV$. A large scintillator mass has nice "calorimetric" properties, high n-capture and γ-ray detection efficiencies. The small number of ν-events releasing detectable energy within the HW containers (see Table I) seems to suggest the use of HW as a "passive" target. This reduces the HW handling problems and allows an optimization of the n-detection efficiency.

3.2. Identification of NC and CC events

NC events, originated in HW, are detected by n-capture (and successive γ-ray interactions) in the gadolinium loaded liquid scintillator surrounding the HW containers. Since CC events are induced, in the same scintillator, by $\overline{\nu}_e$ on free protons, one faces the problem of distinguishing the two event types. An attempt to separate the events into two classes must keep into account the following observations:

1) the n-capture process by Gd is similar, in energy, to a CC-event; detected energies go from $\approx 5\ MeV$ (detection threshold) to $\approx 8\ MeV$. In the case of a CC reaction, the e^+ releases an energy from $\approx 5\ MeV$ to $\approx 30\ MeV$; this primary event is most of the times followed by a delayed n-capture.
2) A partial geometrical separation is possible between events generated in HW and other events, if the HW containers are of smaller longitudinal size than the scintillation counters.
3) A comparison is possible between the total number of events generated in the MACRO SM's containing HW (CC + NC events) and the total number of events generated in the SM's which do not contain HW (mostly CC events).

3.3. Monte Carlo method evaluation of the detection efficiency

The low energy Monte Carlo code [11], developed to evaluate the MACRO detection efficiency for stellar collapse CC $\overline{\nu}_e$-interactions and delayed-n's, was used for this HW proposal, after adaptating it to the different HW apparatus geometry. The efficiency for detecting ν-interactions in HW is: $\epsilon \approx 45\%$ for an energy threshold $E_{th} = 5\ MeV$ and it is rather stable for reasonable parameter choices (like: sandwich "scintillator-HW-scintillator" $60-30-60$ cm; 0.05% gadolinium by weight). Fig.3 shows the positions at which n's produced in HW are captured across the HW and scintillation counter layers ; fig.4 shows a similar distribution for the reconstructed γ-positions (the detected energy centroids).

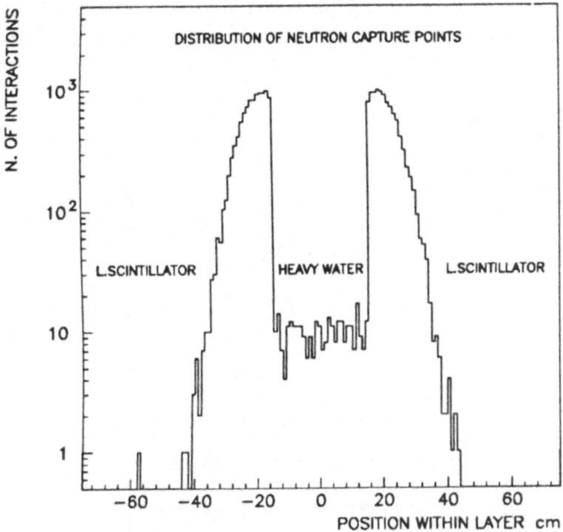

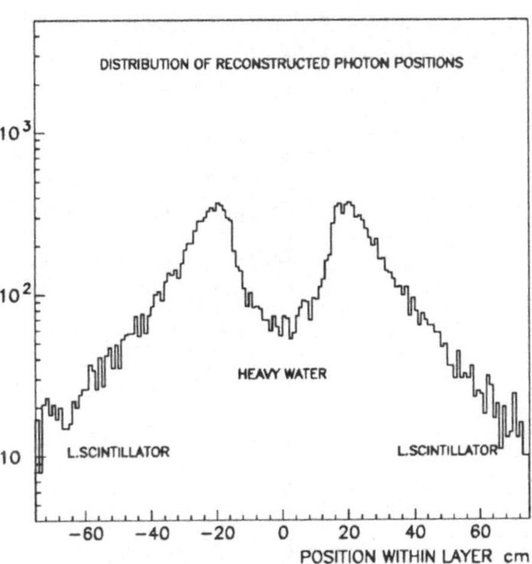

fig.3. Positions of n's-captures.

fig.4. Reconstructed γ-positions (the detected energy centroids).

3.4. Deuterium photodisintegration and radioactivity background shielding

HW is a very good target for ν-interactions because of its relatively large cross section and because of its low threshold; unfortunately ^2D can also be broken by the natural radioactivity background γ-rays (for $E_\gamma > 2.2\ MeV$) present at the experiment location. The rate of n-production by D_2 photodisintegration can be evaluated by the Monte Carlo method and by using the γ energy spectrum, the cross section for D_2 photodisintegration and the geometry of the HW experiment. A preliminary evaluation indicates that the experiment must be shielded against the external natural radioactivity background; a reduction of this background by a factor of 100 (for example: by a water layer $\approx 2\ m$ thick) is needed for a sensitivity to stellar collapse similar to the one of the MACRO experiment (based on CC-interactions).

4. THE DETECTION OF ν NC-INTERACTIONS IN^{12}C

Although the ^{12}C-content of the CH_2-scintillator is quite a good target for supernova ν's, one might take advantage of the "segmentation" of an experiment for finding the best geometrical arrangement of "passive" graphite-layers to be put close to the counters as an extra source of NC events. At the same time one can investigate possible scintillation counter arrangement for improved 15.1 MeV γ-efficiency. The quality of the 15.1 MeV γ-detection is also of interest.

The problems were solved by the Monte Carlo method (by taking into account all detection inefficiencies). An optimized set-up corresponds to "sandwiches" of scintillator-graphite-scintillator layers (30 – 40 – 30 cm) for a total of 1000 *tons* of (no gadolinium) liquid scintillator and $\approx$ 3000 *tons* of graphite. A total number of 35 events is obtained for a stellar collapse at the galactic center; this means an enhancement factor of ≈ 3 over the 10 expected events in the (1000 *tons* of liquid scintillator and 20 *cm*-thick counters of MACRO. The efficiency and the detected 15.1 MeV γ-line are shown in fig.5; efficiency means the fraction of detected events relative to the total number of NC interactions in the liquid scintillator and in the total extra-C. One sees that the 15.1 MeV γ-line is still well resolved although some of events fall in the low-energy tail.

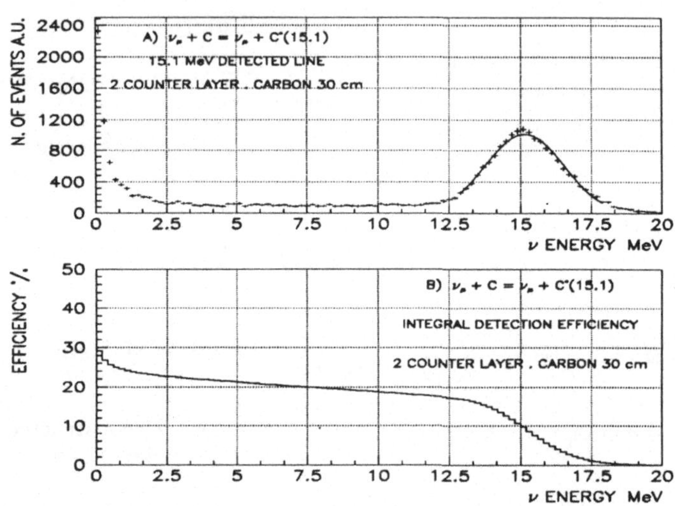

fig.5. The ^{12}C γ-line and the integral detection efficiency.

5. CONCLUSIONS

An experiment containing 200 *tons* of HW would allow the study of stellar gravitational collapse NC ν-interactions. n's from ^2D disintegration can be efficiently detected by MACRO-like counters filled with gadolinium loaded liquid scintillator. An experiment containing scintillator and graphite is also sensitive to NC; the high energy threshold of the ^{12}C reaction makes the experiment less sensitive to the natural radioactivity background.

References

[1] R. Davies., **Neutrino 88** Proceedings (1988)
[2] K.S. Hirata et al., **Phys.Rev.Lett. 65** 1297 (1990)
[3] J.N. Bahcall et al., **Phys.Rev.Lett. 65** 2233 (1990)
[4] B. Schwarzschild et al., **Physics Today Oct** 20 (1990)
[5] S.A.Bludman et al., **Astrophys.J. 326** 265 (1988)
[6] J.N. Bahcall, Neutrino Astrophysics, Cambridge
[7] S. Nozawa et al., **J.Phys.Soc.Jap. 55** 2636 (1986)
[8] J.N. Bahcall et al., **Phys.Rev. D38** 1030 (1988)
[9] S. Ying et al., **Phys.Rev. D40** 3211 (1989)
[10] M. Fukugita et al., **Phys.Lett. B212** 139 (1988)
[11] A. Baldini et al. **Nucl.Instrum.&Meth.A 305** 475 (1991)

NESTOR and SADCO in the Mediterranean

(Tests of modules of deep underwater optical and acoustical neutrino detectors near Pylos, Greece)

presented by I.M. Zheleznykh

I. STATUS OF NESTOR-91 COLLABORATION

I.F. Barinov, A.O. Deineko, V.A. Gaidash, M.A. Markov, A.A.Permyakov, N.N. Surin, D.Yu. Vasilenko, L.M. Zakharov, I.M. Zheleznykh, V.A.Zhukov
Institute for Nuclear Research,
the USSR Academy of Sciences, Moscow, Russia

A.P. Eremeev, V.V. Ledenev, M.N. Platonov, V.Kh. Rucol, N.A. Sheremet
Institute of Oceanology,
the USSR Academy of Sciences, Moscow, Russia

Y.N. Evplov, G.M. Ohrimenko,
Strength of Materials Institute,
the Ukraine SSR Academy of Sciences, Kiev, Ukraine

E. Anassontzis, M. Barone*, P. Ioannou, P. Pramantiotis, L.K. Resvanis, G. Voulgaris
Physics Laboratory, University of Athens, Greece
*** Visitor from INFN, Frascati, Italy.**

Abstract

The results of the Soviet-Greek tests of the 10 PMTs deep undersea detector module for NESTOR - optical neutrino telescope near Pylos, Greece are presented. The vertical muon intensity and the angular distribution of muons at 3000-4000 m depths were measured by that module during r/v "VITYAZ" cruise in July-91. The effective registration area

of the module for vertical muons was evaluated to be close to 400 m . The proposal to construct 10 m NESTOR by the module principle to install a few supermodules or superstrings modules (each superstring consists of 6 modules) is under consideration.

1. INTRODUCTION

The use of the World Ocean as a target for cosmic neutrino detectors had been proclaimed to open good prospects for the high energy neutrino astrophysics and physics of cosmic rays [1,2]. Now Hawaii DUMAND seems to be the Standard Model of the deep underwater neutrino telescope. However some alternative searches were carried out too. In particular in 1989 in the Mediterranean sea new sites were studied and a new DUMAND-type module of 4 PMTs was tested [3]. In July-91 the Soviet-Greek collaboration using R/V "VITYAZ" studied the environmental conditions off the coast of Pylos, Greece (bioluminescence and K 40 background, bottom profile, currents, water transparency, etc.). There is a valley at a depth 3800 m and 7.5 miles to shore which can be chosen as a right site for the optical neutrino telescope. We have deployed and tested a 10 PMTs module of neutrino telescope too.

2. TESTS OF 10 PMTs MODULE

An autonomous module of 10 PMTs (R-2018 HAMAMATSU) in glass BENTHOS spheres has been tested. This is a prototype of NESTOR module. All the module elements including a box of module controller, a power source, a box of pressure gauge are placed on the framework which is made of thin-wall titanium tubes. In the water under an action of positive buoyancy the framework obtains a form of inverted heptahedral pyramid with the base side of 7 m and the height of 3.5 m. Six PMTs are placed at the corners of horizontal hexagon and four PMTs at the pyramid vertex. Assembly and deployment of the module is shown at Figs. 1 and 2.

Figure 1. Assembly of the module module on the board of "VITYAZ".

Figure 2. Deployment of module.

With average optical background of 700 +/- 700/200 photons/sm sq. sec the false starting-up frequency of four central PMTs is less then 1 impulse per minute.

Preliminary analysis for 4 peripheral photodetectors which had been engaged in the time interval +/- 70 ns from the moment of coincidence scheme starting-up, gives the result of less than one false event per year.

3. RESULTS

The differential effective area of particle registration, $S_{eff}(\theta) = S_{eff}(0)*f(\theta)$, where $f(\theta)$ is the dependence function of detector sensitivity from zenith angle; $S_{eff}(0)$ is the effective area, when $\theta=0$.

Table 1
The statistics of muon events at 3338 m, 3697 m, and 4108 m depths

H m	H m.v.e.	Time sec	N events	F sec
3250+/-50	3338	7212	475	(6,59+/-0.30)*10
3600+/-50	3697	8402	356	(4.24+/-0.21)*10
4000+/-50	4108	9897	191	(1.93+/-0.14)*10

The $S_{eff}(0)$ value can be calculated directly from the measurements of muon flux on the H depth, if the vertical intensity of the flux at given depth is known. This intensity is connected with muon registration frequency by such an expression:

$$F = J(\Theta,H)\ S_{eff}(\Theta)\ d\Omega \quad (1)$$

where F is muon statistics at the H depth; Ω is solid angle, in which the muon flux is observed.

So as the detector makes it possible to measure the zenith angle of muon (Fig.3), the index m can be determined. Knowing the dependence of detector sensitivity from zenith angle (f(Θ)) and vertical intensity of flux on the depth of H, the $S_{eff}(0)$ can be obtained from the formula (1).

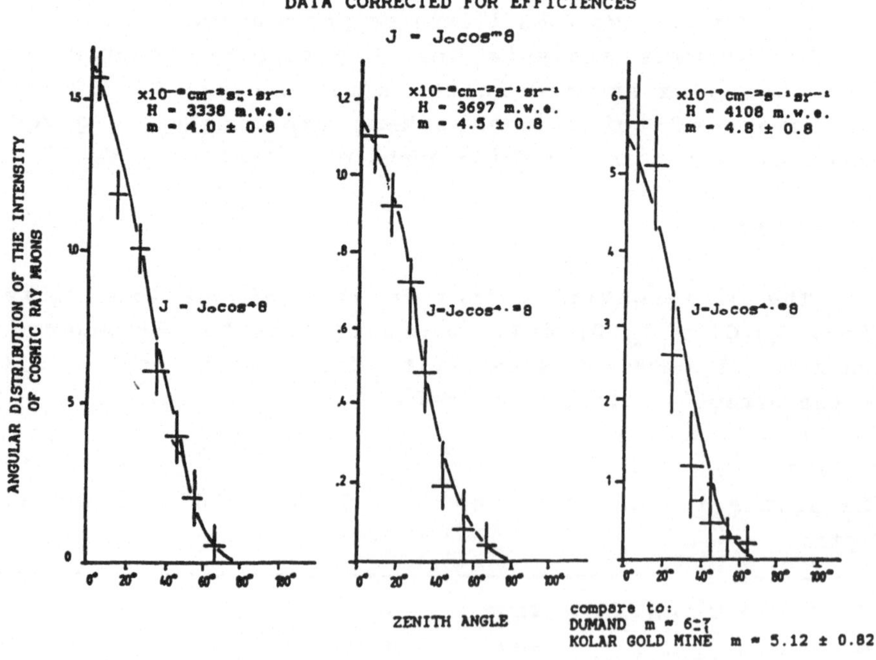

Figure 3. Angular distributions of muons at three depths.

In Vavilov et al (1970) [4] and DUMAND SPS (1987) [5] experiments vertical intensity of muon flux, measured for depth of 3160+/-50 and 3190+/-100 m.w.e. coincides with the one calculated from Miyake formula. That's why for module effective area Seff.(0) calculation we have chosen the results at the depth of 3338 m.

Taking an integral for upper hemisphere (2 sr), we obtained the correlation between I(0,H),F, m, and $S_{eff}(0)$:($f(\Theta)=0.24 + 0.76\cos\Theta$)

$$S_{eff}(0)= \frac{m+1}{2\pi\, I(0,H)} \; \frac{m+2}{m + 1.24} \qquad (2)$$

Substituting in (2) the F value from Table 1 and m=4, we obtained $S_{eff}(0)$=373+/-75 sqr.m., when I(3338m)=1.61*10^{-8} sm s sr. The direct calculation of detector effective area gives $S_{eff}(0)$=373 sq.m., when (550 nm)=0.12, (450nm)=0.05 and PMT threshold equal to 1 ph.e.h.p.

Taking in use the data obtained for $S_{eff}(0)$, it is possible to calculate the vertical intensity of space muons at the 3600 and 4000 m levels. The results are listed in Table 2.

Table 2
The vertical muon intensity at three levels

H m.v.e	m	F sec	I(0,H) sm sec sr
3338	4.0+/-0.8	(6.59+/-0.30)*10^{-2}	(1.61+/-0.40)*10^{-8}
3697	4.5+/-0.8	(4.24+/-0.21)*10^{-2}	(1.13+/-0.40)*10^{-8}
4108	4.8+/-0.8	(1.93+/-0.14)*10^{-2}	(5.38+/-1.40)*10^{-9}

These results coincide with the ones calculated by Miyake formula. The background, which is due to bioluminescence and K40 wasn't over the range of 700+/-700/200 photons/sm sec during the experiment.

4. NESTOR IN THE NEAREST FUTURE (PROPOSAL)

Prospects of construction of a deep underwater muon and neutrino detector - neutrino telescope NESTOR near shores of

Greece are under discussion. It is proposed that NESTOR to be made up by a few (super) strings of 84 PMTs each (type R-2018 HAMAMATSU).

The superstring has such a structure: PMTs are placed in six modules, 14 PMTs per module. Seven PMTs are oriented up by their photocathodes and another seven down. Two of 14 PMTs are placed in the center and others at the corners of horizontal hexagon. Distance between PMTs in one module is 10 m. Distance between modules (planes) is 20 m (see Figs.4 and 5).

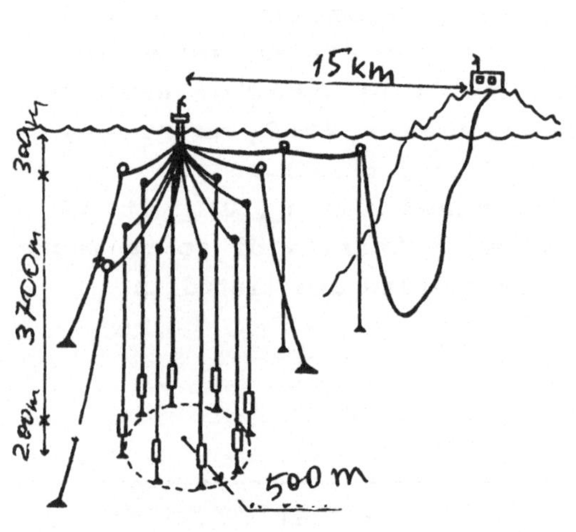

Figure 4. Scheme of NESTOR
(8 superstrings at depth 3.7 km
672 PMTs, V eff - 10^8 m^3 for
Ev=2 TeV,angular resolution 1-4 deg.)

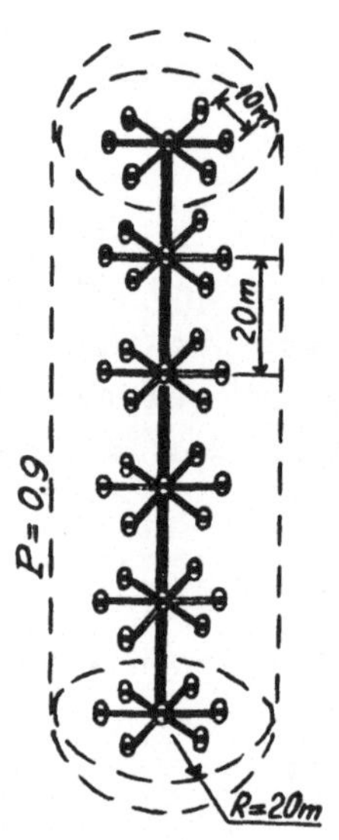

Figure 5. Scheme of superstring of 84 PMTs.

5. CONCLUSION

Joint Soviet-Greek deep underwater experiments in July-91 had shown that hexagonal module can be used as a basic element for NESTOR neutrino telescope.

We consider Pylos as an unique location to build an optical deep underwater neutrino telescope in the Mediterranean.

6. REFERENCES

1 M.A. Markov, Proc. 10 Int. Conf. High-Energy Physics Rochester (1960) 579.

2 H. Blood, J. Learned, F. Reines and A. Roberts, Proc. 1976 Int. Neutrino Conf., Aachen, ed. H Faissner (1977) 688.

3 A.O. Deyneko et al., Proc. #d Int. Workshop on Neitrino Telescopes, ed. Milla Baldo-Ceolin, Venezia, Feb.26-28, 1991, p.p 407-426.

4 Yu.N. Vavilov et al., Bull. Acad. of Sci. USSR, 34 (1970) 1759.

5 DUMAND Proposal, Preprint HDC -3-88.

II. SEA ACOUSTIC DETECTOR OF COSMIC OBJECTS - SADCO (STATUS OF SADCO - 92 COLLABORATION)

A. Butkevich, S. Karaevsky, M. Markov, A. Trenikhin, I. Zheleznykh
Institute for Nuclear Research,
the USSR Academy of Scienses, Moscow, Russia

V. Albul, A. Sinitsky, A. Kurchanov
Institute of Physical, Technical and Radio Measurements, Mendeleevo, Russia

P. Ioannou, L.K Resvanis, G. Voulgaris
Physics Laboratory, University of Athens, Greece

Abstract

The results of tests of deep undersea detector module for SADCO - acoustical neutrino telescope near Pylos, Greece are presented.

1. INTRODUCTION

Acoustic detection of super high energy neutrino ($> 10^{15}$ eV) has been a topic of active research about 15 years ago (Askarian, Dolgoshein [1], Bowen [2], Learned [3]). A search for possible significant fluxes of astrophysical super high energy neutrino from active galaxies nuclei (Stecker) can excite new interest towards the method.Deep underwater acoustical telescope (SADCO) for resonance electron antineutrino detection (threshold energy of an order of $6*10^{15}$ eV is proposed to be deployed off the coast of Pylos (Greece). Effective volume of SADCO telescope is about 10^9 m for $6*10^{15}$ eV with 85% probability of detection and 1 false alarm per month.

Performance of acoustical detector could be enhanced if the experiments are carried out in those regions of World Ocean that have relatively warm water (the larger Grunaisen coefficient the larger the acoustic signal).

2. BACKGROUND NOISE ESTIMATION WITH A SADCO MODULE (IONIC SEA, SUMMER, 1991)

Tests of deep underwater acoustic module at depth near 4000 m were made during the "Vityaz" cruise in July-91.

All electronics and power supply accumulators for the underwater experiment are contained within 1 meter diameter titanium ball. Hydrophones arrays are mounted above the ball. Autonomous bottom station goes down beneath the sea and stops 20 meters up the bottom. An anchor lies on the bottom and can be thrown away by 'emerge' command from the ship. Then the ball comes up to the surface and can be picked up on board. The station can start moving up to the surface on timer commands and in emergency as well. When it's up it transmits acoustic and radio signals to the ship. A lantern which is at the top of the ball (as well as radio antenna) glitters so the ball can be easily seen in the dark.

2.1. Hydrophones arrays

We have got two vertical line arrays of hydrophones. The first one consists of 8 nonuniformly closely spaced sensors with total aperture of 1 meter. The second one is long aperture (30 meters) array of 4 high sensitivity hydrophones. Only single array is mounted up the ball for each experiment. The unit can operate in two modes, noise estimation and signal detection.

2.2. Data acquisition system

Output signals from hydrophones are filtered in 5-35 kHz frequency band, amplified by automatic gain control amplifiers and then pass through 4-bits analog-to-digital converter (sampling frequency is 250 kHz).Intel 8085 based controller is the heart of data acquisition system.It receives digital signals, performs preprocessing and sends data to tape recorder of 10 Mbytes capacity. Dynamic range of data acquisition system is about 60 Db. Level of preamplifiers noise is about 2 nV/Hz .

2.3. Signal processing.Noise estimation

1) Data were tested whether they are random and identically

distributed (run test). It allowed samples to be grouped in blocks of 512 points.

2) Probability density and distribution functions were estimated in each block.We've obtained estimates of four first moments of distribution (mean, variance, skew and kurtosis) and their evolution in time.

3) Gaussian hypothesis were checked against various nonnormal alternatives (chi-square, Kolmogorov tests, methods of skew and kurtosis).

4) Stationarity in time and homogeneity in space were checked (Kolmogorov-Smirnov test).

5) Great blocks of data were formed (up to 500 000 samples) and data were fitted to normal distribution.

6) Autocorrelation and autospectrum functions were estimated.

7) Cross-correlation functions estimates were obtained.

Figure 1. Autonomous deep undersea acoustical module.

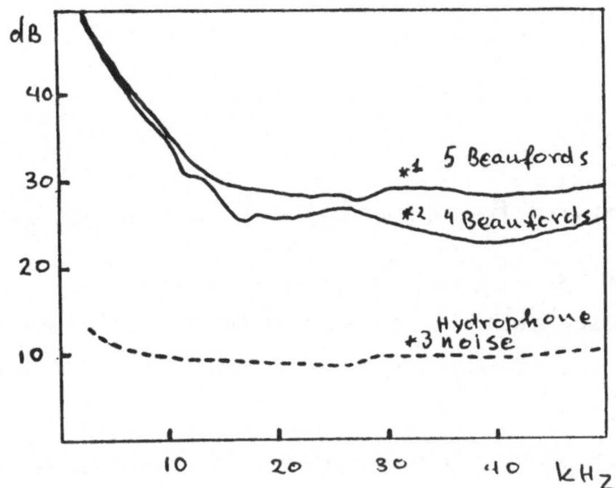

Figure 2. Noise spectrum at 4 km depth at different surface conditions - 1,2. Hydrophone noise -3.

The main result of background noise estimation can be formulated as follows: Input process can well be modeled by multidimensional contaminated normal distribution, its cross-correlation matrix being estimated during up to 5 minutes intervals.

3. FUTURE PLANS

We're going to use 8 modules of that kind deployed on a single vertical string to provide 2-dimensional localization of neutrino cascade. We have found temperature profile of water in the Ionic Sea (14 degrees near a bottom in contrast to 2 degrees in ocean) to be a factor which helps in construction of SADCO.

4. REFERENCES

1 G. Askaryan, B. Dolgoshein, JETP letters, 25 (1979) 277.

2 T. Bowen, Proc. 15 Int. Cosmic Ray Conf., 6 (1977) 277.

3 J. Learned, Phys. Rev. D19 (1979) 3293.

ACKNOWLEDGMENTS

The members of NESTOR and SADCO collaborations are very much obliged to the honorable Mr. John Vretakos, the Major of Pylos for a very fruitful help during experiments near Pylos.

The optical sensor for the lake Baikal project

R.I. Bagduev[b], L.B. Bezrukov[a], B.A. Borisovets[a], A.A. Doroshenko[a], G.N. Dudkin[c], V.B. Kabikov[b], L.A. Kuzmichov[b], G.V. Lisovsky[d], B.K. Lubsandorzhiev[a], P.A. Putilov[d], N.R. Romashkin[d], Z.I. Stepanenko[d], V.A. Toropova[d]

[a]Institute for Nuclear Research of Academy of Sciences, Moscow, Russia
[b]Moscow State University, Moscow, Russia
[c]Tomsk Polytechnical Institute, Tomsk, Russia
[d]Company EKRAN, Novosibirsk, Russia

presented by
B.K. LUBSANDORZHIEV

Abstract
The properties of a new mushroom shape phototube with large hemispherical photocathode are described. We obtained t_{FWHM} = 1,8 ns for transit time distribution of single photoelectron events.

1. INTRODUCTION

The design and development of underwater muon and neutrino telescopes shows the necessity to produce special phototubes with large acceptance, high sensitivity and extremely good time resolution.

Especially for this kind of application the PHILIPS Laboratories have designed the "smart" phototube XP2600 [1,2]. Its main components are an electro-optical preamplifier with a 35 cm diameter hemispherical photocathode and a conventional small PMT. A primary photoelectron emerging from the photocathode is accelerated by 25-30 kV and hits a thin scintillator covered by an aluminium foil. Then, the scintillator generates about 200 photons distributed exponentially in time with $t \sim 35ns$. These photons are registrated by a small PMT matched to the scintillator. the point of photoelectron origin for two azimuthal angles.

2. PHOTOTUBE PROPERTIES

Following the basic design of PHILIPS, we built samples of a soviet tube named QUASAR. Both tubes, XP2600 and QUASAR, were tested in lake Baikal during the winter expeditions in 1988 and 1989, using a trigger telescope to select vertical atmospheric muons (see figure 1) [3]. Figure 2 shows single muon detection efficiencies obtained from these tests.

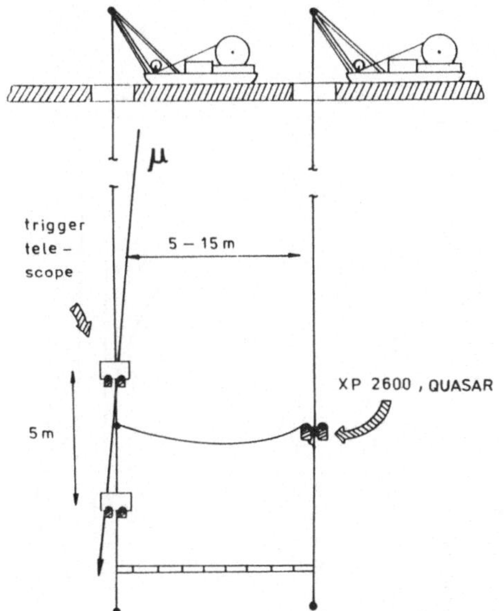

Figure 1. Experimental set–up for measurement of the single muon detection efficiency by "smart" phototubes in lake Baikal.

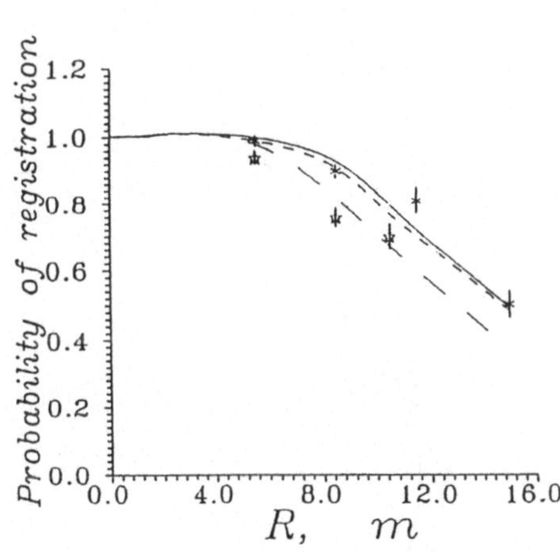

Figure 2. Efficiency of muon detection in lake Baikal as a function of muon distance R for vertical muons (QUASAR data are given for the previous designs with a spherical shape of different sizes)
—— PHILIPS XP2600, M.-Carlo
— — QUASAR ($d = 30cm$), M.-Carlo
- - - - QUASAR ($d = 35cm$), M.-Carlo
* * ** PHILIPS XP2600, exp. data
★★★★ QUASAR ($d = 30cm$), exp.data

Later we understood that for better reconstruction of muon and neutrino events by underwater neutrino telescopes a further improvement of the time measurement is necessary. To achieve this goal the tube shape must be modified and the overall gain of the scintillator-first PMT dynode combination must be increased.

We have developed a new version of the soviet tube,QUASAR-370, which is shown in figure 3. This tube basically follows the original "smart" phototube design, but with a nonspherical (mushroom) shape to minimize the photoelectron transit time differences to less than 1 ns. Figure 4 shows the cathode transit time difference for QUASAR-370 measured at an accelerating voltage of 25 kV, as a function of the point of photoelectron origin.

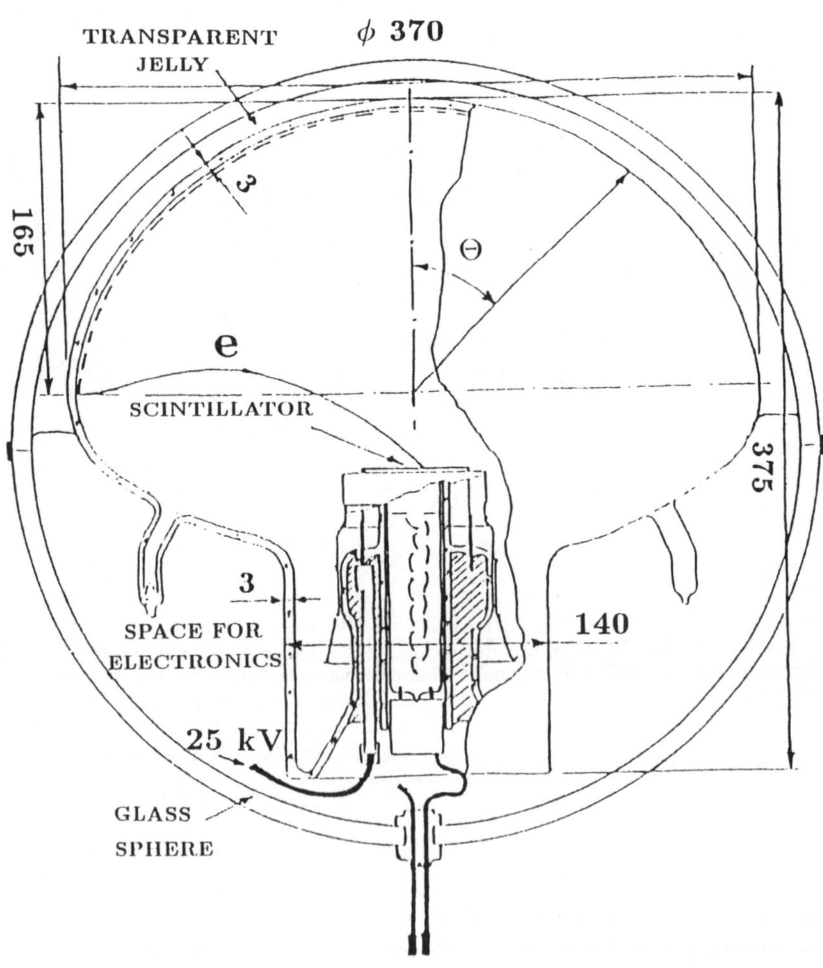

Figure 3. Optical module with QUASAR-370 phototube.

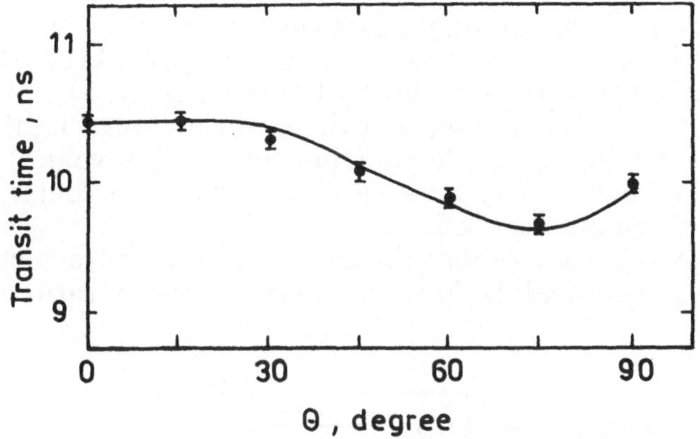

Figure 4. Measured cathode transit time difference for QUASAR-370 versus angle θ (see Figure 3) at an accelerating voltage of 25 kV. Curve – result of calculations.

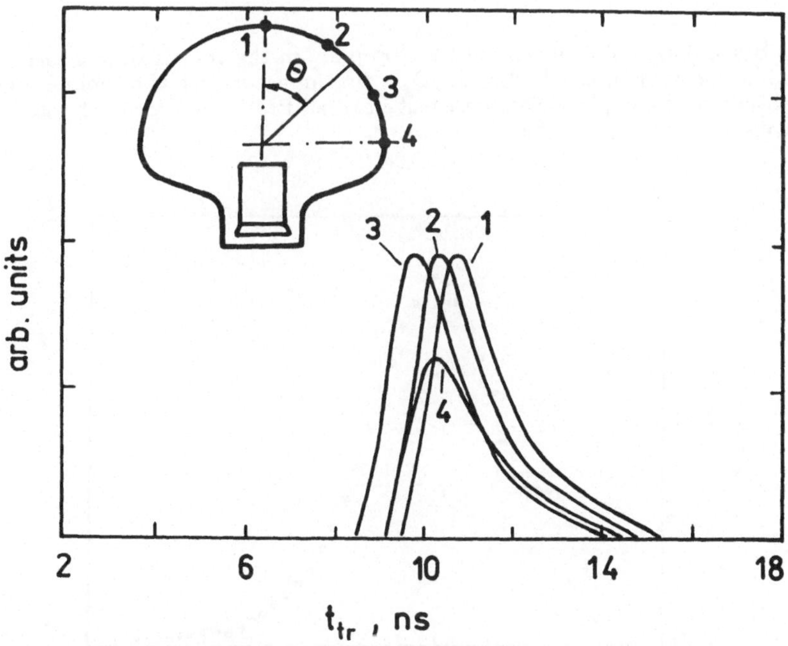

Figure 5. Single photoelectron transit time distribution for point-like illumination of the photocathode of QUASAR-370 at an accelerating voltage of 25 kV obtained by a LED with a flash of $\sim$ 1 ns duration. 1 - $\theta = 0^o$ (t_{FWHM} = 1,8 ns), 2 - $\theta = 30^o$, 3 - $\theta = 60^o$, 4 - $\theta = 90^o$.

Figure 5 shows the single photoelectron transit time distributions for point-like illumination of the photocathode of QUASAR-370, measured with a discriminator with low threshold (see Figure 6). We obtained $t_{FWHM}(\theta = 0^o) = 1.8\,ns$. (Note, that the duration of the light pulse itself ($\sim 1\ ns$) gives a non-negligable contribution to this value!) Moreover, it was observed [4] that the time resolution becomes better when the accelerating voltage is increased.

Figure 7 shows the charge distribution for single photoelectron events. The resolution of the peak is 70%. The peak-to-valley ratio is 2.5.

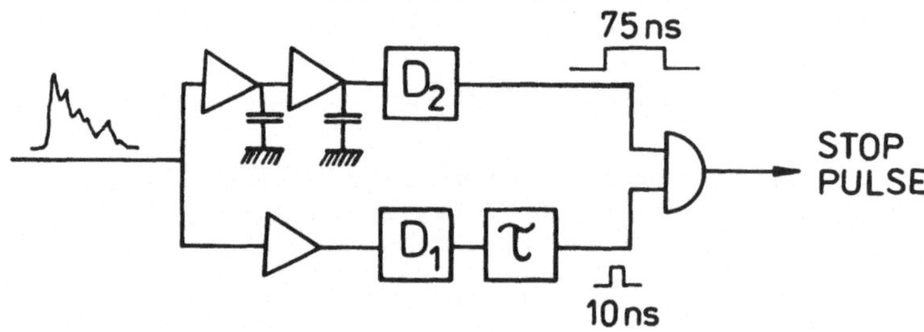

Figure 6. Block-diagram of the circuit for stop pulse in the transit time measurements. D_1 - discriminator with threshold 0,25 a_1, D_2 - discriminator with threshold 0,5 Q_1 (a_1 - mean amplitude of single photoelectron event in small PMT, Q_1 - mean charge of single photoelectron event in preamplifier).

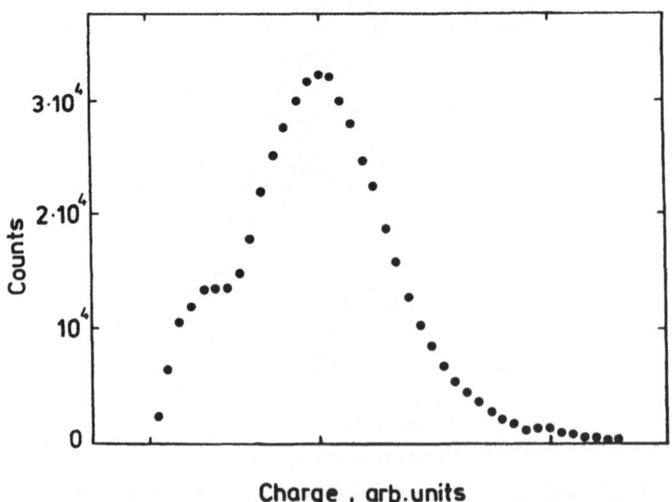

Figure 7. Charge distribution for single photoelectron events in the QUASAR-370 preamplifier at an accelerating voltage of 20 kV.

The counting rate of dark current pulses from single photoelectron events in the preamplifier of the investigated QUASAR-370 was measured to be $2 * 10^4 s^{-1}$ at 20 kV accelerating voltage and room temperature. Note, that this value will be twice smaller at the temperature of Baikal water and that it is smaller than the counting rate due to the natural water luminescence.

Figure 8 shows the relative sensitivity of tube QUASAR-370 as a function of the point of photoelectron origin.

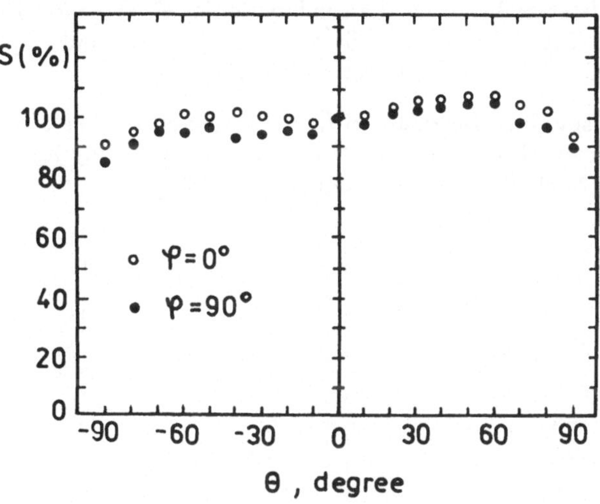

Figure 8. The relative sensitivity of tube QUASAR-370 as a function of the point of photoelectron origin for two azimuthal angles.

3. OUTLOOK

The next step in the improvement of QUASAR-370 parameters is to search for modifications towards new scintillators with brighter and shorter light flashes.

High sensitivity, excellent resolution in time and amplitude for single photoelectrons, rather low dark counting rate, large gain, absence of prepulses and immunity to the earth's magnetic field prove this device to be adequate not only for underwater muon and neutrino telescopes but for a new generation of large scale particle detectors for underground physics and cosmic ray physics. As a first additional application, we currently are investigating the possibility to look for high energy γ-sources by an surface array of QUASAR-370 tubes directed to the night sky.

REFERENCES

1 G. van Aller, S.-O. Flyckt, W. Kuhl et al., A "Smart" 35 cm diameter PMT Proc. French Phys. Society and Swiss Phys. Society Spring Conf. on Physics, Neuchatel 1986, Helv. Phys. Acta, 59 (1986) 1119.

2 S.-O. Flyckt, The "Smart" 15 inch PMT, Proc. DUMAND optical module Workshop, Sendai 1990, Tohoku University (1990) 4.

3 L.B. Bezrukov, B.A. Borisovets, A.V. Golikov et al., Properties and test results of a photon detector based on the combination of electro-optical preamplifier and a small PMT, Proc. Second Int. Symposium "Underground Physics-87", Baksan Valley (USSR) 1987, Nauka, Moscow (1988) 230.

4 I.A. Belolaptikov, L.B. Bezrukov, B.A. Borisovets et al., The lake Baikal deep underwater detector, Proc. 14th Int. Conf. on Neutrino Physics and Astrophysics, CERN, 1990, Nucl. Phys.B (Proc. Suppl.), 19 (1991) 388.

THE HEGRA EXPERIMENT

E. Lorenz

Max Planck Institute for Physics,
Foehringer Ring 6, D 800 Munich 40, FRG

Abstract
HEGRA is a large scintillation counter array for the observation of extended air showers. Details of the detector, its performance and initial results from a source search are presented. A short overview of the on-going upgrade program will be given.

1. Introduction

HEGRA (High Energy Gamma Ray Array) was originally proposed and set up by a group from the University of Kiel. The detector consisted of a 5 x 5 matrix of scintillation counters and 12 distant counters of the same type and was set up on the Roque de los Muchachos, La Palma, Canary Islands. The main aim was to search for gamma induced ultra high energy (UHE) cosmic showers from point sources and was a direct continuation of an earlier experiment at the University Kiel [1]. In the following years groups from the University of Madrid and the Max Planck Institute for Physics, Munich joined with the intention to increase the scintillator array by a significant factor. During 1991 the collaboration was further enlarged by groups from the universities of Hamburg and Wuppertal and the Yerevan Physics Institute, Armenia. The final goal of the collaboration is to build a large area detector, that allows one to measure extended air showers (EAS) between 10^{12} and 10^{17} eV and to measure many shower parameter simultaneously.

A brief summary of the physics program is:

i. Search for galactic γ ray point sources in the UHE region.
ii. Study of the emission parameters of sources such as the energy spectrum, slow time variability and periodical time structure.
iii. Measurement of the muon content of EAS from point sources, e.g. search for 'new' physics like a possible hadronic component of UHE γs or evidence for long lived new neutral particles.

iv. Study of the primary spectrum of cosmic particles around the so-called knee between 10^{15} and 10^{16} eV.
v. Search for diffuse γ radiation from the galactic plane.
vi. Correlated studies of some discrete sources by our detector in the TeV energy region and the Compton gamma ray observatory satellite in the GeV region.
vii. Study of the chemical composition above 10^{14} eV up to a few 10^{16} eV

2. The detector

Plans for a new scintillator array at La Palma were made in 1983 by the Kiel group. A small 5x5 array started data taking in 1988. In 1990 the original scintillator array was increased to 169 counters which are spread over an area of nearly $4x10^4$ m^2. A grid spacing of 15 mtr was chosen as a compromise between a low energy threshold and a large detection area. Each counters is comprised of a light tight containment with a 1 m^2 scintillator which is viewed by two photomultipliers (PM) from 40 cm distance in the original Kiel counters and from 90 cm distance in the later units. A fast 5" diameter PM serves for the trigger generation, the timing signal and the measurement of the charged particle multiplicity up to a factor 5 while the other PM with a 3" diameter and slow rise time covers the particle multiplicity range up to about a factor 1000. A LED light pulser, which is driven by a central common driver, is mounted 20 cm above each fast PM. The purpose of the LED pulser is to calibrate time resolution of the array. The counters are installed in weather protection huts made from zinc coated sheet metal. Each hut is covered by a 4.8 mm lead sheet for the conversion of the much more abundant low energy γs in the tail of EAS. The lead converters lower the energy threshold and improve the angular resolution. Figure 6 shows a vertical section through a counter hut. All huts were surveyed with a precision of $\pm$ 1 cm in the vertical axis and $\pm$ 3 cm in the horizontal axis. Each hut is connected by two 150 m RG 58U coax cables, one for the 5" PM signal and one for the light pulser, a twisted pair cable for the 3" PM signal and supply cables to the central electronic container. At the central electronics container the signals from the 5" PM are split and routed to two constant fraction discriminators (CFD) and a charge sensitive ADC while the signals from the low gain 3" PM are connected to peak sensitive ADCs after appropriate shaping. For the time being both the CFD thresholds are set to $\approx$ 0.2 MIP equivalent threshold. One of the CFDs is connected to a simple majority coincidence which requires ≥ 14 huts firing within 200 ns time. This large overlap time guarantees that we have no trigger bias for showers up to 75° zenith angle. The coincidence signal starts all the ADC gates and the common START for all TDCs, while the second CFDs provide the individual STOP signals after 300 nsec delay. LeCroy CAMAC TDCs of 11 bit range and a 250 ps least count are used.

The current EAS rate is 7.8 hz and the loss due to readout dead-time is about 25%. A Macintosh SE is used as on line computer. The data are written onto WORM optical discs of 800 MBytes capacity. For each event the universal time is also recorded with a least count of 200 nsec. We use as universal time reference a Rubidium clock which is periodically corrected by data from a Cesium standard clock via Navstar satellites.

Further detector parameters are summarized in Table 1.

Table 1: Parameters of HEGRA (summer 1991)

global coordinates	28.80 N, 17.90 W
altitude	2220 mtr above MSL
number of detector huts	169
detector grid spacing	15 m
angular acceptance	360^0 in f, up to 50^0 in theta
energy threshold	$5x10^{13}$ eV (50 % γ detection)
signal for a min. ionizing particle	≈ 30 photoelectrons
time resolution for single particle	σ= 1.3 ns
time resolution of electronics	σ= 0.5 ns
stability of LED pulser time calibration	σ= 0.3 ns
trigger condition	coincidence of ≥ 14 huts
EAS rate	7.8 hz
loss due to readout dead time	25 %
energy range	$5x10^{13}$ - 10^{17} eV
avg. angular resolution	≤ 0.7^0
predicted energy resolution	≈ 30% at 10^{14} eV
short term UTC timing error	200 ns
long term precision of the UTC timing	≈ 20 ms

Figure 1 shows a photo of the scintillator array.

Figure 1: The HEGRA detector on the Roque de los Muchachos, La Palma

3. Main resolution parameters

Two resolution parameters, the angular resolution and to a lesser extent the energy resolution, determine basically the performance of the scintillator array. In the absence of any device that identifies the gammas against the much more abundant hadronic flux one can identify point sources only as a local excess on the otherwise completely uniform illumination of the sky map by charged cosmic particles. The expected signal/noise is directly proportional to the angular resolution (while measuring time and total area enter only as a square root factor).
The calibration of the angular resolution is non trivial because one has no celestial 'test beam' from a strong known source which allows one to measure it directly. Early attempts used the so-called 'chequerboard' method. The array was divided in two interleaved subarrays of alternative counters. The incident direction of the shower was calculated with both subarrays and compared. While the procedure gives satisfactory results for large showers it suffers from low sampling and large detector spacing for small showers. An alternative method was proposed by G.W. Clark [2] to analyse the shadowing of UHE cosmic rays by the moon and by the sun. Figure 2 shows for our experiment the flux reduction of cosmic rays from the

direction of the moon and the sun. If we define the angular resolution σ_θ as the value where 67% off the deficit falls inside the corresponding space cone then we obtain a resolution of $\sigma_\theta = 0.66° \pm 0.1°$. The comparable values for the CASA array [3] and the CYGNUS array [4] are 1.25° and 1.0°, respectively.

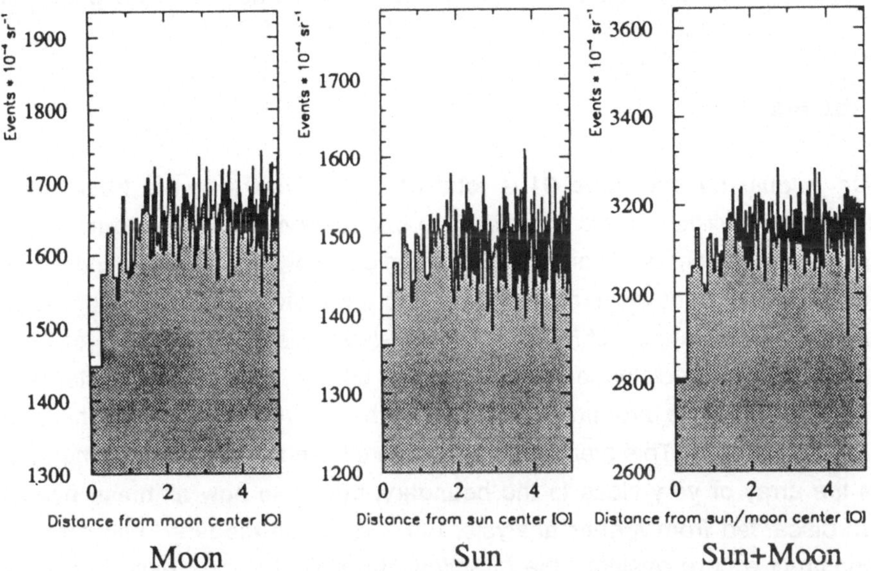

Figure 2 : Shadowing of cosmic rays by the moon, the sun and the combined data

It should be noted that one of the reasons for our high resolution is due to a very stable light pulser calibration of our electronics detector chain. The time calibration is provided by precision LED light pulsers mounted a few cm above the timing PMs. The light pulsers are triggered by a central system every hour in order to cope with the delay changes of the coax cables during daily (yearly) temperature cycles and the slow transit time changes of the pms. Another contributing factor to the stability was the use of constant fraction discriminators that were not affected by amplitude variations of the PMs during daily temperature cycles and the rise time degradation of the fast pulses after the 150 mtr of coax cable. It should be noted that the observation of the shadowing implies also a correct pointing of the array; this cannot be verified by the chequerboard procedure.

The energy of the initial particle is deduced from the shower size and age which are determined from the number of particles that pass the scintillators. We defined the particle number by dividing the actually measured pulse height data by the signal from minimum ionising particles (MIP) and applied appropriate angle corrections. It was found that the pulse height distribution of actual data was sufficient to calibrate the MIP values instead of recording the pulse height distributions of the abundant muons in separate calibration runs. The gain of the 3" pms was found to be very stable; HT voltage adjustment was done only once per year.

4. First results

Between August 89 and June 91 a total of 160 million showers have been recorded. About 95 million of this sample have been reconstructed up to now. We reconstructed the incident direction by fitting a conical shower front to the individual timing signals and the shower size and age from the individual pulse height data which were normalized in units of MIPs. The cone slope was either left free or fixed to 15 nsec delay/100 mtr distance from the shower center. The shower center was determined from the data around the counter with the largest pulse height by a center of gravity method. This method becomes unreliable in the case that the core is outside the array or very close to the boundary, but up to now all these events have been discarded from further analysis. For 98% of all triggered showers we could reconstruct a core position, the direction, the NKG function and fit quality parameters. Finally we required a shower size to exceed $\log(N_e) > 4.0$ and the shower core being at least 2 mtr inside the array bouyndary. About 70% of all triggers fulfilled these basic selection criteria for our global data sample.

Data reconstruction is performed on a DEC station 5000 with a rate of typically 50 hz.

We have searched the data for steady or sporadic directional excess from some possible sources of the northern hemisphere, namely the Crab nebula, Cygnus X-3, Cygnus X-1 and Hercules X-1. For the search we used a cone of 2.24° diameter centered on the source position on the sky globe. These data were defined 'on source'. The size of 2.24° is derived from our angular resolution and maximises the expected signal/background in case of a genuine source. 'Off source' data were derived from a ring extending from 2° to 5° around the potential candidate. The significance of the excesses or deficiencies were calculated by following the prescription of Li and Ma [5]. The flux limits for 90% confidence were obtained by assuming that both the background and the signal obey poisson statistics. The threshold energy of our data has been defined as the energy where the acceptance of our array is 50 %. This is based on the assumption that all incidendent particles were UHE gammas. A threshold energy of 50 GeV has been calculated from Monte Carlo simulations. The socalled median energy of all of our data corresponds to 80 GeV. No significant excess has been found. The result for some sources are listed in Table 2.

Table 2 Flux limits for selected sources

source	Cyg-X3	Cyg-X1	Her-X1	Crab
events 'on source'	22183	22902	22393	20427
events 'off source'	21814	22948	22091	20535
difference	+369	-46	+302	-108
σ(Li&Ma)	+2.42	-0.29	+1.97	-0.73
90% CL Flux [$cm^{-2}sec^{-1}$]	$3.8\ 10^{-13}$	$2.0\ 10^{-13}$	$3.5\ 10^{-13}$	$1.9\ 10^{-13}$

For the search for sporadic emission one has many choices for the time units. We scanned our data in intervals of siderial days. For each potential source we counted these 'days' to begin with the time of the lower culmination point of the source. This guarantees that each of these siderial days contained a full pass over our array. For all the scanned sources we found no excess above the expected fluctuations for the scan time. The only exception was a 3.8 σ excess from the direction of Cygnus X3 on January 20 th of 91. This excess is the largest observed one and coincides with the observation of a large radio flare [6] at the same time. Figure 3 shows the daily fluctuations from the direction of Cygnus X3 around the 20 th of January. The excess occurs basically within one hour. Our observation time is about 6 hours, therefore we cannot make a full phase analysis. Nevertheless the excess data are concentrated at a phase of 0.2-0.3 when we plot them in a 4.8 hour phasogram, using the cubic ephemerides of van der Klis and Bonnet-Bidaud [7]. At the same day (about 8 hours shift in universal time) two other arrays, Cygnus [8] at Los Alamos and CASA[9] at Utah did not observe any excess, while the underground experiment SOUDAN2 [10] claimed an increase of muons from the direction of Cygnus X3.

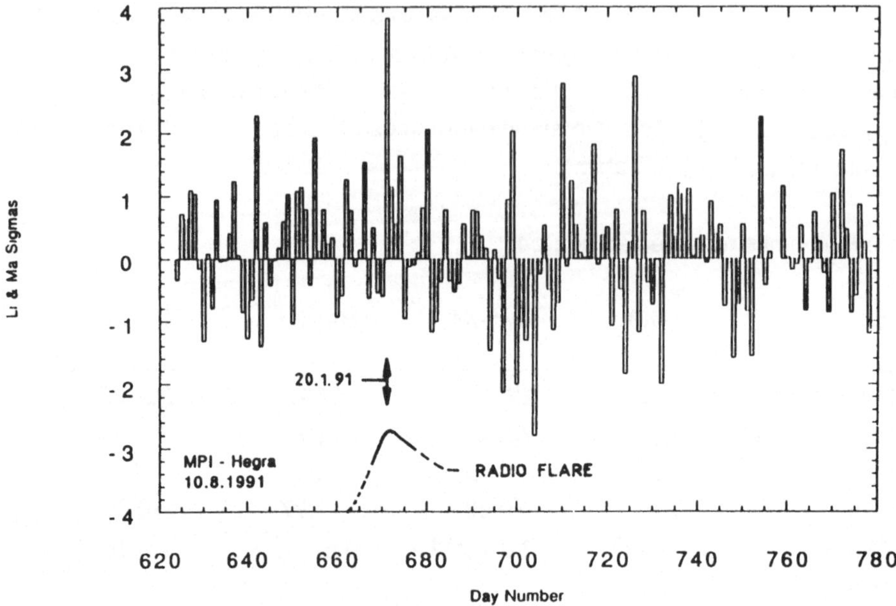

Figure 3 : Daily fluctuation of UHE cosmic rays from the direction of Cygnus X3

5. The upgrade program

For further progress in source searches with EAS we consider it of utmost importance to improve the sensitivity, to lower the detection threshold and to improve the γ/hadron separation (γ/h). Various steps are either already under construction or are in the planning stage:

a) We are currently increasing the scintillation counter density by a factor two in the central array area. This will allow us to lower the energy threshold to $E \approx 2.5 \times 10^{13}$ eV or in connection with trackable muons to $E \approx 10^{13}$ eV, thus a good overlap with air Cerenkov data can be achieved.

b) It is planned to install 34 muon tracking towers of $6 \times 3.4\ m^2$ sensitive area. The towers consist of high density absorbers interleaved with 6 layers of Geiger tubes from the dismantled Frejus proton decay experiment. One can obtain a direction measurement of the muons in one projection only, therefore the towers will have alternative orientations. We expect a track angular resolution of $\approx 1^0$. The observation of tracks gives a very high rejection against errors from breakthrough of energetic electrons. The detection of muons is not only essential for high γ/hadron separation and for the clarification of the existence of so-called muon rich showers from point sources but also important for the study of the chemical composition.

d) In collaboration with the Physics Institute from the Yerevan University it is planned to install 5 air Cerenkov telescopes. These telescopes can either track individual sources or follow one source with very high γ/h separation power due to a stereoscopic view of each shower. These telescopes would mainly track specific sources in a lower energy range from 10 TeV down to 1 TeV, but it is expected that sufficient data can be collected for a subset of events that overlap with data from the scintillator matrix and the muon counters. One telescope of 2.8 m mirror diameter (composed of 18 mirrors with 60 cm diameter each) and a 34 element pixel camera has been completed and should take test data early 92. The other four telescopes will have a mirror diameter of 4 mtr and a 128 pixel camera. These telescopes should be build in 92/93. From Monte Carlo simulations one expects a γ/h separation power of >1:500 for the coincidence mode of operation. Figure 4 shows a photograph of the first telescope.

Figure 4 : A photograph of the first Cerenkov telescope

e) We are currently building a novel air Cerenkov array which allows us to search for sources and to observe transients over one sterad of the night sky. This array should have a very good angular resolution of 0.5-5 mrad and promises to have a high γ/h separation of ≈ 50 in the standalone mode, ≈ 500 in combination with data from the scintillator array and > 10^3 after adding the information from the muon trackers. This separation might allow us to search for diffuse γ production to be expected from the galactic plane. The combined data from the Cerenkov matrix, the scintillator array and the muon trackers should open the possibility to obtain information of the chemical composition of the UHE cosmic radiation.

Figure 5 shows the basic concept of this novel detector. The Cerenkov light disc from EAS has at least a diameter of 200 mtr, has a cone slope of about one third of the corresponding particle shower disc and is much better defined in time. For detection of the light flash one views the night sky by a 7x7 matrix of large diameter hemispherical pms. The light collection of each station is augmented by a Winston cone like light funnel, but at the expense of restricting the observation angle to 1 sr. Special measures have to be taken because of the high background light level from the night sky. Part ob the background is cut out by Schott BG1 blue filters. The DC like background would destroy normal high gain pms. Therefore we use a special pm (EMI D668 K/5) with only 6 dynodes and a series of fast AC coupled amplifiers for the detection of the light flash signal. Figure 7 shows a vertical cut through one station. First test measurements with two prototype detectors mounted side by side confirmed the good time resolution of ≈ 0.6 nsec/station. The array should start data taking in the summer of 1992. We expect a yearly uptime of 15 %. Further details of the air Cerenkov array can be found elsewhere [11].The upgrade program should be completed within the next 18 months.

The improvements will result in a much higher data rate which will rise to > 30 hz during nighttime. Therefore we will replace the CAMAC-Macintosh SE online readout by Fastbus coupled to a DEC 5000 workstation. Figure 8 shows the layaout of the upgraded HEGRA detector with the location of the new elements.

6. Acknowledgements

I want to thank I. Holl for the help to prepare this paper and my colleagues from the HEGRA collaboration for providing me with some of the relevant information and figures.

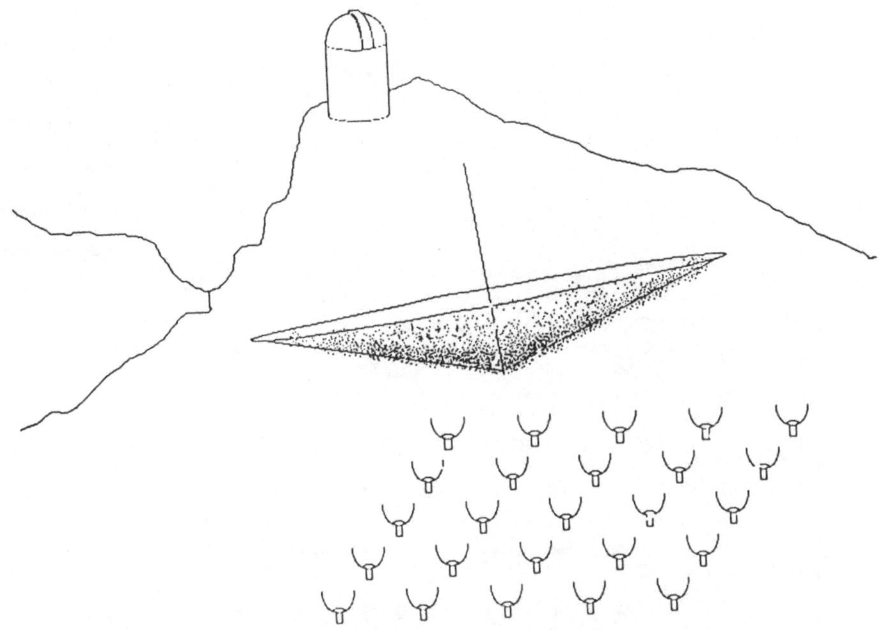

Figure 5 : Basic concept of the novel wide angle air Cerenkov matrix

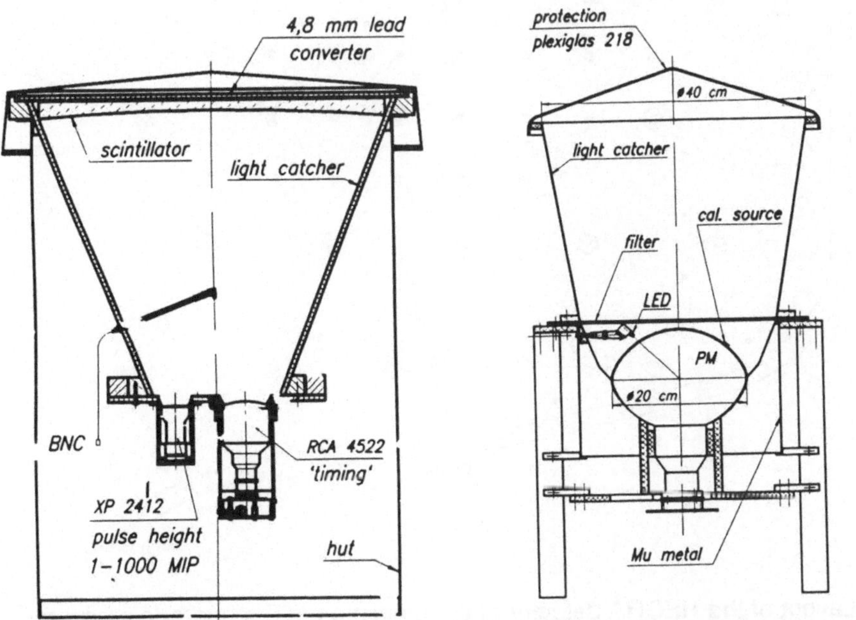

Figure 6 :Cut through a scintillator hut Fig 7 : Cut through a Cerenkov detector

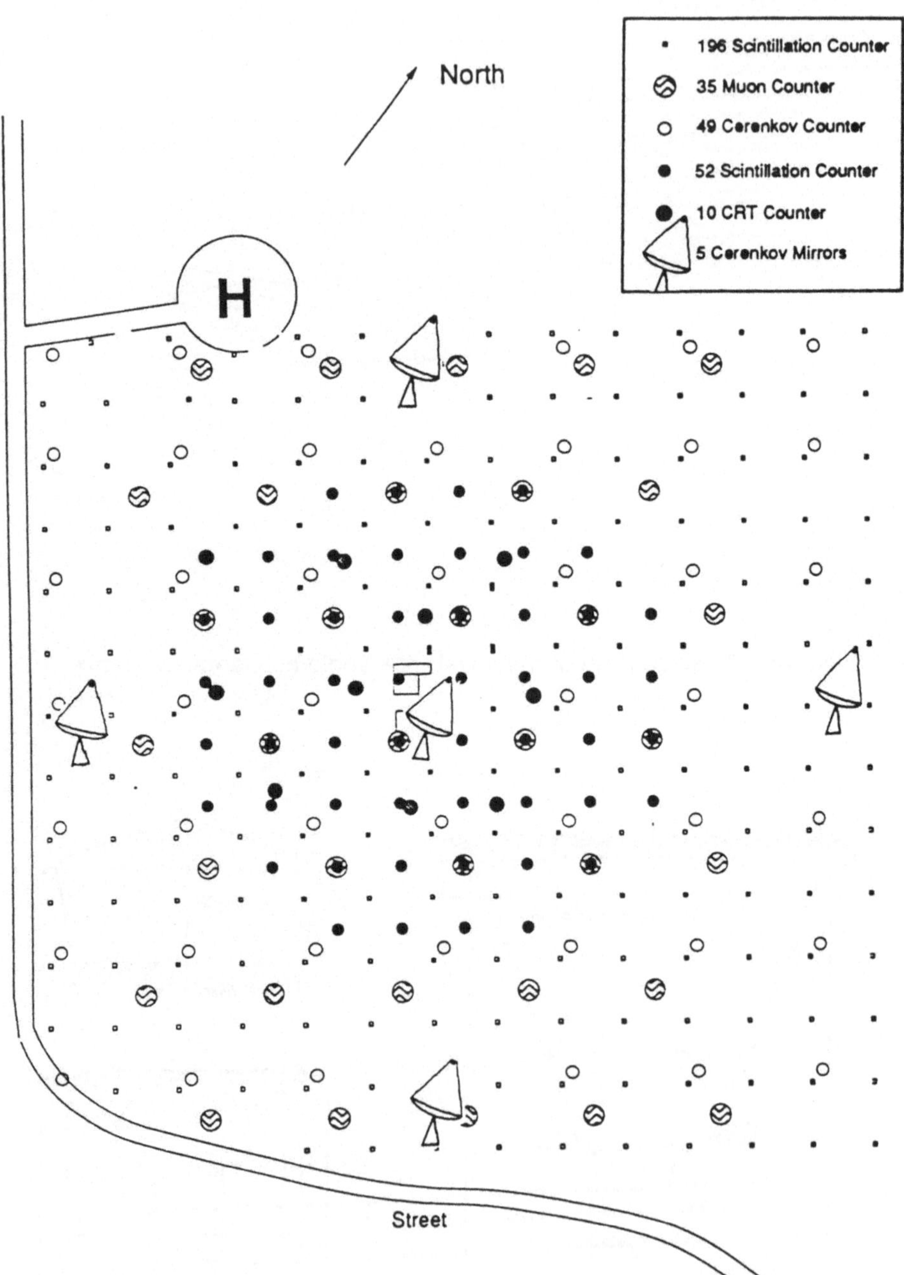

Figure 8: Layout of the HEGRA detector after completion of the upgrade program

7. References

1 M. Samorski and W. Stamm, Ap. J. (1983) L 17.

2 G. W. Clark, Phys. Rev. **108** (1957) 450.

3 B. E. Fick et al., paper OG 10.4.19. 22nd Int Cosmic Ray Conf. , Dublin 11-23, Aug 1991

4 D. E. Alexandreas et al., Phys Rev D43, No 5 (1991) 1735.

5 T.P. Li, Y.Q. Ma, Ap. J. **272** (1983) 317.

6 E.B. Waltman, Naval Research Labs, private communication.

7 M.van der Kliss, J.M. Bonnet-Bidaut, Ast. Ap **214** (1989)203.

8 D.E. Alexandreas et al., paper OG 4.3.18. 22nd Int Cosmic Ray Conf. , Dublin 11-23, Aug 1991

9 R. A. Ong et al., paper OG 4.3.7. 22nd Int Cosmic Ray Conf. , Dublin 11-23, Aug 1991.

10 M.A. Thomson et al., submitted to Phys. Lett. B.

11 M. Bott-Bodenhausen et al.,Proceedings of the Fifth Meeting on Advanced Detectors, NIM **A315** (1992) 236.

Surface water Čerenkov detectors

F. Bobisut

Department of Physics of the University and Sezione INFN, Padova, Italy

Abstract

The various efforts, which are being undertaken to build surface detectors of extraterrestrial neutrinos using the Čerenkov light technique in shallow bodies of water, are reviewed. Other techniques have also been proposed, but the water Čerenkov detectors represent the majority of the present proposals, due to their performance, reliability and cost. They would be able to act both as neutrino and gamma telescopes, improving the present sensitivities by orders of magnitude. Their common main features and capabilities will be discussed and the designs will be briefly presented.

1 MOTIVATIONS

Experiments to detect neutrinos from extraterrestrial origins started more than 25 years ago. In the MeV region solar neutrinos were detected first in the Homestake and later in the Kamiokande experiments at an unexpected low ratio. In 1987 neutrinos from the supernova explosion were observed in several laboratories, mostly in the IMB and Kamiokande detectors, providing a wealth of information on the astrophysics of the supernova and on the very properties of the neutrinos. In the higher energy regions, above $O(1GeV)$, some underground experiments, either dedicated or built primarily to search for nucleon decay, were able to observe interactions induced by neutrinos produced by cosmic rays in the atmosphere. In general the agreement with the calculated neutrino spectra and fluxes is rather good: see Fig. 1. A list of those experiments is given in table I, along with some yet to be completed or in the construction stage. No sources of high energy extraterrestrial neutrinos have been discovered; the limits on the fluxes of neutrino induced muons are: $\Phi_\mu(E_\mu \geq 2GeV) \leq few \cdot 10^{-14} cm^{-2} s^{-1}$. New, more sensitive, and therefore of bigger active area, but still exploratory instruments are now proposed to specifically search for neutrinos with $E_\nu \geq 10GeV$.

On general grounds very high energy cosmic neutrinos are expected, because the energy spectrum of the cosmic rays extends to the EeV ($10^{18}eV$) region and the nuclear interactions producing the high energy cosmic rays should provide also neutrinos.

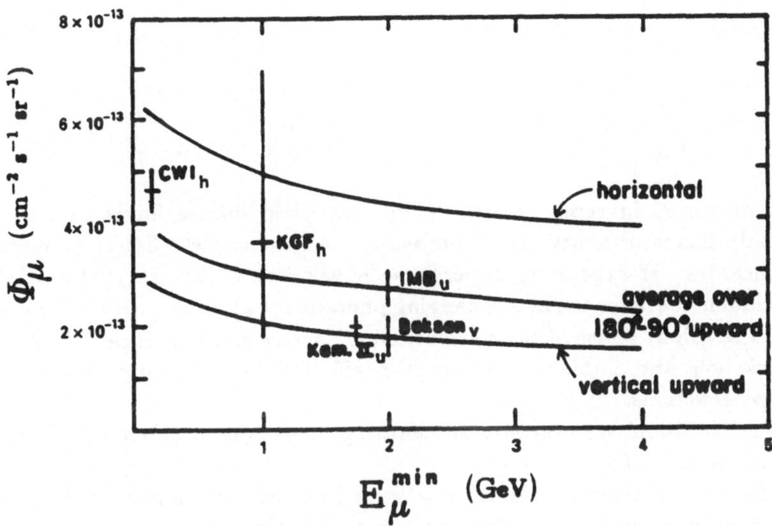

Figure 1: Comparison of measured and predicted fluxes of upward muons. (Reproduced with permission from T.K.Gaisser: Cosmic Rays and Particle Physics, ed. Cambridge Press); CWI [1]; KGF [2]; Kam [3]; Baksan [4]; IMB [5]

Table I

Experiment	Surface(m^2)	Technique
CWI	170	Flash tubes+scint.
KGF	50	Proportional tubes
KGF II	83	" "
SOUDAN	10	" "
NUSEX	12	Streamer tubes
Frejus	75	Flash tubes
IMB	380	Čerenkov
Kamioka	150	Čerenkov
Baksan	300	Scintillator
SOUDAN II	130	Proportional tubes
MACRO	$\sim 10^3$	Streamer tubes+ scint.
LVD	$\sim 10^3$	" " "
Superkamioka	$\sim 10^3$	Čerenkov

Many potential astrophysical sources of neutrinos produced via the processes

$$
\begin{array}{llll}
p+p\rightarrow & \pi^{\pm} & + & X \\
 & |\rightarrow & \mu & \nu_{\mu} \\
 & & |\rightarrow & e \quad \nu_{e} \quad \nu_{\mu}
\end{array}
\qquad
\begin{array}{llll}
p+\gamma\rightarrow & \pi^{\pm} & + & X \\
 & |\rightarrow & \mu & \nu_{\mu} \\
 & & |\rightarrow & e \quad \nu_{e} \quad \nu_{\mu}
\end{array}
$$

have been discussed in recent years. They may yield diffuse fluxes or beams of neutrinos, steady fluxes or short period emissions. An incomplete list of acceleration mechanisms includes: i) expanding supernovae envelopes; ii) binary systems, either "visible" or "hidden" because the accompanying photons are absorbed within the source; iii) galaxies in the bright phase of their evolution; iv) active galactic nuclei (AGN) with a massive black hole and a thick accretion disk, effective in producing TeV and PeV neutrinos via $p\gamma$ reactions.

A complete discussion, also on non accelerating sources of high energy neutrinos, can be found in [6] and [7] .

The identification of these accelerating sources has become controversial in recent years in high energy gamma astronomy. In the exploration of the sky neutrinos are very effective, because not only, like photons, they can travel undisturbed through the galactic magnetic field, but, unlike photons, they do not interact with the microwave background and can traverse large amounts of matter. The detection of these neutrinos is extremely important for two different reasons: i) they would provide unique information on astrophysics, on the origin of the cosmic rays and on the evolution of the universe; ii) if a reasonable number of such high energy ($\geq TeV$) neutrinos are detected, neutrino physics can be studied at energies beyond the present - and foreseeable- accelerator energies (this requires of course some energy resolution of the detector).

Other fundamental questions can be addressed by a detector designed for cosmic neutrinos: the production of neutrinos from the annihilation of cold dark matter candidates (e.g. neutralinos) trapped in the Sun or the Earth; the neutrino oscillations, comparing the expected with the measured rates of events induced by atmospheric neutrinos or by accelerator beams aimed at the detector.

2 DETECTION OF THE NEUTRINOS

The very large target mass needed to produce detectable neutrino interactions can be obtained using the Earth as a target [8],[9],[10]: the neutrinos interacting in the rock via charged current interactions generate muons that can travel over long distances and be detected . The range of the μ increases with E_{ν}, an approximate formula being:

$R_{\mu} \simeq 2.5 \cdot 10^{5} \ln(1 + 2E_{\mu}(TeV))\ m;$

in a similar way increases the rock target mass, which is essentially the product of the detector area and R_{μ}. To improve the present sensitivity by a significant amount a detector with an area $A \sim 10^{5} m^{2}$ is needed, for practical reasons located on surface and therefore recording upward-going muons. Atmospheric neutrinos will also be recorded and they will be interesting in their own, but their contribution to the sample of events should be singled out using their approximately isotropic distribution and their energy spectrum, which is expected to be much steeper than those of the cosmic neutrinos.

The reason why most of the proposed detectors are of the water Čerenkov type is that the technique has been successfully used by the IMB and Kamiokande experiments, a large area can be reliably instrumented and the Čerenkov light directionality and time of flight measurements allow to obtain:

a) sensitivity to high energy cosmic neutrinos;
b) high muon detection and reconstruction efficiency;
c) good angular resolution;
d) energy discrimination capability.

The idea is to instrument a natural or artificial body of water with several planes of large photocathode photomultipliers (PMTs); a light-tight bag encloses the sensitive volume of purified water and an extra depth of water on top acts as a passive shield against the soft component of the cosmic rays showers. Different collaborations are studying these projects: GRANDE in the U.S. [11], LENA in Japan [12], NET in Europe [13] and PAN in Sweden.

3 PROPERTIES OF THE DETECTORS

3.1 Acceptance and angular resolution

The starting point of the extensive studies that have been performed to obtain the best experimental conditions is the MonteCarlo (M.C.) generation of muons that traverse the detector, whose rate is given by

$$N_{\mu}(E > E_{th}) = \int_{E_{th}}^{E_{max}} \frac{dN_{\nu}}{dE_{\nu}} P(E_{\nu}, E_{th}) \, dE_{\nu}$$

where $P(E_{\nu}, E_{th})$ is the probability that a neutrino with energy E_{ν} produces a muon traversing the detector with $E_{\mu} > E_{th}$ [14]. This probability depends on the $\nu N \rightarrow \mu X$ cross section, $\sigma^{\nu}(E_{\nu})$, and on the various interactions the muon undergoes in the rock. $\sigma^{\nu}(E_{\nu})$ increases with the neutrino energy and the muon range with the muon energy: these two properties naturally increase the sensitivity to the higher part of any neutrino energy spectrum.

Kinematics provides the means of obtaining the neutrino direction, because the higher the neutrino energy the more aligned is the μ to the ν direction: for large E_{ν} the average production angle between the muon and the neutrino can be written as

$$\theta_{\mu\nu} \sim 2.6 \; (E_{\nu}/100Gev)^{-1/2} deg$$

(e.g. $\theta_{\mu\nu} \sim 0.8^{0}$ at 1 TeV).

Less than 20% of the neutrinos with energy between 10 and 100 GeV give a muon ($E_{\mu} > 10GeV$) within 1^{0}, but about 80% of them produce the muon within 5^{0}; above few TeV 80% of the neutrinos give a muon within 1^{0}: a small acceptance angle selects the high energy part of a neutrino spectrum. It is generally assumed that the energy distribution of cosmic neutrinos should be represented by a power law :

$$\frac{dN_\nu}{dE_\nu} = CE^{-\gamma}$$

with a spectral index γ smaller than the one observed in the atmospheric neutrino spectrum. Using this formula to calculate $N_\mu(E_\mu > 10GeV)$ the acceptance for a given angle θ_c around the direction pointing to a possible source is obtained as a function of γ. This is reported in Table II: for $\theta_c = 1^0$ the angular acceptance is between 65% and 90%.

Table II

θ_c	$\gamma = 2.1$	2.3	2.5	2.7
1^0	89%	82%	73%	63%
2^0	97%	93%	88%	82%
5^0	100%	99%	98%	97%
10^0	100%	100%	100%	100%

The detection of few events in a year within this angular aperture would indicate a definite signal from a distant source, since the number of upward-going muons from atmospheric neutrinos would be less than one event/year. Both for the acceptance and for the signal/noise ratio the angular resolution should be better than 1^0. This can easily be reached, as shown in Fig. 2, where the angular resolution expected in the GRANDE detector is reported as a function of the zenith angle.

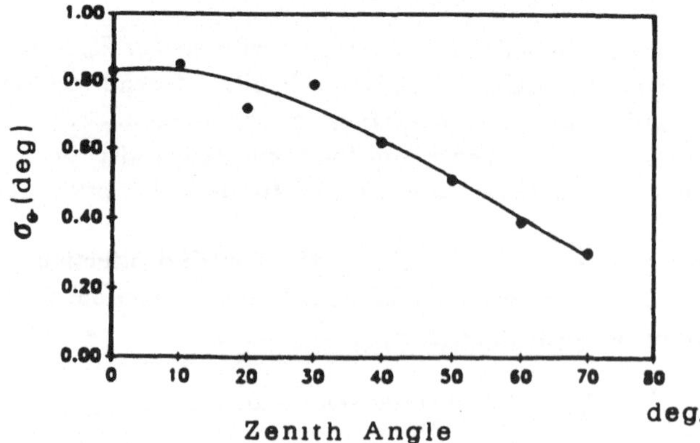

Figure 2: Angular resolution for muon tracks obtainable in the GRANDE detector.

3.2 Reconstruction efficiency

Using the light patterns of the PMTs, the pulse heights (i.e. the number of photoelectrons, n_{pe}, in each PMT) and the relative time a straight line fit can be performed to reconstruct the muon track. In the GRANDE and NET calculations it was recognized that a vertical segmentation of the detector, namely several PMT planes at distance S_p, is essential to obtain both a good tracking and a high rejection of the background induced by downward-going muons. Fig. 3 shows the efficiency, ϵ_1, for reconstructing a track at an angle smaller than 1^0 with respect to the true direction versus the zenith angle θ and the PMT separation s: ϵ_1 will be larger than 0.9 for muons with zenith angles $\leq 60^0$.

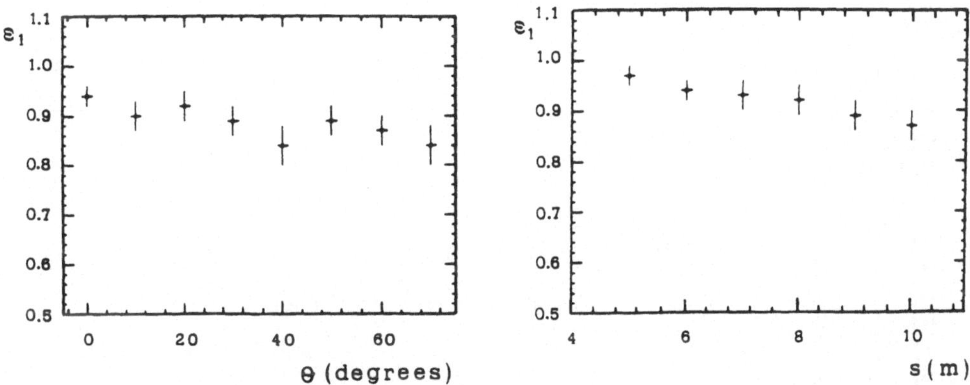

Figure 3: Reconstruction efficiency within 1^0 for muon tracks in the NET detector; a) versus the zenith angle; b) versus the PMT spacing.

3.3 Energy measurement

Although the energy of the interacting neutrinos cannot be directly measured, useful information on the muon energy can be obtained from the amount of Čerenkov light emitted in the detector. At high energy the μ energy loss is approximately proportional to E_μ and the most important processes are pair production, producing e^+e^- pairs mostly with small energy transfers, and Bremsstrahlung, giving rise also to very large energy transfers. The recorded n_{pe} depend on the muon energy. Fig. 4 shows the distribution of the n_{pe} collected by the individual PMTs in different E_μ regions, as calculated in NET. For instance if $E_\mu > 1TeV$, half of the events would have $n_{pe} \geq 600$, whereas for $E_\mu < 100GeV$ practically all events would have smaller n_{pe}. This can be used to select high energy events, decreasing thereby the atmospheric neutrino contribution, to set on- or off-line different energy thresholds and to study the energy composition of the sample.

Recently a M.C. calculation has been performed in the PAN collaboration [15] to simulate in full detail the showers produced by muons with $E_\mu > 1TeV$ and the emitted Čerenkov photons. The muon energy was reconstructed using the photon counts in the PMTs, taking into account not only their number in each plane, but also their densities in two different circles. The results obtained, after optimization of the parameters, is shown in Fig. 5 for a detector with 4 planes, $S_p = 25m$ and $s = 5m$. Even if these calculations have to be repeated with different configurations, it looks very encouraging that with this procedure an energy resolution of about 30% can be achieved.

A very important point concerning the energy determination, still under discussion, is the possibility of some calibration.

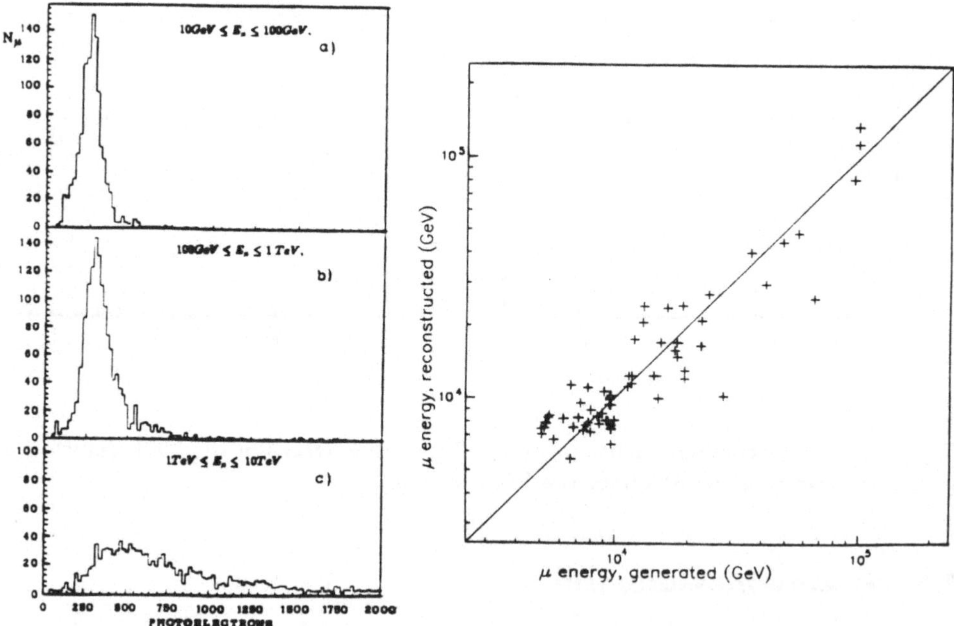

Figure 4: Total number of photoelectrons collected in the NET detector for different muon energy intervals. Figure 5: Generated vs. reconstructed muon energy using the n_{pe} for a particular multi-layer setup: see text.

3.4 Trigger requirements

The electronic trigger to be adopted to record with high efficiency upward-going muons should be very selective against the background due to accidental coincidences produced by downward-going particles, that could give rise to Čerenkov light in the upward direction. The discrimination factor required for such large area detectors is $10^{10} \div 10^{11}$.

The directionality of the Čerenkov light and the coincidence between different PMT planes provide the means to achieve this figure, giving at the same time a high trigger efficiency. The triggering scheme is the following: a number N_{PM} of PMTs in the lower plane C (see Fig. 6a) are required to give a signal in a given time interval, then delayed coincidences are expected in the planes B and A. When the three signals are in coincidence the event is recorded. The trigger efficiency, η_T, for various cuts on the minimum number N_{PM} and on n_{pe} , is shown in fig. 6c. The photons in the Čerenkov cone produce definite light spots in the detection planes. To exploit this feature each plane can be divided, for trigger purposes, in grids (see Fig. 6b) and the events are triggered only when N_{PM} PMTs in adjacent elements of the grid give the coincidence. This requirement still provides an $\eta_T \sim 1$ for $n_{pe} \geq 2$ with at least four PMTs in each plane, see Fig. 6d, and reduces the spurious triggers by a considerable amount.

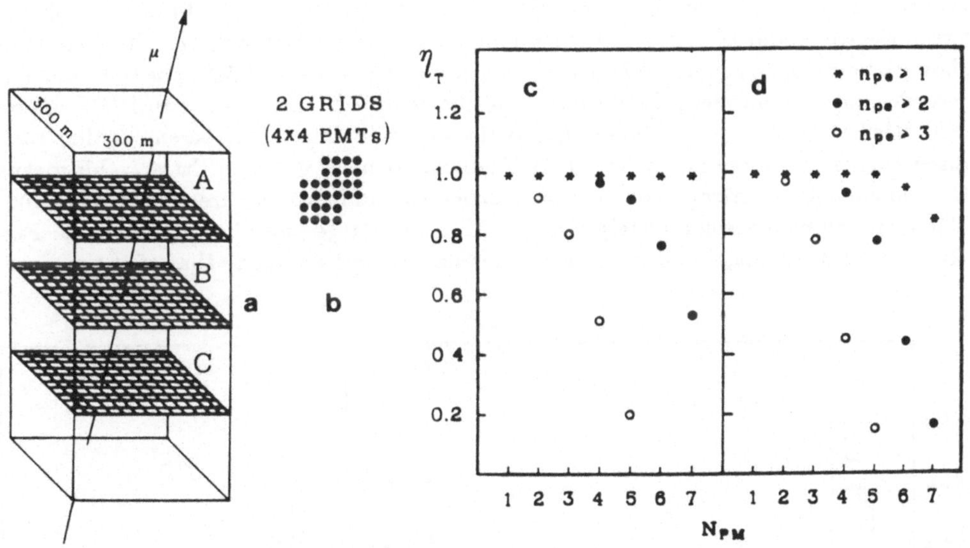

Figure 6: a) A muon crossing a three-layer detector; b) segmentation of the planes for triggering purposes; c) d) trigger efficiency vs. the minimum number of PMTs in the coincidence without and with the segmentation b).

The accidental coincidences due to downward-going muons are expected to give a trigger rate $R_{acc} \simeq 10^{-2} Hz$; the requirement of having a definite light spot, as discussed above, and the angular correlation of the spots in the three planes reduce this rate by a factor 10^{-3}. An upward-facing plane of PMTs, placed in the middle plane position B, acting as an offline veto would further reduce the accidental trigger rate by a factor 10^{-2} - 10^{-3}. In conclusion

$R_{acc} < 10^{-7} Hz = 10^{-2}/day$

which is absolutely negligible, even before any pulse height analysis, track reconstruction and fitting, visual scan .

Another potential source of background are downward-going atmospheric muons scattering in the surrounding rock and energetic enough to cross the detector. Accurate calculations [16] showed that the rate of these events is negligible for $E_\mu \geq 6\,GeV$.

3.5 Electron identification

Electron neutrinos and antineutrinos of very high energy can interact in the detector and eventually produce contained electron induced showers. The electromagnetic cascades would give rise to very characteristic events. Preliminary, and very computer time demanding, M.C. calculations [17] have shown that the number of PMTs with large numbers of photoelectrons is much bigger for electrons than for muons of similar energies. In principle the n_{pe} distribution in the PMTs can be used to distinguish between electrons and muons, but then the question arises : how well can these electron showers be reconstructed? When the correlation between the PMT position and the recorded time is analyzed, the dispersion of the recorded time for the i-th PMT can be calculated : $\Delta t_i = t_i - t_{exp}$, where t_{exp} is the expected time. Fig. 7a shows that such dispersion is very large for the electrons if almost no cut is made on the n_{pe}; this means that no good geometrical reconstuction is expected and hence no direction information. The situation changes completely if only PMTs with large pulse heights are used: Fig. 7b shows that with $n_{pe} > 30$ the dispersion is narrow and similar to that of the muons.

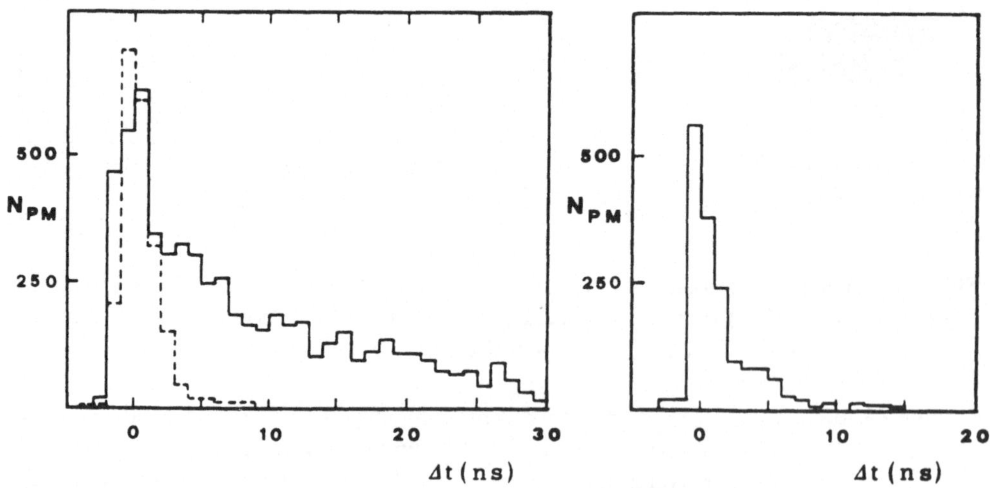

Figure 7: a) Difference between measured and expected time in the PMTs for electron showers (full line) and for muons (dashed) : $n_{pe} \geq 2$; b) electrons: $n_{pe} > 30$.

4 PROPOSED DETECTORS

4.1 GRANDE

The GRANDE collaboration proposed to equip an active area $A \doteq 3.1 \cdot 10^4 m^2$ in an artificial quarry in Arkansas, Fig. 8, to obtain an upward neutrino telescope with a threshold $E_\mu > 6GeV$ and a downward gamma ray telescope with threshold $E_{th} = 2\ TeV$. The gamma detector is designed to detect and distinguish γ, μ and hadrons and to measure the shower direction from the timing pattern of the PMTs with a resolution of 0.3^0. The upper layer acts as a total absorption calorimeter for γ's and hadrons, whereas the deeper upward-facing plane records the muons in the shower. Unfortunately the experiment was not approved and recently a proposal for a scaled down prototype has been submitted: referring to Fig. 8, the diameter of the active surface will be 20 m, $S_p = 5m$, the PMT spacing $s = 3m$. Its purpose is to :

i) demonstrate the up/down discrimination attainable in a detector at the surface of the earth;

ii) demonstrate the performance of the triggering and electronics schemes;

iii) test the plastic bag construction and deployment methods;

iv) obtain actual data to be extrapolated to a full sized detector.

The downward muon rate going through the three layers is expected to be about 11 kHz: a six months running period would test the required discrimination factor of 10^{11}. On the other hand the rate of upward going muons from atmospheric neutrinos is expected to be around 1/34 per day, thus providing 5 events in the same period, useful to test the expectations and the presence of unknown sources of such triggers. The construction time schedule is two years.

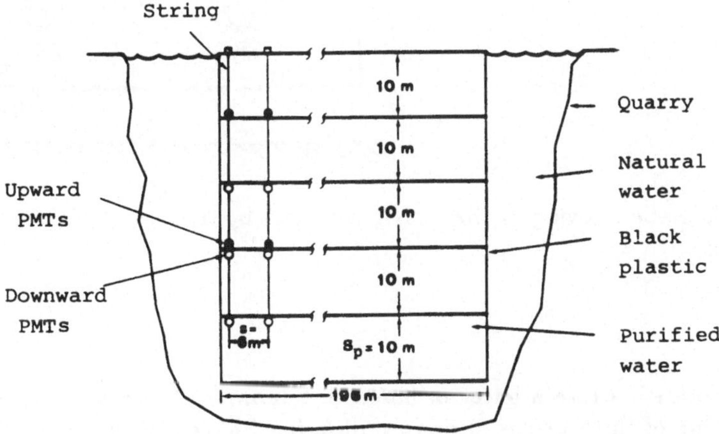

Figure 8: Schematic drawing of the GRANDE I detector

4.2 LENA

The LENA project is shown in Fig. 9a; the area is several $10^4 m^2$ and one active volume with only one downward-facing plane of PMTs is proposed for neutrino induced muon detection and two upward-facing ones for the γ telescope. A very large photocathode area coverage, $\sim 3\%$ of the total, is foreseen and is expected to give an angular resolution of $\sim 1^0$. To test the feasibility and the performance of such scheme a prototype has been built this year in Lake Motosu in Japan, see Fig. 9b. The bag, the cables and the PMTs were deployed last Summer, the detector is now ready and it is planned to take data for a half year.

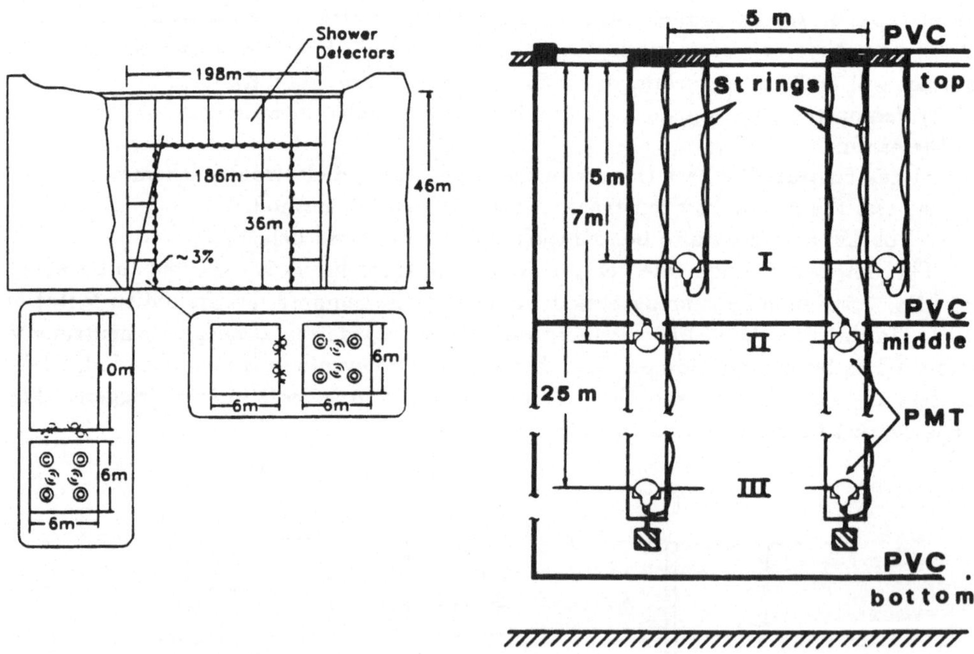

Figure 9: a) Schematic drawing of the LENA detector; b) the prototype deployed in the Lake Motosu

4.3 NET

The NET collaboration wrote a letter of intent in Summer 1990 to build a neutrino telescope consisting of three planes of downward-facing PMTs and one upward-facing one for veto purposes. The main parameters are: $A = 300 \cdot 300 m^2, S_p \simeq 15m$, $\simeq 7000$ PMTs with spacing $s \simeq 5m$. The energy threshold is $E_\mu \geq 10 GeV$. The expected reconstruction and trigger efficiencies were shown in Fig. 3 and 6.

It has been demonstrated that the detector can be placed in an existing quarry, suitably reshaped, and that the necessary water purification can be obtained.

Further studies are in progress to select the best configuration for detecting neutrinos in the PeV region and above, interacting close to or above the detector and to present a proposal in a year from now.

4.4 PAN

A Swedish group is studying the possibility of building a large detector in a lake in the northern part of the country: no written proposal has yet been circulated.

5 CONCLUSIONS

The sensitivity of a $10^5 m^2$ detector for a steady distant source, taking a year of observation time corresponds to a muon flux

$\Phi_\mu(E_\mu > 10GeV) \sim 10^{-16} cm^{-2} s^{-1}$

and, assuming a spectral index $\gamma = 2.1$, to a neutrino flux $\Phi_\nu \leq 10^{-8} cm^{-2} s^{-1}$. The sensitivity will be of course much better for neutrino bursts of short duration.

The predicted rate due to high energy neutrinos emitted in the expansion of supernovae remnants at a distance d is [18]:

$N_\mu(E_\mu > 10GeV) = n\,(L_p/10^{33} Js^{-1})\;(d/10kpc)^{-2}/week$

where $n = 25 - 2.5$ for spectral indexes γ = 2.1 - 2.3.

In conclusion, a water Čerenkov detector on surface is a very appropriate tool to study high energy neutrinos from extraterrestrial origins and γ rays. It will be an exploratory instrument for astrophysics and particle physics: to make a sensible improvement over the present sensitivity and to observe some very interesting phenomena that are theoretically predicted, its area has to be $\leq 10^5 m^2$. All the technical aspects (water transparency, PMTs performance, trigger rates, etc.) have been studied by many groups: the detectors are feasible and the hope is that with the shortest delay at least one of the proposals will enter into the construction stage.

References

[1] M.F.Crouch et al., Phys.Rev. D18 (1978) 2239

[2] M.R.Krishnaswamy et al., Proceedings 15th Int. Cosmic Rays Conference, Plovdiv 1977, Vol. 6, 85

[3] Y.Oyama et al., Phys.Rev. D39 (1989) 1481

[4] M.M.Boliev et al., Proceedings 15th Int. Cosmic Rays Conference, Paris 1981, Vol. 7, 106

[5] R.Svoboda et al., Ap. J. 315 (1987) 420

[6] V.S.Berezinsky in Proceedings of the Third International Workshop on Neutrino Telescopes, Venice 1991, ed. M.Baldo Ceolin, pag. 125

[7] F.Halzen, ibidem, pag. 387

[8] M. A. Markov in Proceedings of the Rochester Conference on High Energy Physics, 1960, pag. 578.

M. A. Markov and I. M. Zheleznykh, Nucl. Phys. 27,(1961), 385.

[9] F. Reines, Ann. Rev. Nucl. Sc. 10, (1960), 1.

[10] K. Greisen, ibidem pag. 63.

[11] A.Adams et al., Proposal to construct the first stage of the GRANDE facility, 1990

[12] M.Koshiba in Proceedings of the International Workshop on Neutrino Telescopes, Venice 1988, ed. M.Baldo Ceolin, pag. 151

[13] L.Moscoso et al., A Neutrino Telescope, 1990

[14] V.S. Berezinsky, C. Castagnoli and P. Galeotti, N. Cim. C8, (1985), 185.

T. K . Gaisser and A. F. Grillo, Phys. Rev. D36, (1987), 2752.

C. Quigg, M. H. Reno and T. P. Walker, Phys. Rev. Lett. 57, (1986), 774.

[15] S.Johansson, Measuring the energy of multi-TeV muons with a water Černkov detector, paper presented at the 21st International Cosmic Ray Conference, Adelaide 1990

[16] J. W. Elbert et al., in Proceedings of the International Workshop on Neutrino Telescopes, Venice 1990, ed. M.Baldo Ceolin, pag. 231.

[17] L.Moscoso, private communication

[18] V.S. Berezinsky and O. Prilutsky, Astr. J. 66, (1978), 325.

T. K. Gaisser and T. Stanev, Phys. Rev. Lett. 58, (1987), 1695.

MACRO at Gran Sasso: results and prospects

The MACRO Collaboration ‡

Presented by Carlo Bemporad

INFN and Dipartimento di Fisica, Università di Pisa, Italy

Abstract

The characteristics of MACRO are reviewed. Results on a search for gravitational stellar collapse and on UHE-astronomy are presented. The MACRO upgrading is discussed.

1. THE MACRO DETECTOR

MACRO (Monopole Astrophysics Cosmic Ray Observatory is located in the Hall B of the Gran Sasso Laboratory, at an average depth of ≈ 3500 $m.w.e.$. The apparatus is optimised for searching GUT monopoles, but it studies also UHE-astrophysics and gravitational stellar collapse, being sensitive to all the penetrating components of the cosmic radiation. For a complete description of MACRO we refer to [1].

MACRO is a large area detector, the acceptance for an isotropic particle flux is $S\Omega = 10^4\ m^{-2}sr$; its final size will be $72 \times 12 \times 10\ m^3$. The apparatus is subdivided into 6 supermodules (SM's), 6 in the lower and 6 in the upper part. The 6 lower SM's are now complete and 3 SM's are taking data; the 6 SM's will all be operational at the beginning of 1992; the construction of the upper part just started and the full MACRO should be completed by the end of 1992.

Each SM of the lower level consists of 2 layers (top and bottom) of liquid scintillation counters, 10 streamer tube layers separated by absorbers, and of a central and passive track-etch detector for highly ionizing radiation. The MACRO sides are closed by 1 scintillation counter and 6 streamer tube layers. The design of the upper part (in the form of a Π placed on the lower part) foresees a layer of scintillation counters and 6 streamer tube extra layers.

1.1. The scintillation counter system

The total liquid scintillator mass (≈ 1000 $tons$) is subdivided into "horizontal" and "vertical counters". The counters of the horizontal layers have dimensions $1200 \times 75 \times 25\ cm^3$, the ones of the vertical layers have dimensions $1200 \times 35 \times 75\ cm^3$; they are all made of thick PVC internally Teflon-lined for light total reflection. A high transparency mineral-oil based scintillator fills the counters for a total active length of 1120 cm. The remaining space, up to the total physical counter length, is occupied by two compartments, filled with non-scintillating mineral-oil and in optical contact with the scintillator. Two (ϕ 20 cm) photomultipliers (only one photomultiplier in the vertical counters), immersed in the oil compartments, collect the light at each counter end.

The energy resolution on events occuring in a large volume container is affected by position dependent corrections; in a counter which has a dominating longitudinal dimension, the event position is well determined by light time-of-flight differences and the resulting energy resolution is at its best. The light attenuation in the long counters is approximately exponential ($\lambda \approx 12\ m$) plus a non-exponential rise near to the PMT's. The energy resolution is $\sigma_E/E = 0.3\sqrt{E}$, the time resolution is $\sigma_t \approx 1.5\ ns$, the spacial accuracy for crossing μ's is $\sigma \approx 15\ cm$.

1.2. The streamer tube system

The streamer tube system consists of ≈ 5000 wires per SM. Eight tubes, each having dimensions $3cm \times 3cm \times 12m$, are combined in a single chamber. The tubes utilize $100\mu m$ anode wires and a graphite cathodes. The tubes operate in a limited streamer regime, with a gas mixture of He (75%) and n-pentane (25%). A two-dimensional readout is performed using signals from the anode wires (X-view) and 26.5° stereo pickup strips (D-view). The overall tube efficiency is $\approx 98\%$. Spacial accuracies for the two views, determined by selecting μ-tracks crossing 10 horizontal planes, are $\sigma_{x'} = 1.1\ cm$ and $\sigma_D = 1.2\ cm$.

2. SEARCH FOR A STELLAR GRAVITATIONAL COLLAPSE

We present in detail the MACRO properties for low energy ν-physics. A running period of about two years was used for clarifying all relevant experimental aspects and for studying background, calibration and set-up procedures needed for the full detector. MACRO analysis methods will be illustrated and results on a galactic supernova search will be presented; a pilot "early warning system" for new supernovas, will be described.

2.1. Supernova theory and $\overline{\nu}_e$-burst detection

Type II supernovae originate from stars of mass greater than $\approx 10\ M_\odot$; the star iron-nickel core compression to neutron star densities releases a large amount of gravitational energy, $\approx 0.1 - 0.2\ M_\odot$. Most of this gravitational energy is emitted in the form ν of all flavours. The average $\overline{\nu}_e$ is $\approx 10\ MeV$, the burst duration is $\approx 10\ s$ and the total energy release is $\approx 3\ 10^{53}\ ergs$ [2].

2.2. Nuclear reactions induced by neutrinos in MACRO

The reaction which by far dominates is: $\overline{\nu}_e + p \rightarrow n + e^+$. This primary reaction is followed, after neutron moderation and capture in the liquid scintillator, by the secondary reaction $n + p \rightarrow \gamma + d\ ,\ E_\gamma = 2.2\ MeV$. Moderation time is $\approx 10\ \mu sec$, capture time is $\approx 180\ \mu sec$.

Deuteron formation 2.2 MeV γ-rays (from now on: γ_2-rays) were never detected in Čerenkov type experiments. The detection of γ-rays from delayed neutron capture in hydrogen is domain of liquid scintillator experiments. This implies energy measurements from $\approx 1\ MeV$, an energy from 10 to 100 times smaller than the one deposited by through-going μ's in MACRO counters.

In Table I the number of detected events from the different reactions in 1000 *tons* liquid scintillator is given for an energy threshold $E_{th} = 7\ MeV$. For the NC and ES reactions ν_x is equivalent to the sum of all ν and $\overline{\nu}$ flavours.

2.3. Gravitational collapse trigger and data acquisition

The problem of background rejection is more easily solved in stellar gravitational collapse than in solar-ν experiments, because of the pulsed character of the supernova explosion. The event cluster generated by a $\overline{\nu}_e$-burst from a supernova explosion might be simulated by rare statistical fluctuations of the average event rate. The two major background components are: cosmic rays and the natural radioactivity background (mostly γ's) from rock, concrete, etc. present at the experiment location [3].

Table I

Number of Events from Stellar Collapse in MACRO (1000 tons of Liquid Scintillator)			
Current Type	Reaction Type	Neutronization Burst	Cooling Stage
CC	$\overline{\nu}_e + p \rightarrow n + e^+$	0	220
NC	$\nu_x +^{12} C \rightarrow \nu_x +^{12} C^*(15)$	0	10
ES	$\nu_x + e \rightarrow \nu_x + e$	0	2

Triggers for the selection of low energy events induced by a supernova ν-burst operate in presence of the cosmic ray and natural radioactivity backgrounds; the first corresponds to events with an energy loss $E \approx 40\ MeV$ (or lower if the counter is only partially crossed) and to a rate $R \approx 2\ mHz$ in a MACRO counter; the second corresponds to lower energies and has $R \approx 5\ 10^3\ Hz$ for $E > 1\ MeV$ and $R \approx 1\ Hz$ for $E > 5\ MeV$ always in a MACRO counter.

Cosmic rays can be largely off-line rejected since they often correspond to coincident events in the scintillation counter system or because identified as tracks by the streamer tube system. The large radioactivity background must instead be hardware rejected; the rejection by an energy threshold must be independent of the event position within the counter. A trigger circuit for stellar gravitational collapse detection must therefore provide such a uniform energy threshold in the range $E = 5 \leftrightarrow 10\ MeV$ ("primary threshold"); moreover, if one wishes to detect the 2.2 MeV photon from n-capture in hydrogen, the energy threshold must be lowered to $E \approx 1\ MeV$ ("secondary threshold") for a time at least comparable to n-capture time in scintillator ($\tau \approx 180\ \mu s$).

The MACRO dedicated stellar gravitational collapse trigger system is based on the circuit PHRASE (*P*ulse *H*eight *R*ecorder *A*nd *S*ynchronous *E*ncoder). A second system: ERP, which provides a generalμ-trigger out of single scintillation counters, has also good stellar gravitational collapse selection capabilities.

The stellar collapse electronics performs the following functions:

1) It provides a trigger for events with an associated energy $E > E_{pth}$, the primary energy threshold, in $\approx$ 80 ns. Typical E_{pth} values are $5 < E_{pth} < 7$ MeV.

2) It lowers the energy threshold for that counter (and for the adjacent counters) to a secondary energy threshold $E_{sth} \approx 1.0$ MeV for a time $\approx$ 1 ms after a primary event in a counter. Secondary events occurring during this time are recorded (a maximum of 14 events), thus allowing the detection of possible 2.2 MeV γ's from delayed neutron capture in hydrogen.

3) It measures with high accuracy ($\sigma \approx 1$ ns) both the time of each event relative to the atomic clock standard time and the time difference between the signals from the two counter ends.

4) It digitizes and stores waveforms (100 MHz) relative to the primary and secondary events. These waveforms and the time information are used for the off-line event energy and position reconstruction.

Stellar gravitational collapse data acquisition systems have problems due to the relatively high rate associated with the low energy thresholds; the raw event rates are $\approx 0.3\ Hz$ for "primary events" and $\approx 3\ kHz$ for "secondary events", for each MACRO SM. The data are read via CAMAC and temporarily stored in two dedicated μVAX's; they are then transferred to the main computer. The stellar gravitational collapse data acquisition can be considered as independent up to the general data storage on disk. In this way the average dead time is very small ($\approx 0.4\ 10^{-4}$ for each MACRO SM); the acquisition system is therefore not dead-time limited even in case of an intense event burst from a supernova explosion.

2.4. "Quality factors" affecting stellar gravitational collapse exps.

The "quality" of the experiments, which intend to look for such an extraordinary event as a stellar gravitational collapse, depends on a series of factors, some obvious, like the sensitive mass, the energy resolution, etc., others less evident, but important for the full success of the experiment. We review some of them:

1) The "energy scale" of the experiment must be well defined and checked.
2) The experiment must be "stable"; this concerns the already mentioned energy scale, but also energy thresholds, rates, etc.; this means to keep under control the low and high voltage power supplies, the ambient temperature, the ventilation (and Radon levels), etc.
3) The power lines must be continuously monitored and any disturbance must be timed and recorded by the data acquisition system; power flickers and electrical noise are the most frequent causes of false event clusters.
4) No interruption of data taking should occur when searching for stellar gravitational collapses; any intervention, repair, calibration of the apparatus, should leave a substancial part of the experiment in normal data acquisition.
5) The data of a stellar gravitational collapse experiment should be analysed online by a "Supernova Watcher". Any suitably defined abnormal event cluster should immediately alert (via computer, modems, computer nets and portable telephones) all interested people. This would allow a prompt and more refined analysis, but also, within a few hours, the optical observation of the early stages of a new supernova by astronomical observatories.
6) Characteristic signatures of bona-fide ν-events, like the observation of the delayed secondary neutron capture reaction $n+p \rightarrow \gamma+d$, $E_\gamma = 2.2\ MeV$ following the primary charged current reaction $\overline{\nu}_e + p \rightarrow n + e^+$, are an important "quality factor"; measurements in the 2 MeV region are particularly difficult due to the high level of the natural radioactivity background at low energies.

It is interesting to examine the solutions offered by the new MACRO experiment to some of the listed items.

2.5. The energy scale determination and the calibration methods

A usual absolute calibration of the energy scale in underground experiments is obtained by crossing cosmic ray μ's (energy loss $\approx 35\ MeV$ in a MACRO counter). For relative calibrations and test purposes, MACRO uses a variable intensity UV-light laser and an optical fiber system. Unavoidable non-linearities over a large energy range and the small cosmic ray rate ($R \approx 1\ m^{-1}\ h^{-1}$) advise an independent absolute energy calibration at low energies ($1 < E < 10\ MeV$); this was obtained by the use of a low intensity Am/Be source (as n and γ-ray emitter via the reaction $^9Be(\alpha,\gamma n)^{12}C$; $E_\gamma = 4.44\ MeV$), externally applied to the scintillation counters [4]. This method is a valuable tool for setting the optimum "stellar gravitational collapse trigger" working conditions both for the detection of the primary $\overline{\nu}_e$-events and for the detection of the delayed neutron captures in hydrogen ($E_\gamma = 2.2\ MeV$; $\tau \approx 180\ \mu s$). The Am/Be source was used to prove MACRO ability of detecting n-capture and to make experimental checks of Monte Carlo calculated detection efficiencies. Fig.1 shows the event energy spectrum when the Am/Be source is applied to a MACRO counter;

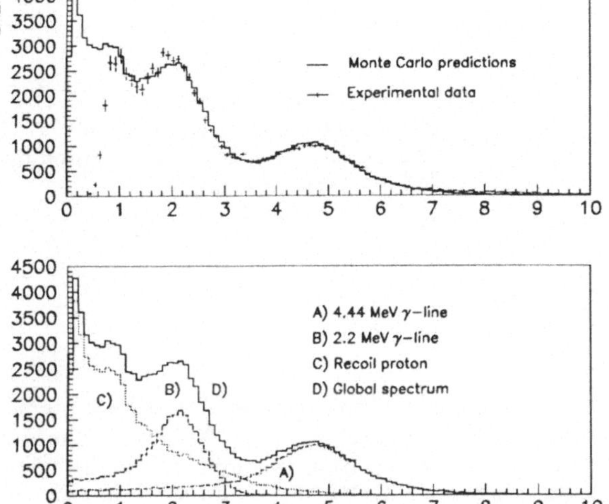

Fig. 1a. Energy spectrum for Am/Be source events. The 4.44 and the 2.2 MeV γ-lines are visible. The full-line histogram is the one of the Monte Carlo simulation.

Fig. 1b. The decomposi tion of the Monte Carlo global result into predictions for different contributing processes.

the 4.44 MeV γ-line and the 2.2 MeV γ-line from n-capture in H_2 are visible. The experimental data are compared (no free parameter !) with the ones obtained by a Monte Carlo calculation which simulates the Am/Be source emission, γ-absorption and detection, n-moderation (by $n-p$ and $n-C$ scattering), n-capture in the liquid scintillator and keeps into account the counter geometry. The contributions of the various processes, taken into account in the simulation, are also separately shown. Fig.2 shows the characteristic time-delay ($\tau \approx 180$ ns) distribution of n-capture 2.2 MeV γ-rays referred to the corresponding prompt Am/Be 4.44 MeV γ-ray. The good agreement between Am/Be-source data and Monte Carlo data supports the evaluation of the efficiency ϵ for delayed n-capture detection, following a primary $\overline{\nu}_e$-event in a MACRO counter; $\epsilon \approx 30\%$.

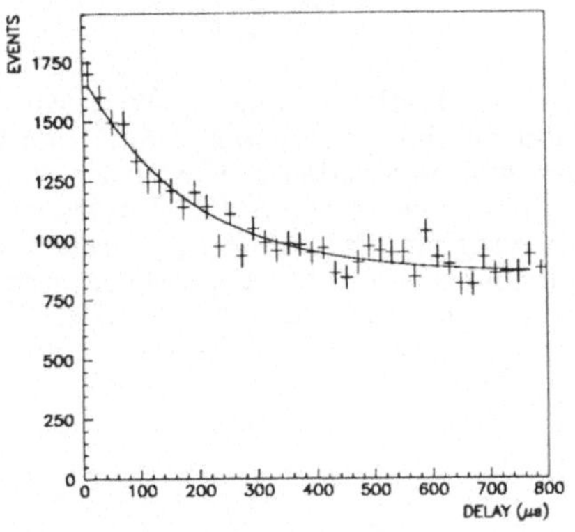

Fig.2. The time-distribution of γ_2-rays, referred to the γ_4 primary.

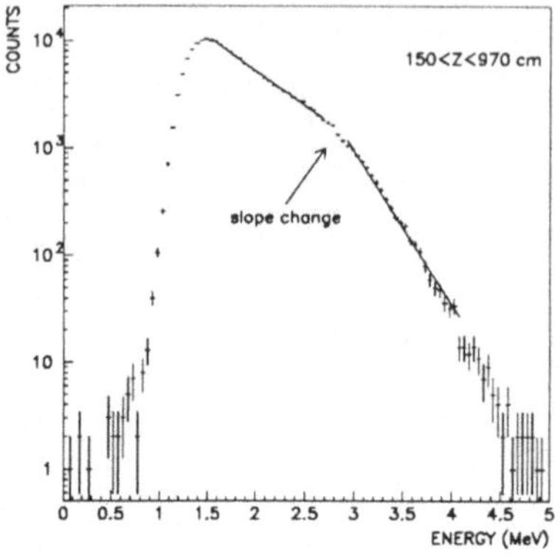

Fig.3. The natural radio activity background energy spectrum measured by a MACRO counter.

Recently we developed a fast method for calibrating in energy all counters; it is based on the natural radioactivity background 2.614 MeV ^{208}Tl-line. The natural radioactivity energy spectrum at the experiment location is complex, but the ^{208}Tl-line produces, even in the energy spectra measured by MACRO counters, a distinct slope-change (see fig.3); the "knee" position corresponding to this slope-change was used as a reference energy.

2.6. The "on-line monitor"

Stellar gravitational collapse data are collected in association with a rather low energy threshold (typically $E_{pth} \approx 7\ MeV$), they are therefore very useful for monitoring the correct behaviour of the scintillation counters and of the associated electronics. An example of such a monitor is presented in fig.4; rates, event multiplicities, etc. are continuously recorded. In case of "anomalies" of whatever nature (stellar collapse or apparatus misbehaviour) an alarm is generated; this allows a prompt and more refined analysis. After gaining sufficient experience on the performance of such a monitor and when the complete MACRO will be active, one will connect this device to an external computer net; the monitor, then a real "Supernova Watcher", would alert, within a few hours, people interested in optically studying the early stages of a new supernova.

2.7. The data analysis

We present the data collected during a period of about 14 months, from March 31, 1990 to June 4, 1991. The live time of SM1 during this period was ≈ 84% due to interventions for regular maintenance; this problem will be avoided in the future when most of MACRO will be active even in case of repairs on one of the SM's. Events with E > 10 MeV were used in this analysis. After applying simple cosmic ray μ rejection criteria which make use of the information from all counters and of the μ-trigger signal from the MACRO streamer tube system, the final rate obtained was ≈ 15 mHz. We have searched for event clusters within sliding 2 s bins beginning at each event time; a "cluster " does not include the event "origin" of the bin.

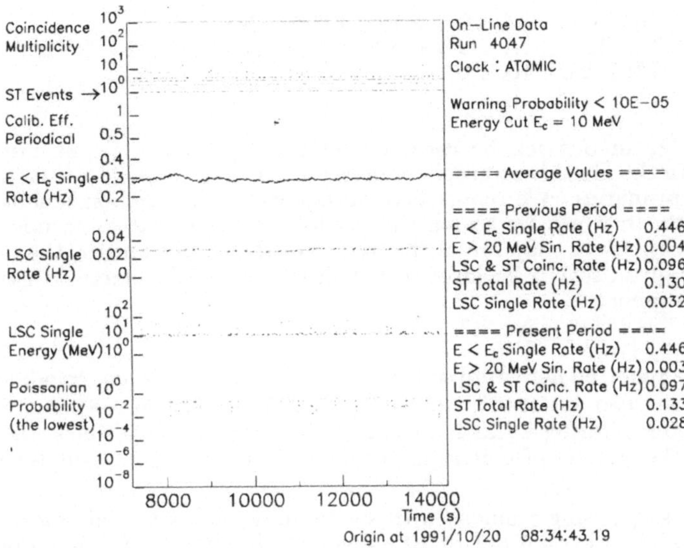

Fig. 4 The on-line monitor. Rates, event multiplicities, etc. are continuously recorded.

The resulting multiplicity distribution is shown in Fig. 5 along with the expected Poisson distribution corresponding to the measured rate (o). No cluster with more than 3 events was found. If a stellar gravitational collapse equivalent to the one from SN1987A had occurred at the galactic center, it would have produced ≈ 8 detected events in a 2 s time window. Fig. 6 shows the number of times in which clusters of multiplicity 1, 2, 3, or 4 occurred vs. the cluster duration. The expectations according to Poisson statistics are also shown.

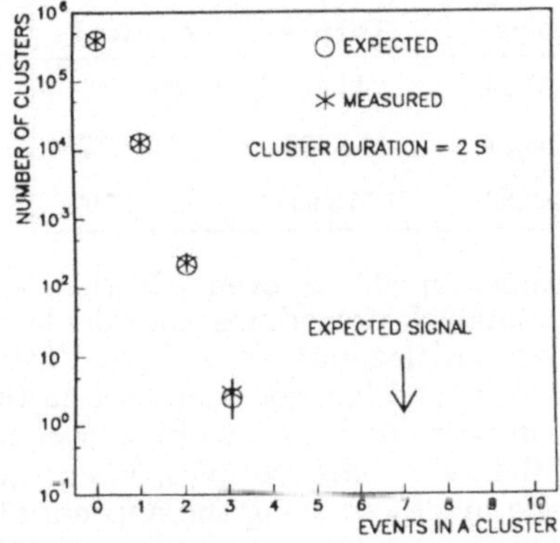

Fig. 5. Number of clusters vs. the number of events in a cluster; the cluster duration is 2 s. Data ($\star$) and expectations ($\circ$) for the 14 month period.

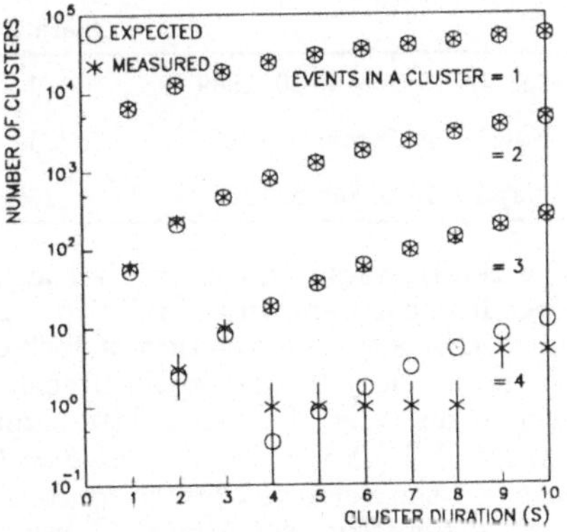

Fig. 6. Number of clusters vs. cluster duration for different number of events in a cluster. Data ($*$) and expectations ($\circ$) for the 14 months period.

3. A SEARCH FOR SIDEREAL ANISOTROPIES WITH UNDERGROUND MUONS AND AN ALL-SKY SURVEY FOR MUON POINT SOURCES

Ultrahigh energy muons can be used to search for point sources of cosmic rays. Muons produced in the Earth's atmosphere preserve the directionality of the source for energetic neutral primaries. We report here an analysis of 1.7 million muons seen by the MACRO detector in which we search the sky for sources of excess cosmic ray flux. Observation of such sources may reveal the nature of the source and/or the mediating particle. Additionally, a search for large-scale anisotropies in the sidereal distributions of event arrivals is reported.

3.1. The data selection

The muon data reported here were collected by the first MACRO supermodule (12 m× 12 m × 4.8 m) during two data runs, and by the first and second MACRO supermodules (24 m × 12 m × 4.8 m) during a third data run. The dates, the live times, the number of μ triggers, and the number of muons analyzed for these three data samples are given in Table II.

In the present study, single muons with successfully reconstructed tracks were retained if they crossed at least 4 out of 10 streamer tube planes. For isotropic events satisfying this criterion, the first MACRO supermodule has an acceptance of $S\Omega \sim 800$ m^2 sr; the first and second MACRO supermodules have $S\Omega \sim 1,600$ m^2 sr. For this study, we have excluded data-taking periods in which the event rate fluctuated by $\gtrsim 2\sigma$ from the mean event rate for the entire run. Individual source fluctuations would not affect rates in this way. In addition, we have excluded particularly noisy events that are typically the result of muon interactions in the apparatus. A total of approximately $1.7\ 10^6$ muons survive these cuts.

Table II Run Parameters for Three Data Samples

Run Period	Supermodules Operational	Live time (hours)	No. of μ Triggers	No. of μ Analyzed
Feb 27, 1989–May 30, 1989	1	1881.14	243,640	223,940
Nov 11, 1989–May 10, 1990	1	2909.37	356,289	319,275
May 10, 1990–Feb 5, 1991	1,2	5032.82	1,378,937	1,185,792

The zenith angle and azimuthal angle distributions for all the events in the local reference frame are shown in fig.7. To verify our data analysis procedures, we calculated the muon intensity as a function of rock depth for events with zenith angle $\theta \leq 60°$; this cut is relevant only to the depth-intensity relationship [5] and has not been used in the subsequent analysis. For each data sample the events were divided into bins of equal solid angle $\Delta\Omega$ ($\Delta\phi = 3.0°$, $\Delta\cos\theta = 0.04$). For the three data samples, intensities corrected to the vertical were computed for each solid angle bin using the appropriate value of the detector acceptance. These intensities were then combined into bins of rock thickness of width 50 hg/cm using an elevation map of the Gran Sasso mountain [6]. The intensities at each rock thickness were combined by weighting every contribution according to its uncertainty. The vertical intensity as a function of rock thickness for muons with $\theta \leq 60°$ is shown in fig.8.

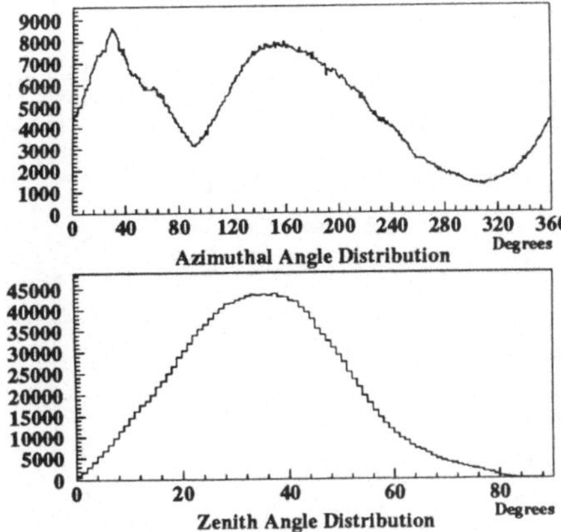

Fig. 7. Azimuthal and zenith angle distributions.

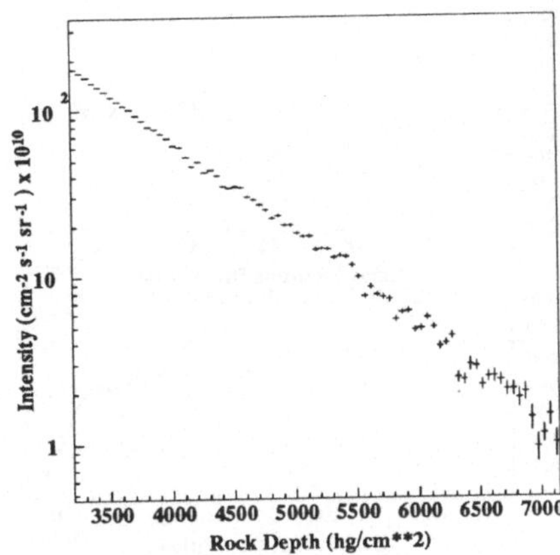

Fig. 8. Intensity distribution for the complete data sample.

3.2. The sidereal anisotropies

In fig.9 are shown the right ascension and declination distributions for the events in the three data samples. The expected right ascension and declination distributions have been calculated by Monte Carlo simulation. Event positions were chosen from the observed two-dimensional distribution of zenith and azimuthal angles. Arrival times were simulated on a run-by-run basis by a Poisson process. For each run the observed event rate was used as the mean. All dead-time gaps were explicitly taken into account. Right ascensions and declinations were computed and analyzed in the same way as the real events. This procedure was repeated 25 times. In fig.9, the average right ascension and declination distributions generated in this manner are shown as dotted lines. The Kolmogorov-Smirnov test yields a probability of 0.999 that the simulated distributions and the data distributions are drawn from the same parent population.

We have searched for sidereal anisotropies in the right ascension distribution using the Rayleigh test. The Rayleigh power of the right ascension distribution is given by $n\bar{R}^2 = \frac{1}{n}\left\{\left[\sum_{i=1}^{n}\cos(2\pi m\,\alpha_i/24)\right]^2 + \left[\sum_{i=1}^{n}\sin(2\pi m\,\alpha_i/24)\right]^2\right\}$, where α_i is the right ascension of the ith event, n is the total number of muons, and m = (1,2) for the first/second moment, respectively.

Since event times in our simulations are chosen from a Poisson distribution, the simulations will not exhibit sidereal anisotropies. We have computed the expected distributions for the first and second moments of the Rayleigh power that reflect the live time distribution from 652 Monte Carlo simulations. Fig.10 shows these distributions. Values for the two moments computed from the MACRO data distributions fall in the shaded bins. The data samples we have investigated are consistent with a right ascension distribution with no sidereal anisotropies.

3.3. The all-sky survey

Using our Monte Carlo distributions we can make an all-sky search for point sources of muons in excess of the expected background. First the data were binned in equal solid angle bins $\Delta\Omega$ ($\Delta\alpha = 3.0°$, $\Delta\sin\delta = 0.04$). An average of 25 Monte Carlo data runs was then used as the expected background. These background events were binned in the same way as the data.

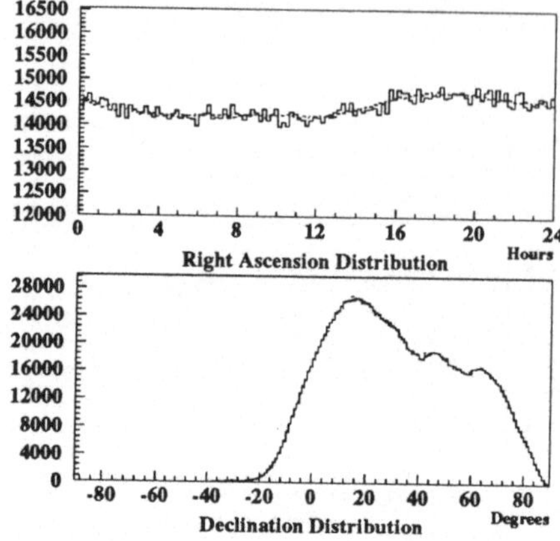

Rayleigh Power Distribution

2nd Moment Rayleigh Power Distribution

Fig. 9. Right ascension and declination distributions. Monte Carlo backgrounds are superimposed.

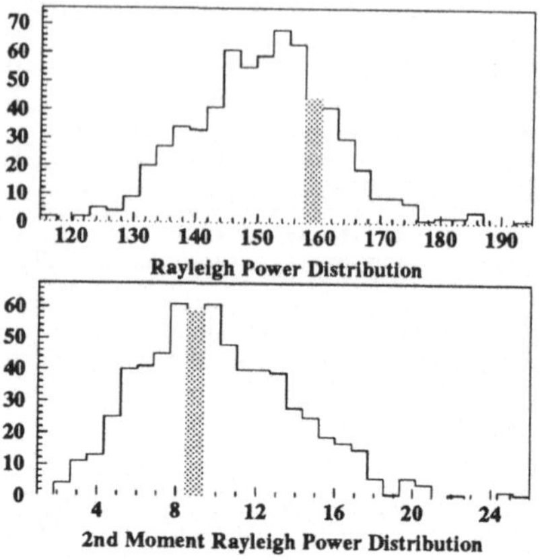

Fig. 10. Rayleigh power distributions for the first and second moments. Data values are in shaded bins.

We computed the deviation from the mean for every bin, $\delta = (\mathrm{n} - \mathrm{e})/\sqrt{\mathrm{e}}$, where n is the number of events in the bin and e is the expected background. In fig.11 we show the distribution of these deviations. There are no deviations greater than 3.8σ. Superimposed onto this distribution is the best-fitting Gaussian $\chi^2/\mathrm{DoF} = 85/100$. The mean of this Gaussian is 1.4×10^{-3} and the rms is 1.01, as is expected from a random distribution.

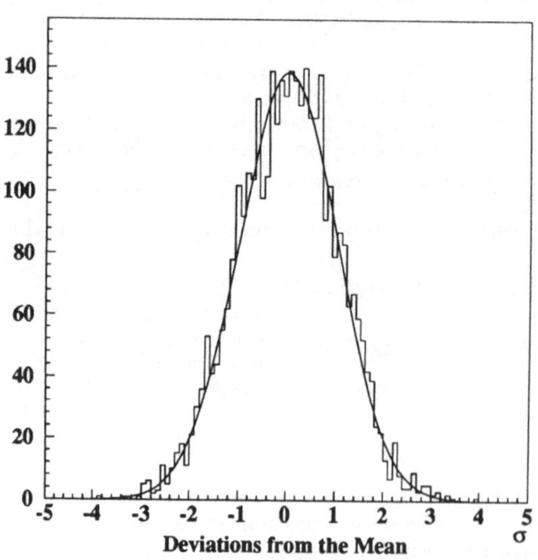

Fig. 11. Distribution of deviations from the mean in the all-sky survey.

4. SEARCH FOR ASTROPHYSICAL POINT SOURCES OF MUONS

Due to its high tracking resolution, MACRO is well suited to search for muon excesses in the direction of celestial point sources of VHE or UHE photons. In this paper we analyze the muons from the direction of the reported sources of VHE and UHE photons: Cyg X3, Her X1, 1E2259+59 and the Crab. The threshold energy for single muons due to the rock overburden is ≈ 1.5 TeV.

4.1. The data selection and analysis

The data analyzed consist of 3 data samples (see section 3.1.) which total $\approx 1.7\ 10^6$ muons obtained over a live-time Δt of 9823 hours. We have selected for this analysis single muon events that cross at least 4 of 10 horizontal streamer tube planes.
Steady Muon Signals: We have searched for an excess of muon tracks pointing back to a 1.5° half-angle cone centered on Cyg X3, Her X1, 1E2259+59 and the Crab. The average pathlength traversed by muons from each source window and the number of events, n, in each window are listed in Table III. The expected number of background counts determined by Monte Carlo simulation, n_{bkd}, is also given in Table III. These Monte Carlo simulations explicitly account for the detector live-time distribution throughout the data taking periods. There is no evident signal from any source at the 1σ level. /break/indent The upper limit to the steady muon flux at the 90% C.L. was computed by $F_{stdy} = 1.28\sqrt{n_{bkd}}/(\bar{\epsilon} A_{eff} f \Delta t)$, where $\bar{\epsilon} = 0.87$ is an estimate of the live-time weighted efficiency; A_{eff} is the live-time weighted effective area for the window; and f is the fractional time the source window is $\geq 10°$ above the horizon. The values for A_{eff}, f, and F_{stdy} are given in Table III.

Table III Search for Steady Muon Signals from Point Sources

Source	<depth> (MWE)	n	n_{bkd}	A_{eff} (m^2)	f	F_{stdy} ($cm^{-2}s^{-1}$)
Cyg X3	3960	491	485	190	.66	$< 7.4 \times 10^{-13}$
Her X1	3885	455	489	194	.64	$< 7.5 \times 10^{-13}$
1E2259+59	3795	584	608	183	1.0	$< 5.7 \times 10^{-13}$
Crab	3560	520	523	200	.52	$< 9.3 \times 10^{-13}$

Periodic Muon Signals: In the search for periodic muon signals in a 1.5° half-angle cone, a correction was applied to the arrival time of each event to account for the earth's motion with respect to the solar system barycenter.

For Cyg X3, we searched for a muon signal modulated by the 4.8 hr X-ray period using the parabolic ephemeris of van der Klis and Bonnet-Bidaud [7]. The phase diagram for these data are shown in fig.12 as a solid line and the average background is shown as a dotted line. This figure shows that the largest deviation above background is at the 2σ level.
For Her X1, we searched for a muon signal which was modulated by either the 1.7^d orbital period [8] or the 35^d (presumed precessional) period [9]. For the 1E2259+59 data, we searched for a muon signal modulated by the first and second harmonic of the 6.98^s pulsar period [10]. (A large fraction of the X-ray power is emitted in the second harmonic.) For 1E2259+59 the X-ray period has been determined accurately enough to maintain phase coherence only over ≈ 30 days. We therefore broke the data into 17 separate 30^d segments, analyzing each segment individually. We searched for evidence of a statistically significant modulated signal in these three sources using both the Rayleigh and Protheroe tests. In Table 2 we list for Cyg X3 and Her X1 the probabilities,

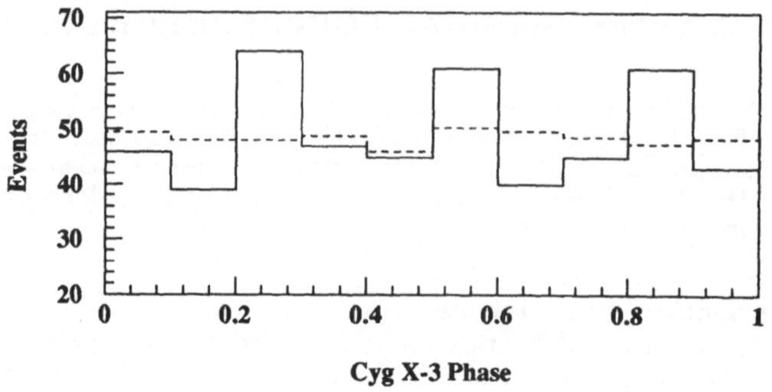

Fig. 12. Phase diagram for Cyg X3 in a 1.5° half-angle cone.

$W(>n\bar{R}^2)$ and $W(>\Upsilon_n)$, that the Rayleigh power, $n\bar{R}^2$, and the Protheroe power, Υ_n, have values at least as large as we found for events distributed uniformly in phase—a test of the null hypothesis. These probabilities were computed by Monte Carlo simulation. Both tests imply there are no statistically significant modulated muon signals at the periods investigated. There is also no evidence in 1E2259+59 for a muon signal modulated with the pulsar period. In Table IV we have estimated the upper limit to the modulated flux in the 1.5° half-angle cone from the relation $F_{mod} = (n - n_{bkd} + 1.28\sqrt{n})/(\bar{\epsilon}A_{eff}f\Delta t)$, using the phase bin with the largest excess above background. For 1E2259+59 a similar calculation yields $F_{mod} \lesssim 2 \times 10^{-12}$ $cm^{-2}\,s^{-1}$ as a typical value for an individual 30 day run.

Table IV Search for a Modulated Muon Signal from Point Sources

Source	P_0	$W(>n\bar{R}^2)$	$W(>\Upsilon_n)$	F_{mod} $(cm^{-2}s^{-1})$
Cyg X3	4.8^h	0.90	0.52	$< 6.9 \times 10^{-13}$
Her X1	1.70^d	0.52	0.68	$< 4.1 \times 10^{-13}$
Her X1	34.9^d	0.64	0.32	$< 6.8 \times 10^{-13}$

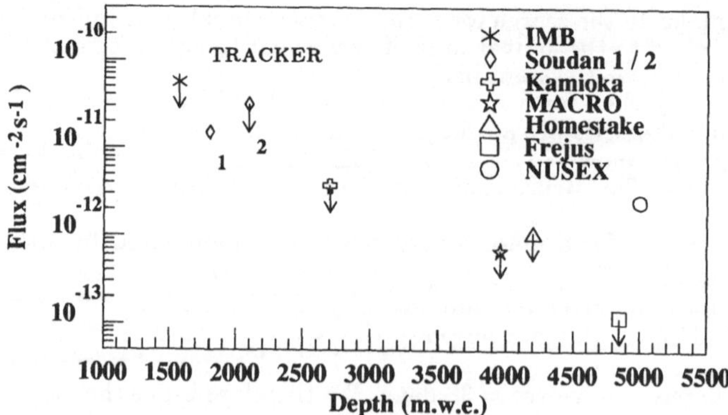

Fig. 13. Measurements of the modulated muon flux from Cyg X3.

In fig.13 we present our flux limit from the direction of Cyg X3, compared with the limits from other underground detectors.

Period Search: We have searched through period space for a modulated muon signal with a period displaced from the fiducial periods P_0. We chose for the increment in period space 10% of an Independent Fourier Spacing (IFS), $\Delta P = 0.1P_0^2/T$. For Cyg X3 and Her X1, T is the run time for a data sample. For 1E2259+59, T is equal to 30 days. We have searched through 30 IFS on either side of P_0.

Let χ_{max} represent the maximum Rayleigh or Protheroe power we have found for any source in our period search. As a test of the null hypothesis, we have calculated the probability W that χ_{max} represents a random fluctuation in the rate of cosmic ray muons, $W(>\chi_{max}) = 1-(1-p_1)^N$. In this expression, p_1 is the probability of obtaining χ_{max} from events distributed randomly in phase as determined by Monte Carlo simulation, and N is the effective number of trials. For Cyg X3 and Her X1, $N \approx 61 \times 3 \times 3$ – we have searched through 61 IFS; 3 represents the penalty for oversampling the IFS [11]; and 3 is the number of run periods searched. For 1E2259+59, $N \approx 61 \times 3 \times 17$. In all cases investigated, $W(>\chi_{max}) \gg 0.99$, which suggests that there is no modulated muon signal with a period displaced slightly from P_0.

Search for Short Term Variability in Cyg X3: The muon events from the 1.5° half-angle cone around Cyg X3 for data sample 3 have been analyzed for short term variability on a time scale of one day. During this period there were three reported radio outbursts, a large burst on 21 January 1991 and two lesser flares on 14 August and 5 October 1990 [12] .

The points in fig.14 show the deviations $(n - e)$ for every day during this period, where n is the number of events in the window and e is the expected number of events on that day determined by Monte Carlo simulation. We have found no deviations in excess of 3.9σ. In periods consisting of five days on either side of the radio bursts, no deviations above 3σ were found. The solid curve in Fig.14 shows the predictions of Poisson statistics. Clearly the simulated distribution matches the data distribution well, reinforcing the conclusion that MACRO saw no statistically significant outburst during the period of data sample 3.

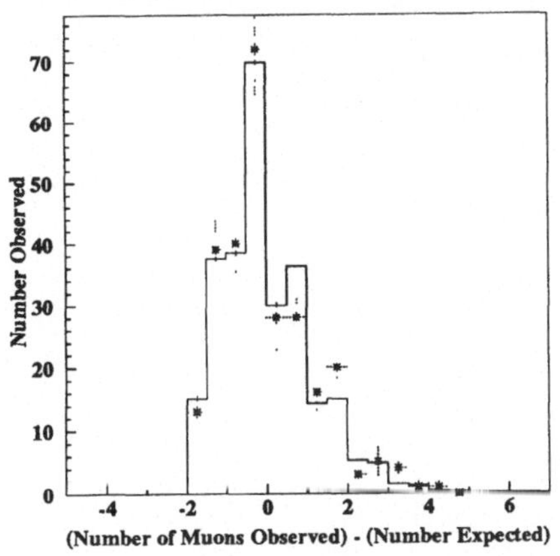

Fig. 14. Daily deviations from expected for CygX3.

‡ **Bari:** R. Bellotti, F. Cafagna, M. Calicchio, G. De Cataldo, C. De Marzo, O. Erriquez, C. Favuzzi, P. Fusco, N. Giglietto, P. Spinelli; **Bartol:** J. Petrakis; **Bologna:** S. Cecchini, G. Giacomelli, G. Mandrioli, A. Margiotta-Neri, P. Matteuzzi, L. Patrizii, F. Predieri, G.L. Sanzani, E. Scapparone, P. Serra Lugaresi, M. Spurio, V. Togo; **Boston:** S. Ahlen, R. Cormack, E. Kearns, S. Klein, G. Ludlam, A. Marin, C. Okada, J. Stone, L. Sulak, W. Worstell; **Caltech:** B. Barish, S. Coutu, J. Hong,E. Katsuvounidis, S. Kyriazopoulou, G. Liu, R. Liu, D. Michael, C. Peck, N. Pignatano, K. Scholberg, J. Steele, C. Walter; **Drexel:** C. Lane, R. Steinberg; **Frascati:** G. Battistoni, H. Bilokon, C. Bloise, P. Campana, P. Cavallo, V. Chiarella, C. Forti, A. Grillo, E. Iarocci, A. Marini, V. Patera, F. Ronga, L. Satta, M. Spinetti, V. Valente; **Gran Sasso:** C. Gustavino, J. Reynoldson; **Indiana:** A. Habig, R. Heinz, L. Miller, S. Mufson, J. Musser, S. Nutter; **L'Aquila:** A. Di Credico, P. Monacelli; **Lecce:** P. Bernardini, G. Mancarella, D. Martello, O. Palamara, S. Petrera, P. Pistilli, A. Surdo; **Michigan:** E. Diehl, D. Levin, M. Longo, C. Smith, G. Tarlé; **Napoli:** M. Ambrosio, G. C. Barbarino, F. Guarino, G. Osteria; **Pisa:** A. Baldini, C. Bemporad, F. Cei, G. Giannini,* M. Grassi, R. Pazzi; **Roma:** G. Auriemma,** S. Bussino, C. Chiera, P. Chrysicopoulou, A. Corona, M. DeVincenzi, L. Foti, E. Lamanna, P. Lipari, G. Martellotti, G. Rosa, A. Sciubba, M. Severi; **Sandia Labs:** P. Green; **Texas A&M:** R. Webb; **Torino:** V. Bisi, P. Giubellino, A. Marzari Chiesa, M. Masera, M. Monteno, S. Parlati, L. Ramello, M. Sitta

*Univ. di Trieste
**Univ. della Basilicata

References

[1] The MACRO Collaboration, **Nucl.Instrum.&Meth.A 264** 18 (1988)

[2] for a review see: W.D. Arnett et al., **Annu.Rev.Astron.Astrophys. 27** 629 (1989)

[3] A.. Alessandrello et al., **Nuovo Cimento 103A** 1617 (1990)

[4] A. Baldini et al., **Nucl.Instrum.& Meth. A305** 475 (1991)

[5] T.K. Gaisser, **Cosmic Rays and Particle Physics**, Cambridge University Press, 34 (1990)

[6] The MACRO Collaboration, **Phys.Lett.B 249** 149 (1990)

[7] M. van der Klis et al., **Astr. Astrophys. 214** 203 (1989)

[8] J.E. Deeter et al., **Astrophys.J. 247** 1003 (1981)

[9] H. Ögelman, **Astr.Astrophys. 172** 79 (1987)

[10] K. Koyama et al. **Pub.Astron.Soc.Japan 41** 461 (1989)

[11] O.C. de Jager, Potchefstroom University, Ph.D.Thesis (1987)

[12] The Soudan2 Collaboration, M.A. Thomson et al., preprint (1991)

Muons in γ-ray Air Showers and the Photoproduction Cross Section

R. S. Fletcher[a], T. K. Gaisser[a] and F. Halzen[b]

[a]Bartol Research Institute, University of Delaware, Newark DE 19716
[b]Dept. of Physics, University of Wisconsin, Madison WI 53706

Abstract

We review the status of muon production in γ-ray induced air showers, and discuss the high-energy behavior of the photoproduction cross section. Even if the photoproduction cross section increases with energy above a TeV, the muon-poor criterion remains good for selecting photon-induced showers. Conversely, true signals, with high muon content from point sources would imply new physics.

1. Introduction

"Muon-Poor Astronomy" is an awkward phrase that refers to gamma ray astronomy at ultra-high energies with air showers. Air showers initiated by photons have relatively few muons as compared to background showers generated by cosmic ray protons or nuclei, because the cross section for photoproduction is relatively small. Thus, the absence of muons can be used as a tag to discriminate candidate photon showers from ordinary cosmic ray cascades.

Ultra-high energy gamma ray astronomy ($\gtrsim$ 100 TeV) is the realm of air shower experiments because likely fluxes are much too low to be studied with small detectors flown above the atmosphere in balloons or spacecraft. A pertinent example is the flux of diffuse photons from decay of neutral pions produced when cosmic ray nuclei interact with interstellar gas. Even from the direction of the galactic center, the ratio of photons to cosmic rays where it is measured in the GeV range is small,

$$\phi_\gamma/\phi_p < 10^{-4}\,. \qquad (1)$$

This ratio remains at about this level up to air showers energies ($\gtrsim$ 100 TeV)[1]. The small value of this ratio is largely a consequence of the fact that the photons travel straight out of the galaxy and so have only one chance to be observed. In contrast, the protons (and other charged nuclei) may cross the galactic disk many times before diffusing out of the galaxy.

Since the cosmic ray flux falls quickly as energy increases, the intensity of photons soon becomes too low for detectors flown on spacecraft or in balloons at the top of the atmosphere. For example, above 100 TeV, the cosmic ray flux is about 3×10^{-9} cm^{-2}s^{-1}sr^{-1}.

A photon flux four orders of magnitude smaller would give only one or two events per day in an area the size of a football field.

The idea of using muon-poor air showers for UHE gamma ray astronomy was originally suggested nearly 30 years ago[2, 3]. Modern versions of this technique are presently exploited at several air shower arrays, including the Cygnus experiment at Los Alamos[4], the Akeno experiment[5] and the Chicago-Michigan-Utah experiment[6].

A related technique at lower energy has been developed by the Whipple Observatory group[7] for the TeV energy range. In this energy range, air showers die out before the particles reach ground level. The showers are detected by the atmospheric Cherenkov light emitted by shower electrons high in the atmosphere. According to simulations, Cherenkov images of cosmic ray showers, show a structure with relatively large fluctuations characteristic of the hadronic subshowers. Images of showers initiated by photons, on the other hand, are relatively symmetric about the axis. This calorimetric technique (which uses a parameter called "azwidth") was proved by application to the Crab Nebula[7].

In both cases, the techniques for discriminating against the cosmic ray background of hadronic showers depends on the small value of the photoproduction cross section relative to the cross section for pair production in an electromagnetic shower. The relative probability of photoproduction is

$$R_\gamma = \frac{\sigma(\gamma + \mathrm{air} \to \mathrm{hadrons})}{\sigma(\gamma + \mathrm{air} \to e^+e^-)} \approx 3 \times 10^{-3}$$

up to 150 GeV, the highest energy at which the photoproduction cross section is measured at present. As a consequence of cascading, the ratio of the muon content of a photonic shower to that of a hadronic shower is about an order of magnitude larger than R_γ. This result was obtained long ago by semianalytic estimates[8] and confirmed by Monte Carlo calculations[9, 10, 11]. The logic can be introduced by using Heitler's approximation[12] to an electromagnetic cascade. Suppose a 100 TeV photon strikes the atmosphere within view of a detector. In every radiation length (λ_R) a photon will produce a e^+e^- pair and every electron or positron will radiate a photon. This model should be on average correct as high energy pair production and bremsstrahlung cross sections are roughly equal. The atmosphere is filled with sequential layers of depth λ_R (actually $\ln 2\lambda_R$) in which particles subdivide by pair production or bremsstrahlung in 2 particles with 1/2 the initial energy. The cascade is explicitly constructed below:

Layer (n)	Energy ($E_0/2^n$)	Multiplicity (2^n)
0	100 TeV	1 (γ)
1	50 TeV	2 (e^+e^-)
2	25 TeV	4 ($e^+\gamma e^-\gamma$)
⋮	⋮	⋮
6	1.6 TeV	64
7	0.8 TeV	128
⋮	⋮	⋮
12	25 GeV	4100
13	12 GeV	8200 $\left(\frac{1}{3}\gamma,\ \frac{2}{3}e^\pm\right)$

Experiments observe muons of GeV energy. They are abundantly produced in layer $n = 13$ which contains 8200 particles of which 1/3 are photons of 12 GeV energy. This is also the last layer where one can produce muons of a few GeV energy. The number of muons is given by $R_\gamma \times 1/3 \times 8400 \simeq 10$, where R_γ is the probability that the photons photoproduce a π (which decays into a muon), rather than an e^+e^- pair. In the Heitler model all previous layers together contain as many photons as the last one, therefore we obtain a total of 20 muons. What is dramatically illustrated by the example is that the result is inescapable as photoproduction of muons is almost exclusively by γ-rays with energies explored in accelerator experiments; there is no room for new physics. Also, GeV muon production by other processes such as $\gamma \to \mu^+\mu^-$ and $\gamma \to$ charm $\to \mu$ is negligible. For a proton-initiated shower the muon content is 30 times larger as a result of the abundant hadronic production and decay of charged π's. Thus, as long as the photoproduction cross section is small compared to the pair production cross section, no Monte Carlo calculation is required to convince oneself that muon-poor astronomy works.

2. The Photoproduction Cross Section at High Energy.

To make firm predictions for the small, but finite, muon content of γ-ray showers, we need to know the photoproduction cross section at all energies. At the low energies relevant for Cherenkov astronomy, this cross section is known experimentally[13], but at high energies, we must rely on calculations or extrapolations. In this section we review the expected behavior of the photoproduction cross section at high energy[14, 15], and show that photoproduction is small compared to the pair production cross section at all energies.

At high energies the hadronic (pp, πp, etc.) scattering cross sections rise. This rise can be understood in the QCD based mini-jet model[16, 17]. At high energies, there are a large number of gluons in a hadron which can scatter, producing jets. The rise of the total cross section is a direct consequence of this rapid increase of the number of gluons in the proton. At high energies the photon also develops a gluon structure function and therefore a similar increase of the photoproduction cross section is anticipated.

In photon-proton collisions "quarks and gluons inside the photon" are the origin of jet production with large high energy cross sections[18, 19]. The inclusive jet cross sections eventually exceed the typical values of the total photoproduction cross section measured in accelerator experiments[20, 21]. This signals the onset of multi-jet production which can be included by exploiting the eikonal models designed to link the rise of the total cross section to the rising jet cross section in proton-antiproton collisions. The predicted increase in the photoproduction cross section should be measurable at high energy electron-proton colliders[15, 22].

The total cross section for jet production in photon-hadron interactions has two components; see Fig. 1. The "direct part" is given by

$$\sigma_{\text{direct}} = \sum_{i=g,q} \int_{\hat{t}_{\min}}^{\hat{s}-\hat{t}_{\min}} d\hat{t} \int f_{i/p}(x) \frac{d\hat{\sigma}_i}{d\hat{t}} dx \,. \tag{2}$$

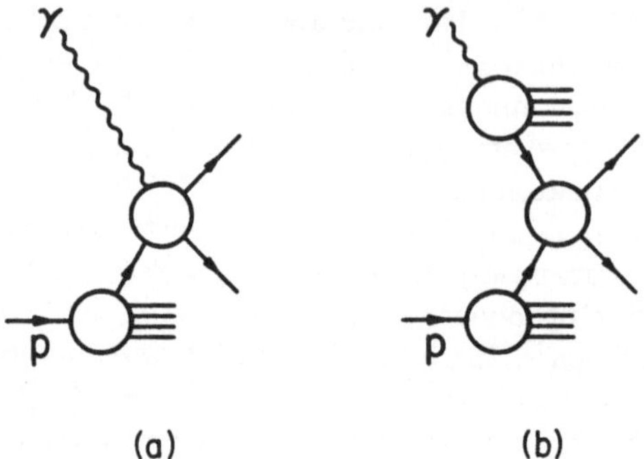

Figure 1: Direct (a) and resolved (b) contributions to dijet production.

Here the photon couples directly to a parton in the proton. The parton flux is given by the parton distribution $f_{i/p}$ and x is the fractional momentum of the parton in the proton. $\frac{d\sigma}{dt}$ is the parton level γ-parton cross section[24]. The "resolved" part is, in the dominant pole approximation[25], given by the parton model expression

$$\sigma_{parton}(s) = \int_{\hat{t}_{\min}}^{\hat{s}/2} d\hat{t} \int F_\gamma(x_1) F_p(x_2) \frac{d\hat{\sigma}}{d\hat{t}} dx_1 dx_2 \,. \tag{3}$$

Here, the photon interacts as a cloud of partons. F_i is the flux of quarks and gluons in the photon or proton. The latter we approximate by

$$F(x) = g(x) + \frac{4}{9}\sum_i [q_i(x) + \bar{q}_i(x)] \tag{4}$$

and, in this approximation, the parton cross section is

$$\frac{d\hat{\sigma}}{d\hat{t}} = \frac{9\pi}{2}\frac{\alpha_s^2}{\hat{s}^2}\left[\left(\frac{\hat{t}}{\hat{u}}\right)^2 + \frac{\hat{t}}{\hat{u}} + 1 + \frac{\hat{u}}{\hat{t}} + \left(\frac{\hat{u}}{\hat{t}}\right)^2\right] . \tag{5}$$

At high energies the calculated jet cross section can exceed the typical photoproduction cross section of $100 - 120\mu$b measured in accelerator experiments. This is the signal that the multiple scattering, or eikonal, corrections to the cross section need to be considered.

Following the formalism developed for $p\bar{p}$ interactions[16], we introduce the average number of parton interactions as a function of impact parameter

$$n(b,s) = \sigma_{parton} A(b) = P_{had} \times \frac{\sigma_{parton} A(b)}{P_{had}} , \tag{6}$$

where σ_{parton} is the mini-jet cross section given by Eq. 2. The last expression in Eq. 6 should be understood as follows. The probability for a photon to interact like a hadron, *i.e.* like a cloud of colored partons, is P_{had} which should be proportional to α_{em}. Then the cross section for that "hadron" to interact with the proton is σ_{parton}/P_{had}.

The simplest way to approximate the parameter P_{had} is to use the standard vector meson dominance model result[14]

$$P_{had} = \frac{4\pi\alpha}{f_\rho} \approx \frac{1}{300} \,. \tag{7}$$

An alternative estimate can be made based on a simple parton model inspired ansatz

$$f_{q/\gamma} \approx P_{had} \times f_{q/\pi} \,, \tag{8}$$

where the π is chosen because it is the only meson for which the structure functions are known. Because the shapes of the photon and π parton distributions are rather different, this expression cannot be exactly satisfied, but for x in the vicinity of 0.1 this ansatz gives $P_{had} \approx 1/150$ for the distributions of Refs. [26] and [27]. Alternatively, instead of normalizing the two distributions at a single point, we can compare the momentum integrals of the parton distributions. It is well known that for a hadron,

$$\sum_i \int_0^1 dx\, x\, f_{i/\pi}(x) = 1 \,. \tag{9}$$

For the photon

$$\sum_i \int_0^1 dx\, x\, f_{i/\gamma}(x) \approx 1/170 \,. \tag{10}$$

From Eq. 8, this suggests $P_{had} = 1/170$, which is in the range discussed above.

$A(b)$ is the overlap function in impact parameter space of the colliding photon and proton. We model it as the Fourier transform of the product of the pion and proton electromagnetic form factors[23, 29]

$$A(b) = F.T. \left[\frac{1}{(1+q^2/\mu^2)^2}\frac{1}{(1+q^2/\nu^2)}\right] \tag{11}$$

$$= \frac{\nu^2}{2\pi}\frac{\mu^2}{\mu^2-\nu^2}\left[\frac{\mu^2}{\mu^2-\nu^2}\Big[K_0(\nu b) - K_0(\mu b)\Big] - \frac{\mu b}{2}K_1(\mu b)\right] \,. \tag{12}$$

In computing the total hadronic cross section we first add a constant term, $n_0(b)$, to $n(b,s)$ in order to parameterize the energy independent, low energy photoproduction cross section. Once a hadronic interaction occurs, the probability that there is no hard collision is $\exp(-n(b,s)/P_{had})$. The total inelastic cross section is the probability that there is one or more collision, therefore

$$\sigma_{inel} = \int d^2b\, P_{had} \times \left(1 - e^{-(n(b,s)+n_0(b))/P_{had}}\right) \,. \tag{13}$$

For P_{had} equal to 1 we recover the expression for $p\bar{p}$ interactions.

The interpretation of the expression for the inelastic cross section is clear. The photon interacts as a hadron with a probability $P_{had} \sim \alpha_{em}$. The hadronic cross section is then given by the normal mini-jet expression for the inelastic cross section, where the average number of interactions per *hadronic* interaction is $n_{had}(b,s) = n(b,s)/P_{had}$. This result preserves the expansion in the total number of jets produced, as required if the model is to make sense. Because the multiplicity of jets per hadronic event is larger for P_{had} small, the total number of hadronic events, *i.e.* the photoproduction cross section, will be smaller.

Figure 2 shows the inelastic photoproduction cross section as a function of energy for the two values of P_{had} given above. The cutoff in $\hat{t}$ is chosen in the 1 GeV^2 range as suggested by phenomenological fits of similar models to proton-proton cross sections. A significant increase of the cross section is anticipated as a result of the gluonic structure of the high energy photon. Better predictions will be possible once the parameter t_{min} can be extracted from data. At the HERA collider the photoproduction cross section should be measurable up to $\sqrt{s} = 300$ GeV, *i.e.* $E_\gamma \approx 20$ TeV. As seen in Fig. 2, the cross section is very sensitive to t_{min} in this threshold region. Once new data becomes available from fixed target experiments with higher energy secondary photon beams at Fermilab, and from the HERA collider, better predictions for the high energy behavior of photoproduction will be possible.

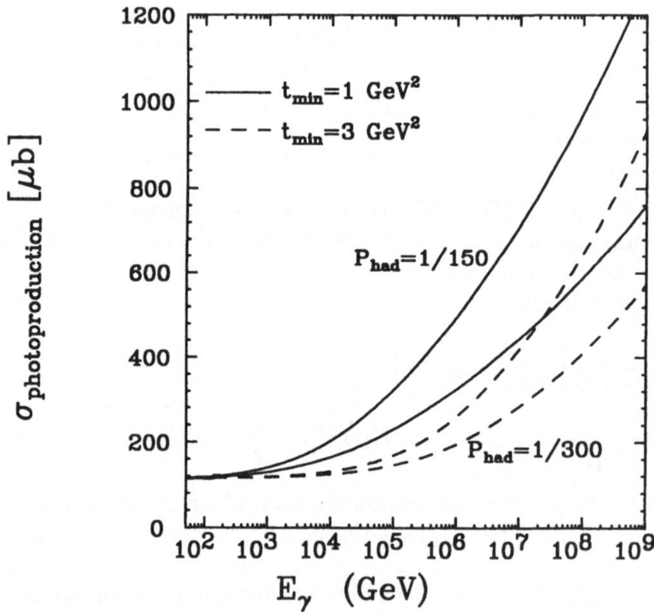

Figure 2: The photoproduction cross section versus energy for two values of t_{min}, and two values of P_{had}. All calculations use the Duke-Owens proton distribution functions[28], set 1.

All of the predictions in Fig. 2 are of the same order of magnitude as traditional extrapolations of the low energy photoproduction cross section. Thus, the predictions of muon number in air showers will not be strongly affected by the increasing photoproduction cross section.

3. Conclusions

We have seen that photon air showers should not contain many muons. Despite this, claims that Cygnus X-3 beams muons at us have now persisted for many years and were boosted by recent observation during a spectacular radio-burst lasting 5 days in January 1991[31]. Several experiments reported signals of order 4σ during this period, while others failed to observe the flare. As usual there are no definite inconsistencies, as one can unfortunately invoke time dependence of the source, varying thresholds of detectors and the like to avoid conflict. On the other hand, it must be said that the observations, while still low in statistics, are several and difficult to criticize. After the controversies of a few years back, experimentalist have introduced procedures to extract signals from background that have an unambiguous and calculable statistical meaning.

What if muons from Cygnus X-3 and other sources are confirmed? In the context of the Standard Model there is one possible way out and it does not work. This would be to increase the value of the photoproduction cross section and therefore raise the value of R_γ. This cannot be done in the last layer where muons are produced by 12 GeV photons because the cross sections are known experimentally. A large cross section at $10-10^2$ TeV escapes direct experimental scrutiny but 1) is not expected theoretically, and 2) is useless because there are too few photons of that energy in the shower to make a difference; see above table. Again, this confirms results obtained by Monte Carlo simulation. Even though increased photoproduction can increase the average number of muons, especially high energy muons, this is achieved by the occasional occurrence of an event where a high energy photon interacts hadronically in it's first interaction and produces a very large number of muons. The average is raised by these special events but a typical photon shower has the normal muon content. In observations, on the contrary, most showers are apparently anomalous.

We therefore must conclude that "muons from Cygnus X-3" would represent a revolution in particle physics, if not astronomy and astrophysics. One would be observing some totally new interaction or neutral particle , with a scale not far above those presently probed by accelerator experiments.

Acknowledgements

This research was supported in part by the U.S. Department of Energy under Contracts No. DE-FG02-91ER40626 and No. DE-AC02-76ER00881, in part by the Texas National Research Laboratory Commission under Grant No. RGFY9173, in part by the University of Wisconsin Research Committee with funds granted by the Wisconsin Alumni Research Foundation.

References

[1] T. K. Gaisser F. Halzen, T. Stanev, Radiation from Cosmic Ray Interactions in the Galaxy, to appear in *Proc. of the 22nd Int. Cosmic Ray Conference*, Dublin (1991).

[2] K. Suga et al., *Proc. of the 8th Int. Cosmic Ray Conf.* (Jaipur), **4** (1963) 9; J. Phys. Soc. Japan, **17** Supplement A-III (1962) 138.

[3] R. Firkowski et al., J. Phy. Soc. Japan, **17** Supplement A-III (1962) 123.

[4] Cygnus Coll., D. E. Alexandreas et al., LA-UR-91-2167, LA-UR-91-2142, submitted to Nuc. Inst. and Meth.

[5] M. Nagano et al. (Akeno), J. Phys. **G10** (1984) 1295.

[6] J. W. Cronin et al., EFI-91-35 (Sept. 1991), submitted to Phys. Rev. **D**; K. G. Gibbs et al., Nuc. Inst. and Meth. **A264** (1988) 67.

[7] T. Weekes et al., Ap. J. **342** (1989) 379.

[8] S. Karakula and J. Wdowczyk, Acta Phys. Pol. **24** (1963) 231; O. Braun and K. Sitte, in *Proc. of the 9th IntCosmic Ray Conf.* 1965, London, (1966).

[9] T. Stanev, T. K. Gaisser and F. Halzen, Phys. Rev. **D32** (1985) 1244.

[10] P. G. Edwards, R. J. Protheroe and E. Rawinski, J. Phys. **G11** (1985) L101.

[11] T. K. Gaisser, T. Stanev, F. Halzen, W. F. Long and E. Zas, Phys. Rev. **D43** (1991) 314.

[12] W. Heitler, *Quantum Theory of Radiation*, (Oxford, 1944) 2nd edition.

[13] D. Caldwell et al., Phys. Rev. Lett. **40** (1978) 1222.

[14] J.C. Collins and G.A. Ladinsky, Phy. Rev. **D43** (1991) 2847.

[15] R. S. Fletcher, T. K. Gaisser and F. Halzen, Phys. Rev. **D45** (1992).

[16] L. Durand and H. Pi, Phys. Rev. Lett. **58** (1987) 303; Phys. Rev. **D40** (1989) 1436.

[17] M. M. Block et al., Phys. Rev. **D41** (1990) 978.

[18] M. Drees and R. M. Godbole, Phys. Rev. Lett. **61** (1988) 682.

[19] M. Drees and R. M. Godbole, Phys. Rev. **D39** (1989) 169.

[20] M. Drees and F. Halzen, Phys. Rev. Lett. **61** (1988) 275.

[21] M. Drees, F. Halzen and K. Hikasa, Phys. Rev. **D39** (1989) 1310.

[22] R. Gandhi and I. Sarcevic, Phys. Rev. **D44** (1991) 10.

[23] R. Gandhi, I. Sarcevic, A. Burrows, L. Durand and H. Pi, Phys. Rev. **D42** (1990).

[24] R. Gastmans and T. T. Wu, *The Ubiquitous Photon*, (Oxford Univ. Press, Oxford, 1990).

[25] V. Barger and R. Phillips, *Collider·Physics*, (Addison-Wesley, New York, 1987).

[26] M. Drees and K. Grassie, Zeit. Phy. **C28** (1985) 451.

[27] J. F. Owens, Phys. Rev. **D30** (1984) 943.

[28] D. Duke and J. F. Owens, Phys. Rev. **D30** (1984) 49.

[29] S. Tilav, U. of Delaware Ph.D. Thesis, unpublished.

[30] M. Samorski and W. Stamm, Ap. J. **268** (1983) L17.

[31] M. A.Thomson et al., Phys. Lett. **B** (in print) and various contributions to the *Proc. of the 22nd Int. Cosmic Ray Conference*, Dublin (1991).

JULIA

Joint Underwater Laboratory and Institute for Astro-Particle-Physics

*Peter C. Bosetti**

III Physikalisches Institut, RWTH Aachen, Aachen, Germany

Abstract

This is a summary of the motivations and plans for and the status of JULIA, a proposed multidisciplinary underwater laboratory. The concept of JULIA allows for progress in Physics and Astrophysics, Oceanscience, Geology, Environmental studies as well as Detector-Development, Informationtechnology and Computer-Aided System Design and Surveillance.

* Now at Vijlen Institute for Physics, Vijlen, The Netherlands

Introduction

During the last few decades it became more and more obvious that the interplay of quite different branches of science can lead to substantial progress in the fields of research not anticipated a priori. One example is the now apparent connection of Elementary Particle Physics, Astrophysics and Cosmology. Other examples include the increasing interplay between both Theoretical Physics and Meteorology on one side and Computational Science on the other side. JULIA, **J**oint **U**nderwater **L**aboratory and **I**nstitute for **A**stro-Particle-Physics, is a proposed laboratory in which the multidisciplinary concept is inherent already in the design phase.

One of the main objectives of the JULIA experiment is obviously the detection of comic ray particles in the deep ocean by observing their emitted Cherenkov light with light sensors. Besides this a permanent laboratory in the deep ocean offers a variety of further applications. Amongst others, longterm measurements of the radioactivity, transparency (and thus pollution) of the water, bioluminescence, variation of the earths magnetic field, benthic currents and seismic activities should be mentioned.

Aims of JULIA

The main components of the JULIA experiment are based on the detection of Cherenkov light in the deep ocean. and will be described in the next section. In the following, a brief summary of the fields of research and development is given, where JULIA eventually could contribute.

- Physics and Astrophysics

Detection of atmospheric neutrinos and search for Neutrino-Pscillations
Detection of solar neutrinos and search for neutrinos from supernovae
Search for galactic and extragalactic high energy neutrino point sources
Acoustic detection of ultra-high energy particle interactions in the ocean

- Ocean Science

Measurement and longterm monitoring of bioluminescence
Measurement of radioactivity in the ocean
Monitoring of benthic currents

- Geophysics

Measurement of variations of the earth's magnetic field
Monitoring seismic activities
Determination of the earth's density profile from atmospheric neutrino interactions

- Environmental Studies

Online monitoring the clarity of the water by measuring the light absorption

- Detector Development

Improvement and tests of ultra-sensitive light detectors
Improvement on acoustic detectors for interactions of ultrahigh energy cosmic ray particles

- Informationtechnology

Gigabit data transfer from the deep ocean via monomode fibre optic cable
Online data analysis
Parallel Processing

- Computer Aided System Design and Surveillance

Expertsystem development on performance and surveillance of the detector system
Underwater remote control robotics with advanced capabilities

The major items that will be adressed and progress can be expected are:

Detector Development
Central element of the optical modules are large area photomultiplier. A traditional design, essentially an extension of the technology used in smaller tubes, does not offer the requirements of single electron resolution and transit time jitter of not more than a few nanoseconds. A substantial progress has been done with the design and development of "Smart" photomultiplier. These photomultiplier are essentially a system of a "photon-preamplifier" with large photocathode and a small photomultiplier with good energy- and timing resolution.

Data Handling
Eventually, the laboratory will consist of a large amount of detectors and monitoring devices leading to a huge amount of data accumulated in the deep ocean. Data rates in the gigabit range have to be transferred to a shore station and analyzed there. In a first step an online analysis has to be performed to separate genuine data from backgrounds in order to keep the data that has to be saved permanently at a manageable rate. This involves the development of "intelligent" trigger logics, large data-transfer rates via monomode optical fiber cable and a sophisticated system of parallel processors.

Software Development

Both for the design phase and the operation of the laboratory sophisticated computerprograms have to be developed on the basis of expertsystem methods. During the design phase, an optimisation of the composition and arrangement of the detector components for the applications shall be done using Monte Carlo simulations.

During the operational phase, an expert system should allow for an optimisation of the performance as well as minimisation of maintenance and repair time.

Robotics

Due to the nature of the laboratory, a sophisticated robot has to be involved both during the deployment and the operational phases. The robot has to be able to move accurately in three dimensions and perform small repair work in situ as well as install and remove parts to be repaired or exchanged from shore.

A conceptional design for the JULIA Detector

One of the primary purposes of the JULIA detector is the detection of high, medium and low energy neutrinos from galactic and extragalactic point sources. The neutrinos will be detected by the cherenkov light from the charged secondaries produced by their interaction in the seawater. For low energy neutrinos, this will be electrons, for high energies, mostly muons will be detected.

In figure 1, a possible design for the JULIA detector is shown[1]. It consists of three different parts nested into each other. The outer most part consists of a series of strings with large area photomultipliers (35 cm diameter) with 50 m spacing both horizontally and vertically. This results in a threshold of around 1 TeV for muons produced by neutrinos. The effective volume, depending on the energy of the muon, is of the order of 10^8 m^3.

Inside this widely spaced array the medium energy detector is placed. Photomultiplier of smaller size (9") are spaced rather closely (1 - 2 meters horizontally, 5 - 10 meters vertically) to be sensitive for athmospheric neutrinos in the 1 GeV energy range. This part of the detector is in particular sensitive to neutrino oscillations by comparing the flux of up- and down going muons. By using the outer part of the detector neutrino interactions inside this volume can be selected. The energy of the muons can be determined by their range.

The inner most part is designed for the detection of low energy neutrinos, i.e. those from supernova burst. In the design, special emphasis is given to a large detection volume by minimizing the costs. Therefor the design is made is visualized in fig 2. A cylinder of diameter of 10 m height and 10 m diameter is taken as example. The inner part of the cylindic sides is covered with reflecting material and only the upper area is equipped with photomuliplier tubes for economic reasons. Even though the detection efficency for supernova neutrinos is reduced by not having the whole surface equipped with PMTs, the reflections make up for a substantial amount of this loss[1]. A typical interaction of a 25 MeV neutrino inside this detector is shown in figure 2. To reduce the background from athmospheric neutrinos, a small plane of material different from water can be integrated to produce total reflection of particles coming from the upper hemisphere. In fact, simulations suggest that a detector of this type could even work at sea level[1] and thus be multiplied to increase the sensitivity.

The central element of the high energy part of the detector is the large area (15‘) photomultiplier Philips XP2600 (Smart PMT"), embedded in a standard pressure housing for deep ocean applications[2]. This photomultiplier consists of the combination of an elctro-optical preamplifier with a conventional small phototube. Photoelectrons are accelerated with a high voltage (25kV) to a scintillator placed in the center of the glass-bulb. This scintillator is read out by a small fast 11-stage phototube (XP2982). The photomultiplier is shown schematically in fig. 4 and as working example in fig 5.

The result of the high acceleration voltage is a high gain in the first stage and essentially 100% collection efficiency over the whole cathode area. As a typical value, 30 photoelectrons (PE) are converted in the smalll phototube for each primary PE. Due to the good statistics in the first stage, the tube provides an excellent timing and energy-resolution on the low photoelectron- level [3].

The energy-resolution on the 1PE-level is better than 50% FWHM [4] (see fig. 6), resulting in the possibility of a clear separation between one, two, and more PE. This is demonstrated in figure 7, which shows a typical distribution of the integrated charge, when the tube was illuminated with a blue LED of constant low intensity. A fit with lienarly spaced Poisson distributed Gaussians describe the distribution well[4].

An accurate gain calibration of this photomultiplier can be done in situ any time using the dark current signal originating from thermal emission of photoelectrons from the cathode as is demonstrated in figure 6.

The time-jitter for this tube is ~5ns (FWHM) for 1 PE and decreases as $1/\sqrt{n}$ with n the number of PE, as can be seen from figure 8, where the measurement of point illumination with a fast green LED with an intrinsic time-jitter of 1.8 ns is shown. The contribution of transit-time differences resulting from different origins of the photoelectron on the cathode are estimated to be ~1.4 ns [4]

The pulse structure of this PMT can be seen from figure 9, where a typical pulse for a low PE signal is shown as recorded with an 300 MHz FADC. The signal is determined by the decay time od the scintillator, which has been measured to be ~60 ns. This requires the readout- electronics to integrate the complete charge of the induced signal to obtain an optimized energy resolution. Due to the long decay-constant of the scintillator conventional charge to time converters require long conversion times. Therefore we have designed the system to convert the charge parallel to its collection. As a result, the time of conversion of a pulse is proportional to the number of photoelectrons of the signal. This leads to small deadtime (~180ns) for 1 PE- signals as such as background light in the ocean (K40-decays) [5].

Two different discriminator thresholds are implemented. The first (low threshold) determines the time of the signal, triggering on the leading edge of the PMT-pulse. The second (high treshold) requires the pulse-height to be above a preset level. The circiut measures the integrated charge of PMT-pulses with good linearity and time accuracy. The output signal is an ECL-pulse which can be easily handled in further data aquisition [5].

The module is controlled by a 68301-CPU based computer with an OS/9 operating system. This enables not only downloading software via a 300 baud modem-signal superimposed on the 48 V DC power supply line, but also allows for the regulation of relevant OM parameters. These parameters include the monitoring and setting of the two internal readout-thresholds and the high voltage of the PMT. It also monitors the PMT count rates, the temperature within the module at three different locations, possible water leakage and the internal reference voltages and their currents. The implemented software also constantly checks on the performance of the module and is capable of reporting on alarm conditions [6].

Besides these optical detectors acoustic detection of highest energy interactions can as well be implemented as well as all other oceanographic and geological sensors. Also, it should be noted, that such an optical module can not only detect natural radioactivity in the water but any additional radioactive source of the same magnitude online. Furthermore the light originating from natural radioactivity gives a measure of the absorption length for light in water, enabling to also measure any pollution online, independent of the source of the water pollution. This makes such a module an online detector for environmental observations in water of a very large volume.

JULIA Test Cruise

Early 1991 we have performed in collaboration with the Instituts for High Energy physics, Zeuthen, and the Institute for Lagerstättenlehre, Aachen, a first feasibility study wuithin the JULIA project. The site selected was south of the Canary Island of Gomera (27'45 N, 15'10 W), as the ocean depth at this site is 3378 meters only 17 Km off shore. Measurements have been performed with the German research vessel SONNE. The vessel is shown in fig. 10, it is equipped with adequate mechanical and electric workshops.

The mini-string, designed for this test consists of three optical modules arranged with 2 meters vertical distance, connected to a junction box which supplied them with electric power via a coax cable. The optical modules were a prototype version of the final light sensors. In a 17'pressure housing (Benthos) a Smart 35 cm PMT

(PHILIPS) was installed. The readout as well as the power supply and the electrical/optical signal transfer were also part of the module. The halfshere opposite to the photocathode was blackened.

Pulses from the PMT's were tranferred via a monomode fibre optic cable to the vessel. The cable was a special design by AEG. It allowed a direct data transfer with an accuracy of better than 0.1 ns for the timing of the signal. The "raw" data from each PMT were transmitted into the on-board NIM electronics, were they could be discriminated individually. Any coincidence signal could be selcted for any time window via soft- and hardware. Thereafter, the data were fed into a CAMAC crate and read in to a computer, were grafics displays could be obtained.

In fig 11 the watering of the mini-string is shown. Several technical difficulties turned up, however, we finally succeeded in measuring over a longer period at a depth of 1000 meters the light intensity with one of the modules. This measurement is shown in fig 12. Comparing this spectrum with a reference spectrum obtained on board of the vessel a significant excess of entries at higher energies in the data from deep underwater is obtained, proving that external light sources have been detected.

In addition to testing the JULIA concept, we have performed several measurement in situ to investigate the possibility for a permanent underwater laboratory at that site according to the specifications mentioned above. Via echolot measurements a detailed bottom profile in that area was obtained. It allows in conjunction with the photographs and videos taken at the site that the location appears to be very well suited for such a laboratory with respect to flatness and bottom quality.

Furthermore, we have taken data on the salinity, temperature, sound velocity and density as a function of water depth. These data are shown in fig. 13 to 16 and show the expected values and certainly are in accordance with the requirements for a permanent underwater laboratory.

Conclusions and Outlook

It has been shown that the general scheme for the detection of external light sources as proposed for the JULIA experiment is feasible. We have developed a special read out and remote control electronics for the so called Smart large diameter Philips PMT's enabling to easily distinguish between natural radioactivity in the ocean on the one photoelectron level from higher intensity Cherenkov light origininating from cosmic ray particles. Bioluminescence can be separated as well due to the different pulse time structure. The concept allows for a most flexible setting of triggers and selction of data to be analysed.

For a first generation experiment the optical modules will be developed further and improved to fulfill the requirements for the DUMAND experiment. In late 1993 several of them are expected to be deployed as part of the DUMAND II experiment.

Acknowledgement

I like to thank all collaborators for many fruitful discussions and their help in the preparations of this talk.

References

[1] G. Wurm. Zum Nachweis kosmischer Neutrinos mit Wasser-Cherenkovdetektoren. Diploma thesis, RWTH Aachen, in preparation.

[2] G. van Aller, S. O. Flyckt, W. Kühl, P. Linders and P. C. Bosetti, A "smart" 35 cm Diameter Photomultiplier. Helvetica Physica Acta, 59, 1119ff., 1986.

[3] P. C. Bosetti, Neutrino detection using "smart" large diameter photomultiplier. Workshop on Neutrino Masses and Neutrinoastrophysics, Telemark IV, Ashland, Wisconsin, USA, 1987.

[4] C. H. Wiebusch. Zum Nachweis schwacher Lichtquellen im Ozean mit Hilfe eines neuartigen großflächigen Photomultipliers. *PITHA 91/20.* Diploma thesis, RWTH Aachen, 1991.

[5] F. Beißel and V. Commichau. A fast charge to time converter V04. Internal report HD04, RWTH Aachen, February 1991.

[6] U. Berson, Entwicklung und Test eines Optischen Moduls zum Nachweis schwacher Lichtquellen in Wasser-Cerenkov-Detektoren. Diploma thesis, RWTH Aachen, in preparation.

FIGURE CAPTIONS

Fig. 1. Conceptual design of the JULIA detector. It consists of three different parts, sensitive for high, medium and low energy neutrinos, respectively.

Fig. 2. The low energy part of the JULIA detector, sensitive to superneutrinos. Two examples of interactions are shown, the upper figure shows the simulation of an upcomig electron (i.e. an accepted signal), the lower one a background interaction, showing the total reflection of the produced photons.

Fig. 3. Simulation of a 25 MeV electron in the inner part of the JULIA detector.

Fig. 4. Schematic drawing of the "smart" PHILIPS 35 cm photomultiplier.

Fig. 5. Photograph of the "smart" 35 cm diameter PHILIPS photomultiplier.

Fig. 6. Distribution of the dark current of the "smart" PHILIPS 35 cm PMT.

Fig. 7. Distribution of integrated charge of PMT pulse resulting from the illumination with an LED of constant low intensity.

Fig. 8. The pulse structure of two typical low PE pulses.

Fig. 9. Timejitter as a function of the number of photoelectrons. The dashed curve shows the indicated parametrisation.

Fig. 10. The german research vessel SONNE.

Fig. 11. Watering of the three optical modules during the first JULIA test experiment.

Fig. 12. Energy spectrum obtained with a "smart" PHILIPS PMT in 1000 meters depth.

Fig. 13. Salinity as a function of depth at the site of the JULIA test cruise off the Canary Islands.

Fig. 14. Temperature as a function of depth at the site of the JULIA test cruise off the Canary Islands.

Fig. 15. Sound velocity as a function of depth at the site of the JULIA test cruise off the Canary Islands.

Fig. 16. Density as a function of depth at the site of the JULIA test cruise off the Canary Islands.

400m
50m
50m
400m
10m
2m
16m
80m
High Energy
Array ~1TeV
Medium Energy
Array ~1GeV
Low Energy
Supernova Detector
~10MeV
Mulitplexing
Shore Station

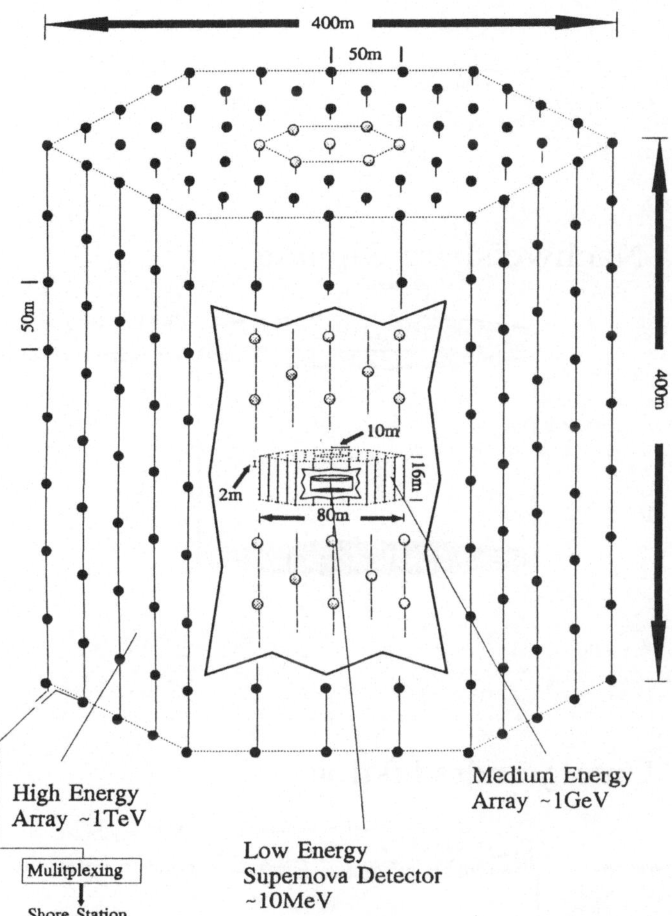

Figure 1.

Nachweisbare Signale
PMT Puls
PMT - Ebene
Wasseroberfläche
γ
γ
e
ν

Untergrundreduktion
μ
PMT - Ebene
Totalreflektion
Wasseroberfläche
μ
γ
γ
γ
γ

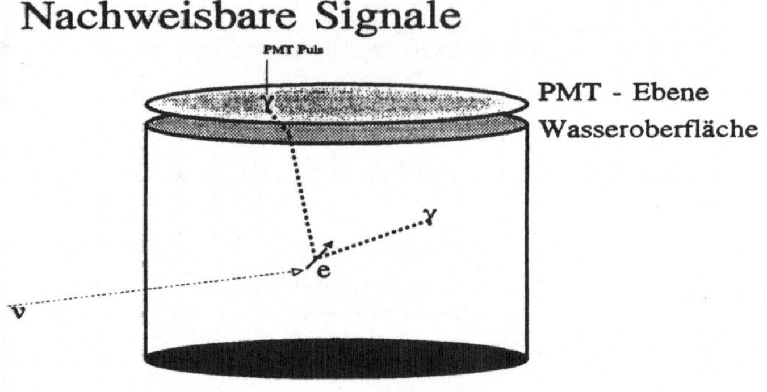

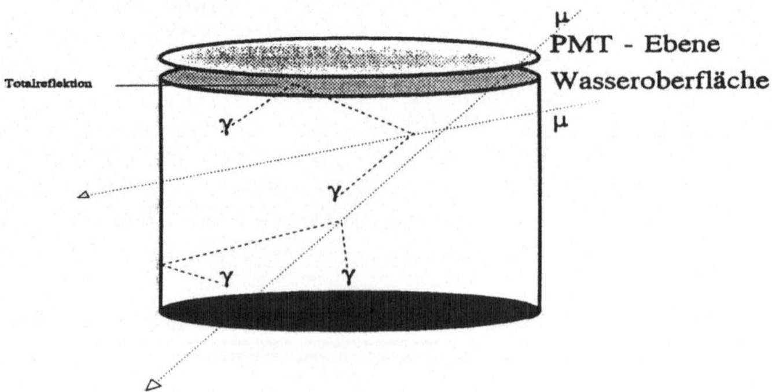

Figure 2.

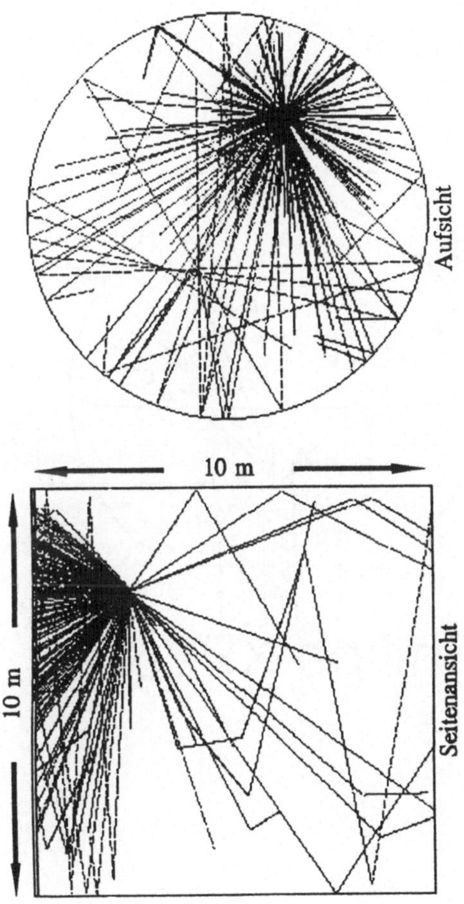

Figure 3.

34,5cm
Photokathode
SbKCs
25kV
4,6cm
e⁻
γ
Szintillator
$YSiO_5$:Ce
Konventioneller
Photomultiplier
z.B. XP2982A
Philips XP2600

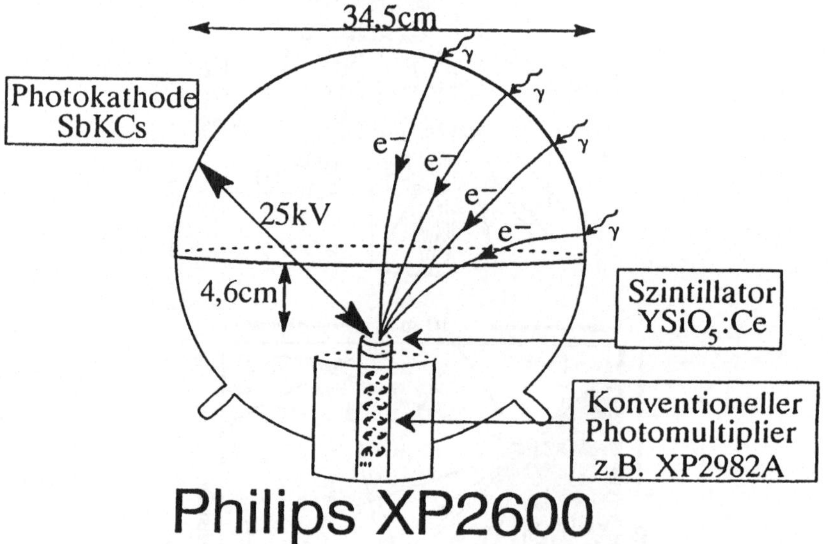

Figure 4.

Figure 5.

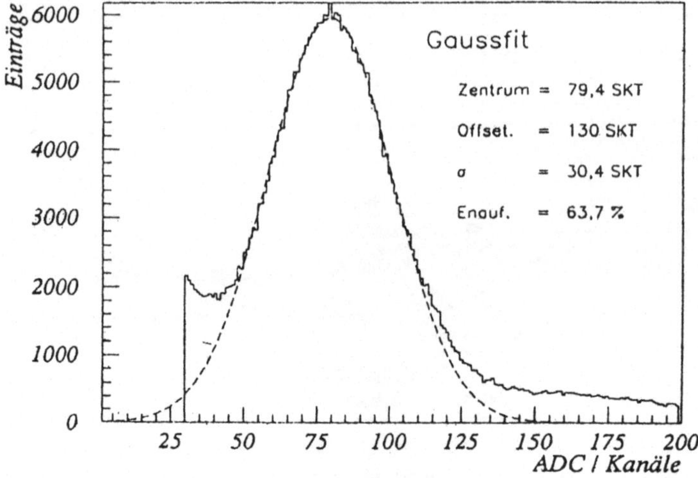

Figure 6.

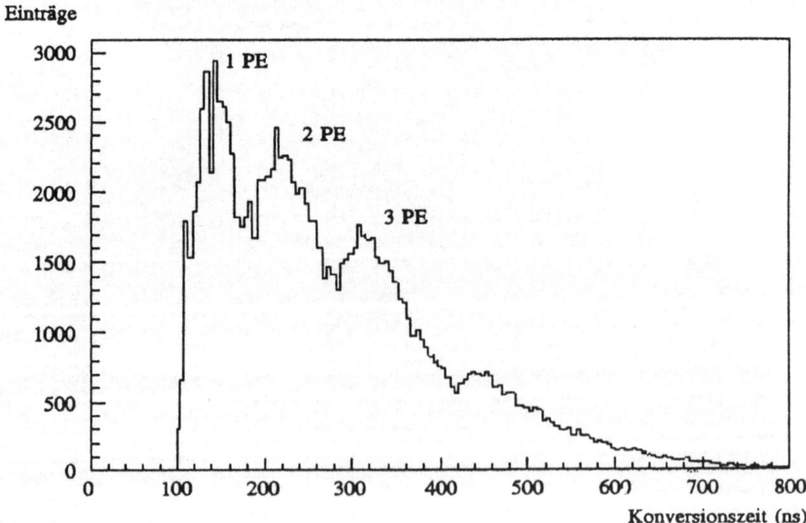

Figure 7.

TOT = 100ns
100mV
Integral = 200Skt ≈ 1PE [ns]
TOT = 160ns
100mV
Integral = 200Skt ≈ 1PE [ns]

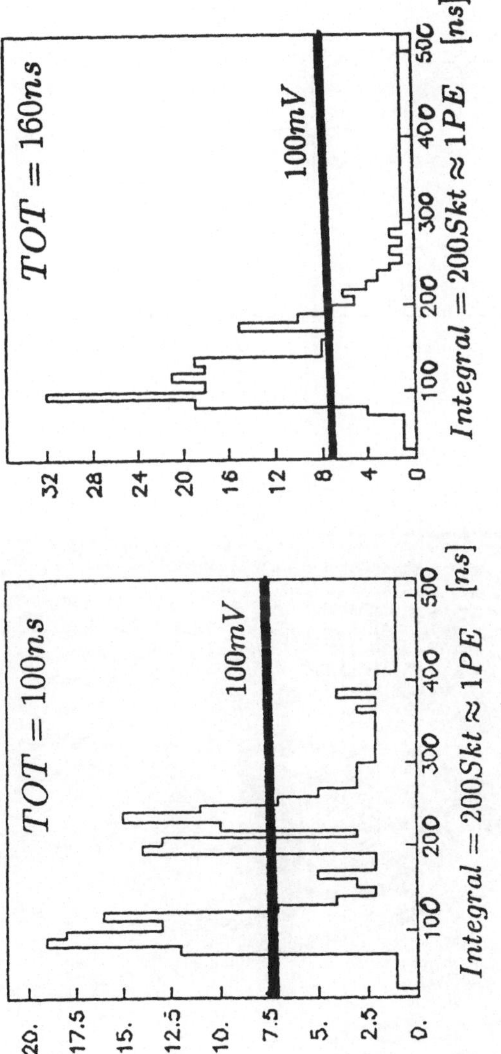

Figure 8.

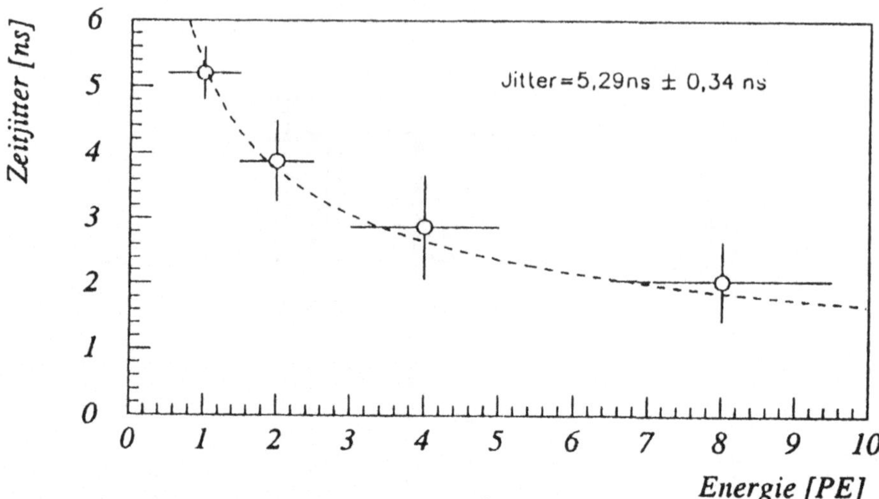

Figure 9.

Figure 10.

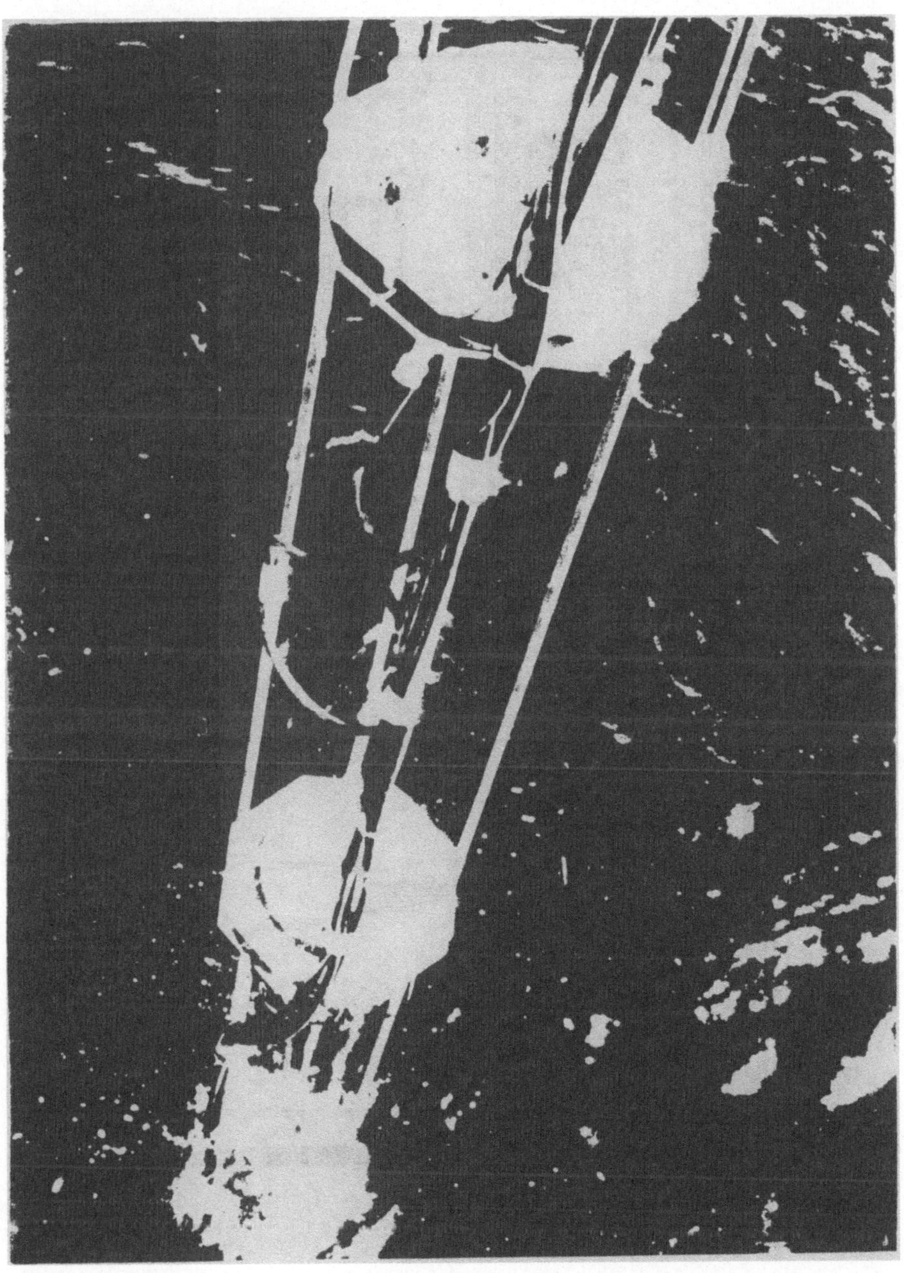

Figure 11.

Desk Romeo CAMAC File I/O HISTOGRAM FIT HELP 15:17:08
Histogram
Text
-- File Header --
File :jr1b.DAT
Run Number :jr
Date and Time :14:35--26: 2:1991
Format ID : 2
Active PMs : 1
PM ID : 0
Threshold : 30.000
Window [nsec] : 2000.000
No. of coinc. : 1
Remote HV : 955.000
Experim. depth : 1000.000
Experim. Temp : 0.000
No. of Columns : 1
column names :EN-PMT2
ID 1
INT 953
MEAN 107.309
SIGMA 67.670
E 1
4.00
3.00
2.00
1.00
0.00
0
100
200
300
400
500
Histogramm 1

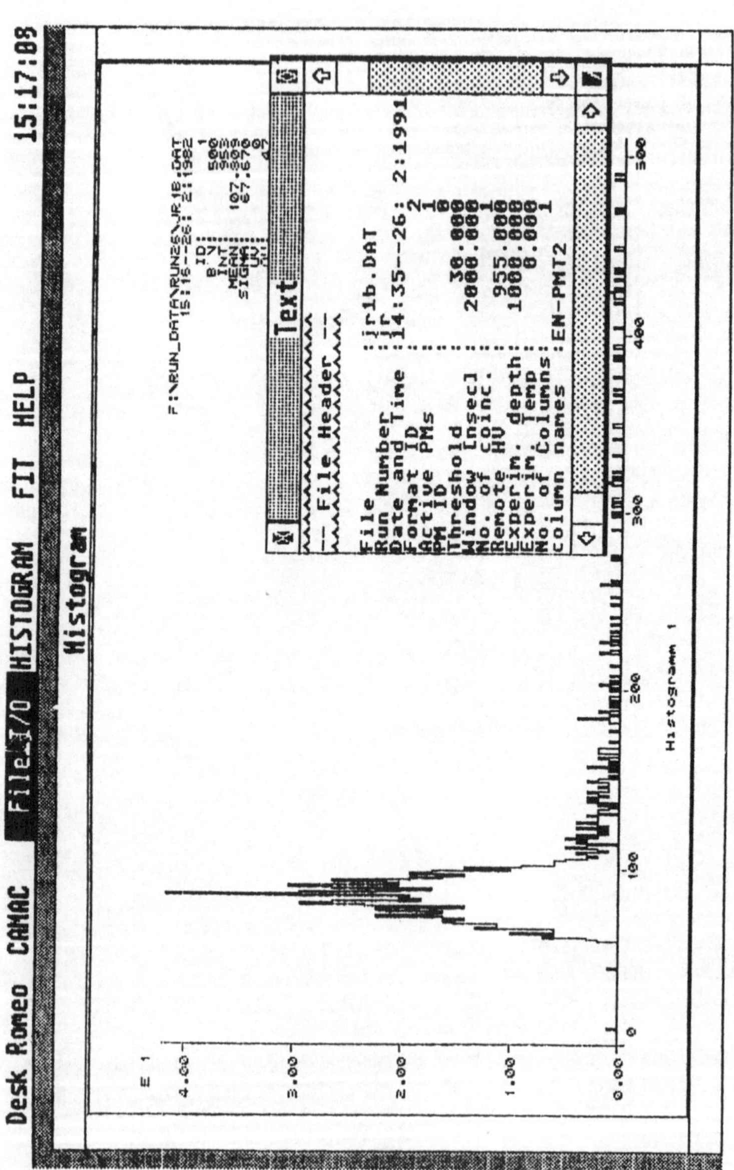

Figure 12.

SALINITY [PPT]
34.0
35.0
36.0
37.0
3399m
0
200
400
600
800
1000
1200
1400
1600
1800
2000
2200
2400
2600
2800
3000
3200
3400
[m]
SO74 JULIA RWTH AACHEN STATION NO.: 1
MULTI-SONDE

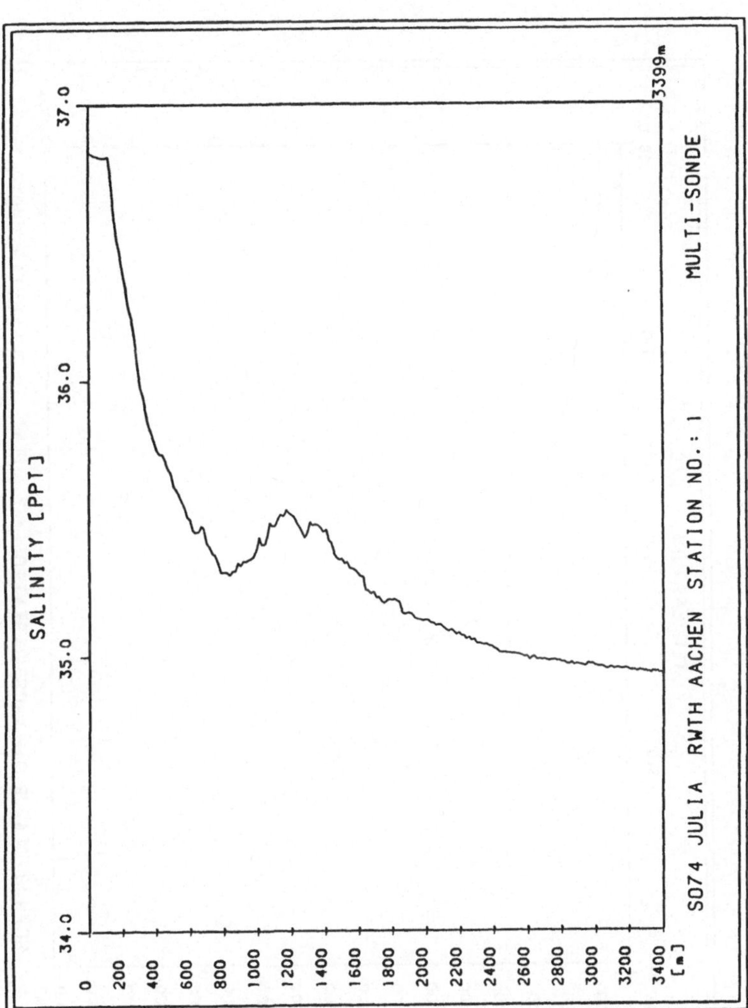

Figure 13.

TEMPERATURE [°C]
0.0
5.0
10.0
15.0
20.0
0
200
400
600
800
1000
1200
1400
1600
1800
2000
2200
2400
2600
2800
3000
3200
3400
[m]
3399m
SO74 JULIA RWTH AACHEN STATION NO.: 1
MULTI-SONDE

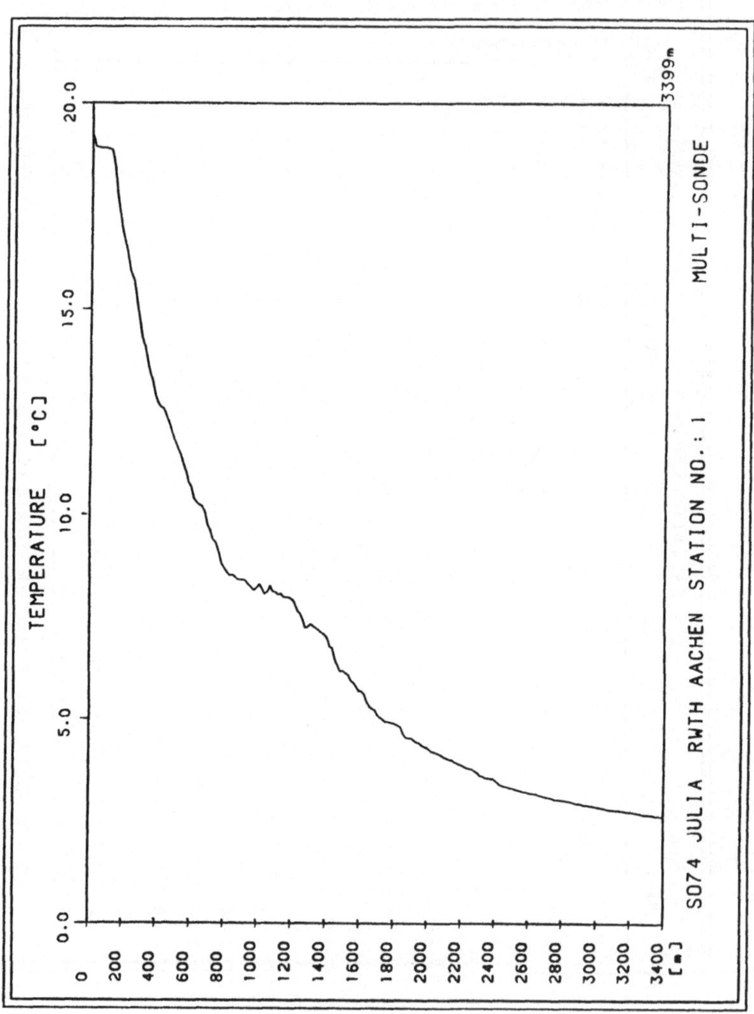

Figure 14.

SOUND VELOCITY [M/S]
1480.0
1490.0
1500.0
1510.0
1520.0
1530.0
3399m
0
200
400
600
800
1000
1200
1400
1600
1800
2000
2200
2400
2600
2800
3000
3200
3400
[m]
SO74 JULIA RWTH AACHEN STATION NO.: 1
MULTI-SONDE

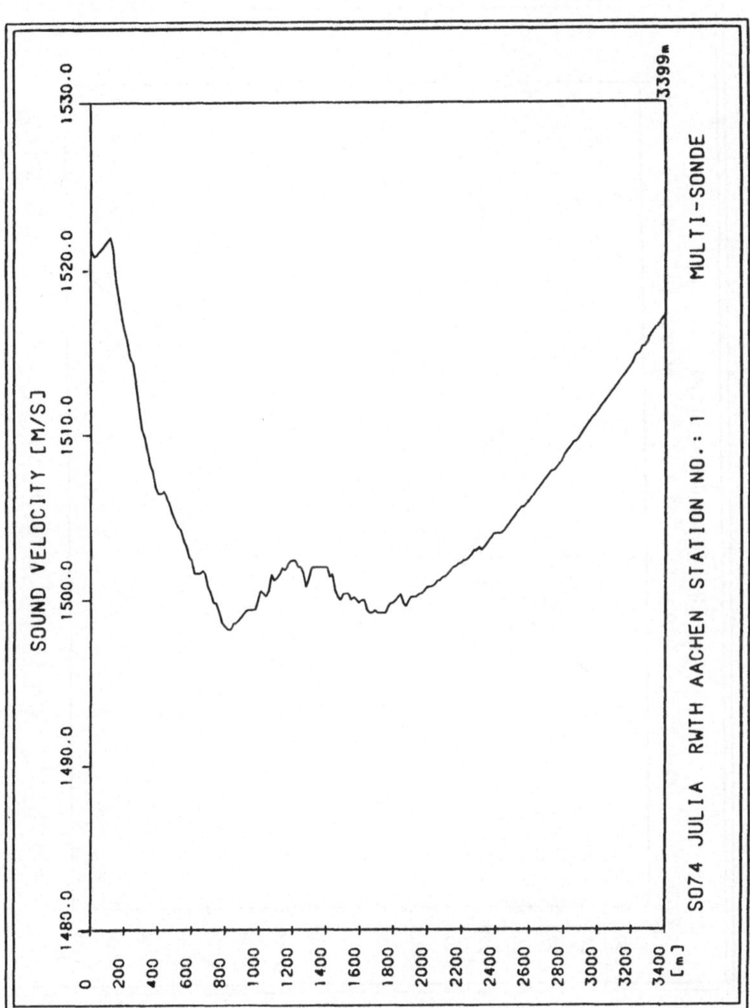

Figure 15.

DENSITY SIGMA (S.T.P)
20.0
25.0
30.0
35.0
40.0
45.0
50.0
0
200
400
600
800
1000
1200
1400
1600
1800
2000
2200
2400
2600
2800
3000
3200
3400
[m]
3399m
SO74 JULIA RWTH AACHEN STATION NO.: 1
MULTI-SONDE

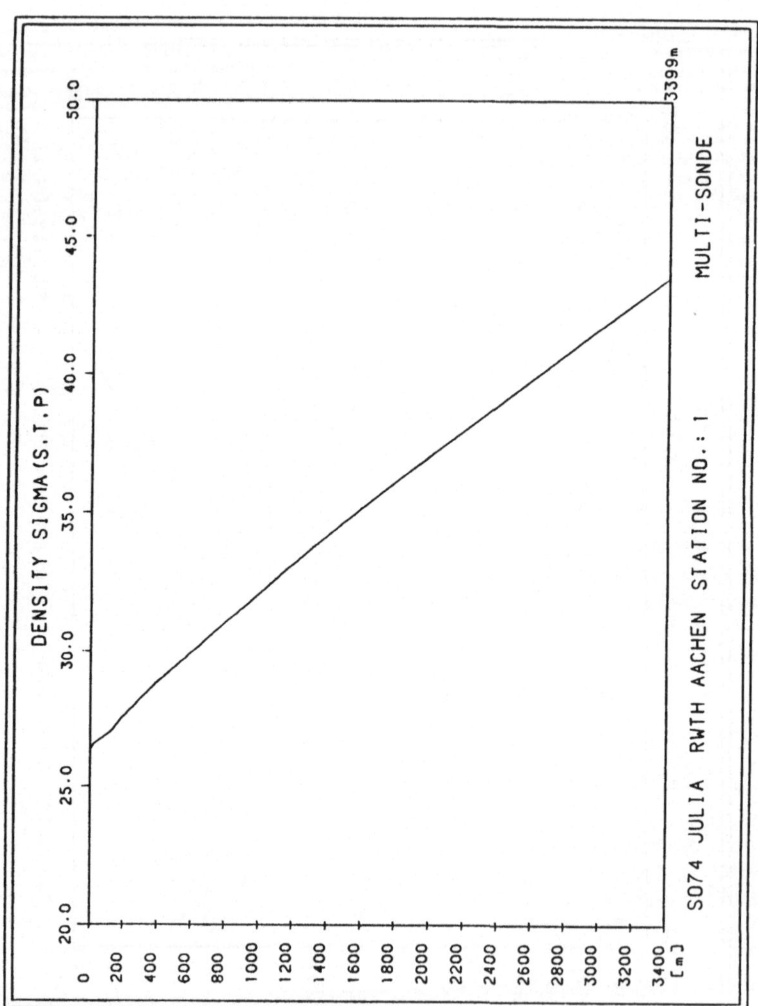

Figure 16.

AMANDA: Antarctic Muon And Neutrino Detector Array

S. W. Barwick[a], F. Halzen[b], D. M. Lowder[c], J. Lynch[a], T. Miller[c], R. Morse[b], P. B. Price[c], A. Westphal[c], G. B. Yodh[a]

[a]Department of Physics, University of California, Irvine, CA 92717, USA
[b]Department of Physics, University of Wisconsin, Madison, WI 54930, USA
[c]Department of Physics, University of California, Berkeley, CA 94720, USA

Abstract

Enormous detectors with unprecedented sensitivity will be required to search for astrophysical sources of neutrinos. If the transparency of deep polar ice is similar to that measured in the laboratory, then AMANDA (Antarctic Muon And Neutrino Detector Array) may be the most cost-effective way to reach the detector volumes required to search for astrophysical sources of high energy neutrinos or WIMP annihilation within the sun.

We describe a series of tests that will be conducted near the South Pole Station during the '91–'92 Antarctic campaign. The primary goal of these tests is to extend the optical transparency measurements of polar ice to a depth of one kilometer by measuring the rate of downward-moving muons with a prototype string of 4 optical modules. Additional objectives include the measurement of up/down discrimination, background light levels, and timing resolution. The design of the prototype string and its expected performance based on laboratory calibrations are discussed.

1. Introduction

The experimental challenge of high energy neutrino astrophysics is to conceive detectors of large sensitivity in an environment shielded from the severe cosmic ray backgrounds at the surface of the earth. The AMANDA concept is based on a new idea which suggests that optically transparent ice found deep under the Antarctic ice cap may be used in much the same way that the DUMAND facility uses ocean water as an integral part of a large volume particle detector [1]. Evidence is mounting that the optical attenuation length of deep polar ice is comparable to pure polycrystalline ice grown in the laboratory. It is known that ice cores retrieved from a depth of approximately one kilometer become bubble-free and photographs of the cores show them to be optically transparent. Gow and his colleagues [2] have measured the distribution of bubble diameters and bubble density as a function of depth in Antarctic ice. At depths between 800–1100 meters, trapped air is forced into the ice lattice forming nitrogen and oxygen hydrates (molecules of $N_2O_2 * 6H_2O$). Approximately 0.06% of the ice changes from the usual hexagonal structure into a cubic crystal, called a clathrate hydrate, to accommodate the hydrate molecules. The remaining ice retains its structure and becomes bubble-free [3]. Recent measurements of in-situ polar ice at depths of 220 meters suggest that the maximum attenuation length of ice exceeds 18 meters, comparable to measurements of polycrystalline ice grown in the laboratory ice (shown in Figure 1). Moreover, our studies of in-situ ice leads us to conclude that gas bubbles trapped locally in the ice during the installation of the photodetectors should not significantly impact the sensitivity of the optical sensors because the scattering of light from air bubbles in ice is primarily forward-directed and negligible additional absorption is introduced.

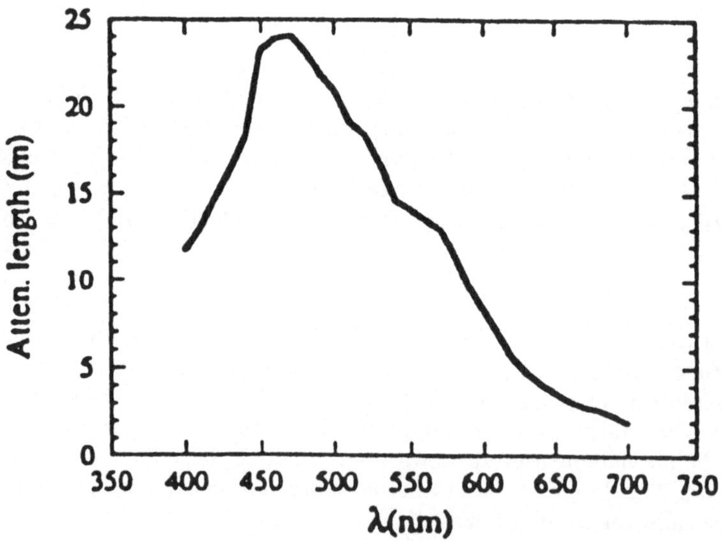

Figure 1: Optical attenuation length of polycrystalline ice grown in lab [4]

A series of tests are scheduled for the upcoming '91–'92 Antarctic campaign which are designed to evaluate the usefulness of deep Antarctic ice as an integral component in a large-volume particle detector. A string of photomultiplier tubes will measure the optical transparency of ice located between 500–1000 meters beneath the South Pole. The rate of downward moving muons seen by the string is related to the effective sampling volume and hence related to the optical transparency of the ice. Additional goals of the protostring include the measurement of up/down discrimination and timing resolution. The string will provide experience with gain and timing calibration techniques, reliability assessment, quality control, remote monitoring and data acquisition.

The Polar Ice Coring Office (PICO) has implemented a number of different technologies to drill boreholes in the ice. One relatively inexpensive and quick method of drilling uses hot water to melt through the ice. Surface snow is collected, heated to near boiling in a pressure tank, and distributed via a high pressure hose. The hot water melts the ice within a cylinder whose radius is determined by the rate of descent. Numerous holes of modest diameter (6"–7" diameter) have been drilled to depths in excess of one kilometer and a large diameter hole (14" diameter) has been drilled to 500 m [5]. It takes about 30 hours to drill a large diameter hole to 1 km. After a borehole is drilled, a string of photomultiplier tubes is lowered to depth.

2. Why Antarctica? Physics advantages

It is no secret that Antarctica, located on the far end of the globe, has a rather unforgiving environment. Experimental campaigns must be carefully planned to accommodate the formidable logistical requirements. We consider an Antarctic-based experiment because the AMANDA concept has several compelling science advantages:

(i) instrumenting polar ice with photomultiplier tubes allows the experiment to follow science. The detector can be gradually increased in resolution or (alternatively) expanded in coverage over the essentially unlimited expanse of the stable icefields;

(ii) ice is a quiet, sterile medium. Studies of Antarctic ice cores have shown that the concentration of radioactive β-emitters such a 40 K is nearly four orders of magnitude below concentrations in seawater. In addition to the lack of sources of background light, intrinsic dark noise is typically only $\sim$ 1.5 kHz (0.3 pe discrimination) in the constant $-$ 55° C temperatures. Since the background rates are small, extremely simple trigger schemes can be implemented with off-the-shelf, inexpensive electronics;

(iii) a cosmic source is observed continuously and remains at fixed angle, making the subtraction of comic ray backgrounds extremely simple;

(iv) a large fraction of the detector electronics can be located at the surface. Moreover, timing can be maintained at a few nanosecond resolution over kilometer lengths of cable because the ice temperatures below a few meters are very stable throughout the year;

(v) the existing Bartol-Leeds South Pole air shower array (SPASE [6]) can be used to determine the absolute pointing accuracy of AMANDA by looking at events which simultaneously trigger both detectors (although the hemispherical PMTs are pointed down, they have some sensitivity to down-going muons);

(vi) most new air shower arrays are located in the northern hemisphere (CASA, CYGNUS II, AKENO, etc.). Thus, AMANDA can monitor the neutrino signal and air shower arrays can monitor gamma emission from any northern source at the same time.

In addition, there are a number of practical advantages to considering South Pole Station as a base for AMANDA. It is manned throughout the year, providing food and lodging. As much as 10 kilowatts of power are available to experimenters which is adequate for AMANDA. There are communication links via satellite which are capable of transmitting $\sim$ 5 MBytes of data per day.

3. Project Description

We have constructed two prototype strings of 4 modules containing 8" hemispherical PMTs from Thorn EMI (9351KB). The PMTs are separated from each other by 5 meters and housed in glass pressure vessels specially designed to withstand pressures up to 150 atmospheres (supplied by Benthos, Inc.). The string will be used to measure the flux of penetrating muons between 500–1000 meters depth. This type of experiment inherently integrates over many of the systematic uncertainties of the full-scale neutrino detector, such as the module/medium optical coupling after refreezing, the wavelength-dependent attenuation properties of ice, or possible de-focussing of the Cherenkov cone by the ice medium. Since the muon flux can be reliably estimated to within 10% [7], the coincidence rate determines the effective area of sensitivity. As the muon rate is comfortably large, a few hours of data collection at a given depth is sufficient to determine the transparency properties of the ice to an accuracy of 10%. The total separation between the top and bottom module is 20 meters which is large enough to determine muon trajectories from timing and pulse height information, allowing the study of up/down discrimination.

The light collection efficiency of the string will be monitored as it freezes into the borehole to make sure that no difficulties are encountered during the refreezing process. Long term studies will be made to assess module reliability and mechanical integrity. It is hoped that the string will remain in operation until the next campaign season.

4. Optical Module and String Design

The optical detection system was designed to be electrically and mechanically robust. In addition, the sensors must tolerate pressures as large as 100 atmospheres, operate at the $-55°$ C ambient temperatures, and consume < 1 watt of power per detector.

The optical module contains a single 8" hemispherical PMT, divider chain, and a pair of green LEDs for calibration. We chose the Thorn-EMI 9351KB PMT because it has excellent single photoelectron resolution (peak-to-valley ratio > 2.0), good timing at the single photoelectron level (typical timing resolution was measured to be 3.5 ns FWHM, scaling as 1/(Npe)0.5 up to Npe $\sim$ 10), high gain (1.5×108 at 1400 V), and low cost ($1050 ea. in quantities of 10). The PMTs are embedded in a clear gel (RTV 6156) and housed in a clear glass pressure vessel supplied by Benthos, Inc. The contour of the glass envelope of the PMT almost exactly matches the contour of the 10" diameter pressure vessel. The RTV serves to match the indices of refraction between the pressure vessel and PMT, and the RTV mechanically insulates the PMT from external shock and vibration. The high voltage and PMT output are routed over a one kilometer length of RG122 via commercially-available underwater connectors, supplied by Brantner and Assoc. (San

Diego, CA). It is expected that the optical modules will be highly reliable since they contain only a few, simple electronic components besides the PMT. The modules are slightly buoyant in water to insure that they maintain their separation as they are lowered to depth. The depth of of the lowest module will be determined by measuring the water pressure with a sensor supplied by Paine (Seattle, WA).

Signals from the PMT are separated from the high voltage and routed into a frequency-dependent amplifier which partially compensates for the distortion of the pulse shape introduced by the long coaxial cable. Table 1 shows the affect of both the cable and amplifier on the signal characteristics.

Table 1
PMT pulse characteristics, −55° C, 1400 V

	Pulse Height (V)	risetime (ns)	FWHM (ns)
Before 1 km cable	0.7	4.0	10.0
After 1 km, before amp	0.003	110	270
After amp	0.03	30	80

The timing resolution of the optical module was determined using light from a pulsed dye-laser, pumped by a Nitrogen laser (pulse width = 300 ps), which is routed into a fiber optic splitter. Most of the light impinges on a photodiode (PD). The signal from the PD is discriminated and delayed by 5.2 msec, providing the start pulse for the TDC (Lecroy 2228A) and gate for the ADC (Lecroy 2249A). The remaining light passes through an attenuator before illuminating the photocathode of the PMT. The output from the PMT propagated through one kilometer of RG122 coax, amplified, and discriminated at the 0.3 pe level. The discriminator output provided the stop pulse. The optical module is placed in a refrigerator which maintains a temperature of −55° C to an accuracy of 1° C. The results of the timing tests are given in Table 2. The timing resolution was measured as a function of the number of photoelectrons generated by the laser pulse. The test were performed using two configurations: (1) using a 10m cable between optical module and signal/HV pickoff electronics, (2) using a 1 km cable between module and electronics. Timing degrades in (2) because the signal-to-noise ratio becomes significantly worse. Finally in configuration (3) a frequency-dependent amplifier was used. We obtained a 5 nsec FWHM for 1 pe over a 1 km cable.

Table 2
FWHM Timing resolution (ns), −55° C

configuration (see text)	1 photoelectron	2 photoelectrons	4 photoelectrons
1	3.5	2.5	1.9
2	6.5	5.5	3.5
3	5.0	3.8	-

5. Status

Two strings of optical modules have been designed, constructed, and tested in the laboratory. The modules perform well at low temperatures. Timing resolution has been shown to be adequate for the transparency tests and should improve with the use of the frequency-dependent amplifier.

References

[1] F. Halzen and J. Learned, *Proceedings of the Inter. Symp. on VHE Cosmic Ray Interactions*, the Univ. of Lodz Publishers, edited by M. Giler (1989).

[2] A. J. Gow, J. Geophy. Res. **76** (1971) 2533; A. J. Gow and T. Williamson, Cold Regions Research and Engineering Laboratory, Research Report 339 (Oct. 1975).

[3] S. Miller, Science, **165** (1969) 489.

[4] T. C. Grenfell and D. K. Perovich, J. Geophys. Res. **86** (1981) 7447.

[5] J. Sonderup, director of PICO, private communication.

[6] N. J. T. Smith, et al., Nucl. Inst. and Meth. **A276** (1989) 622.

[7] S. Miyake, *Proceedings of the 13th International Cosmic Ray Conference*, Vol. 5, (Denver, CO, 1973), 3638.

Signal Processing With JULIA

Christopher Henrik V. Wiebusch

III.Physikalisches Institut, Rheinisch–Westfälisch Technische Hochschule Aachen, Hyskensweg, 5100 Aachen, Germany

Abstract

A new project, JULIA , is developing new technologies for next stages DUMAND–like underwater experiments. These experiments use large area lightsensors for the detection of faint Čerenkov–light, produced by neutrino induced myons. During a cruise in Feburary 1991 a small testdetector was deployed in the Atlantic ocean near the Canary Islands. The technical concept of this test–experimet is discussed here. The most important aspects were tests of a new type light–sensor (*"smart" photomultiplier*), and an analog transmission of data to ship via single mode optical fibers. First results show, that it is possible to increase experimental accuracy and reliability, and also to decrease the price of detector components.

1 Introduction — Why JULIA ?

Major goals for next–stage under–water–detectors are on the one hand bigger detector areas to increase the luminosity, on the other hand smaller distances between optical sensors to include the detection of low energy neutrinos. In order to achieve this, the number of necessary optical sensors has to increase. This seems reachable, if cheaper and more simple technologies are used.

For the moment the most important aim of the JULIA project is the development of new technologies for signal processing and ocean–technologies, in order to achieve a better accuracy as well as lower costs for future detectors.

JULIA means: Joint Underwater Laboratory and Institute for Astroparticlephysics. The JULIA experiment is supposed to be a next–stage underwater Čerenkov–detector, but opposite to current stages of DUMAND or BAIKAL the main emphasis is on medium and low energy neutrinos.

The pilosophy can be stated in a small sentence: *As simple as possible.* Simple technology provides a high reliability as well as a low price. As we will see, it is possible, to increase accuracy as well. From a technological point of view the momentary status of JULIA is more a laboratory, than a new experiment.

This talk provides information on the technical concept of a testcruise carried out in febuary 1991 in the Atlantic ocean near the Canary–islands with the german research vessel *"Sonne"* [1, 2].

2 A technical overview

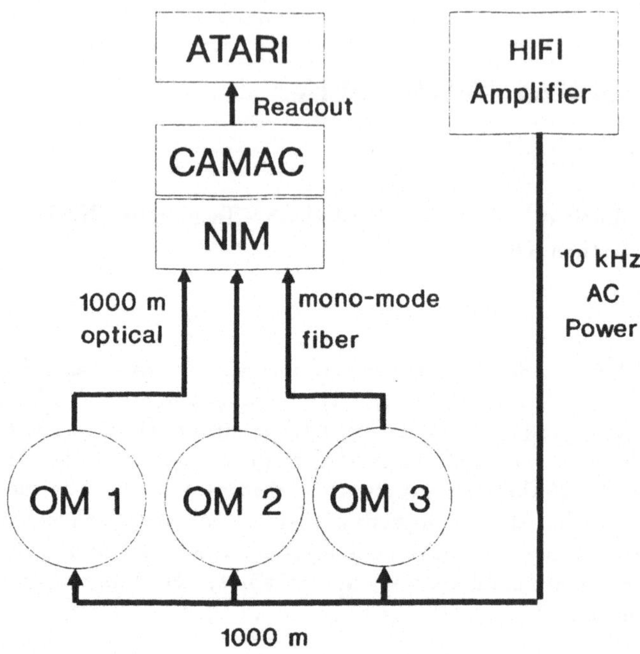

Figure 1: Schematic overview of the test–detector

The test–detector consists of three optical sensors. These were assembeled into one vertical string with spacings of two meters. Central part of each sensor is a large area ($\emptyset 35cm$) photomultiplier. Together with a glass–pressure housing and electronics these photomultiplier are usually called optical modules (OMs).

The AC electric power was supplied with two Hifi–amplifiers ($100W$) via $8km$ Coax–cable, which also carried the whole detector.

The signals from the OMs were directly transmitted from the read–out electronics inside the OM's to the ship via three single mode optical fibers (without digitization). These signals could be processed on board with conventional NIM and CAMAC electronics. The data were read out via ATARI Mega ST computers and written to disk [3].

Due to the limited size of the optical cable of $1km$, larger dephths were not reachable. An overview on the testdetector can be obtained from figure 1.

3 Optical modules

All OMs in different experiments have a glass–pressurehousing and a large area photomultiplier[1] in common. Due to a small power budget and the physical require-

[1] There are ideas for OMs consisting of some small phototubes, but none is built so far.

ments the electronic is almost similar, too. The electronic parts can be divided into slow and fast devices. Whereas the slow devices fulfill control-functions, the fast electronics process the incoming data, without digitizing.

In the JULIA fast circuit (see below) the analog pulses from a "smart" Philips-PMT are charge-integrated by a readout-unit called DMQT.

Philips PMT	analog $\longrightarrow$ pulse	Readout DMQT	ECL $\longrightarrow$ pulse	Laser Diode	optical $\longrightarrow$ fiber

The ECL-pulse coming out of the DMQT is feed into a circuit which drives a laser-diode. From this diode the signal is transmitted[2] via a single mode optical fiber to ship.

Usually the central part of an optical module's slow electronics is a remote control unit (**RCU**) wich is essentially a microcomputer. The RCU enables the comunication to a module via a modem[3].

During the JULIA testcruise the OMs were only passive, what means that there was no computer inside and no comunication possible. By changing the frequency of the supplied power between 8 and $12kHz$ Ac (with an oszillator before the HiFi-Amplifier) it was possible to regulate the high voltage of the PMT and thus the gain of the PMT and its pulseheight.

This concept was only practicable for a short cruise. Permanently detectors require a RCU to know, " what is going on deep in the ocean".

4 The "smart" Phototube

The ("smart") 15' Photomultiplier XP2600 by Philips [4] consists of the combination of an electro-optical preamplifier with a conventional small phototube (see figure 2).

Photoelectrons are accelerated with high voltage ($25kV$) to a scintillator placed in the center of the glass-bulb. This scintillator is read out by a small conventional 11-stage phototube. This principle is sketched in figure 3.

The result of the high accelleration voltage is a high gain in the first stage. As a typical value, 30 photoelectrons (**PE**) in the small phototube are converted for each PE from the cathode. Due to the good statistics in the first stage, the tube provides an excelent time and energy-resolution on the low photoelectron-level. The energy-resolution on the 1PE-level is 50% FWHM. A good seperation between one, two or more PE is possible and can be obtained from figure 4. This figure shows a typical distribution of the integrated charge. The tube is illuminated with a LED of constant intensity. A fit with equidistant, poisson distributed gaussians fits the distribution well. The time-jitter lies below $5ns$ for 1PE and decreases as $1/\sqrt{n}$ with n, the number of PE. Due to a collection efficiency of 100% and the high acceleration voltage, the tube behaves uniform over the whole cathode. The spherical cathode-geometry is not the best shape to optimize transit-time-differences. The transit-time difference between 0° and 90° is about $4ns$ for single point ilumination.

[2]Without digitisation of the time-information.

[3]Signals are superimposed on the powerline.

Figure 2: Photo of the "smart" tube

It is proven[4], that this value can be reduced to less than $1ns$. As a consequence of the decay time of the scintillator ($\sim 45ns$) this PMT has long pulses ($\sim 150ns$). It should be possible to reduce the pulselength as well as the time–jitter with the choice of a faster scintillator [5].

5 "Smart" readout — DMQT

Requirements for a readout circuit fitting to the "smart" Photomultiplier are:
1. It should be self–triggering because there is no external trigger in the ocean).
2. In respect of the good PMT–energy resolution it should work lineary to reflect this advantage. Due to the scintillator decay the pulses are not smooth, so linearity can be achieved only by integrating the pulse's charge. This integration should be fast to avoid a long dead–time.
3. This circuit must have a low time–jitter.
4. The outputs should be norm–pulses to enable easy further processing.

The readout circuit, DMQT, measures the integrated charge with good linearity and time accuracy [6]. The integration of the pulse–charge takes place parrallel to its collection. As result the time of conversion is proportional to the integrated charge. Contrary to conventional charge–integrating–circuits with a fixed integration–window the DMQT

[4]The russian tube: *Quasar* (see this proceedings) and calculations [5].

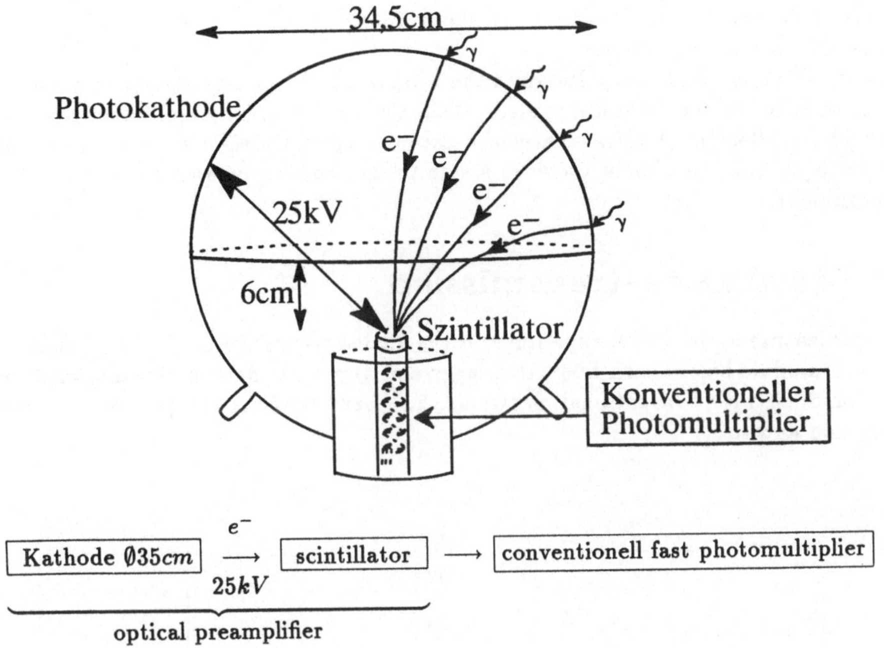

Figure 3: Schematic diagram of the "smart" tube

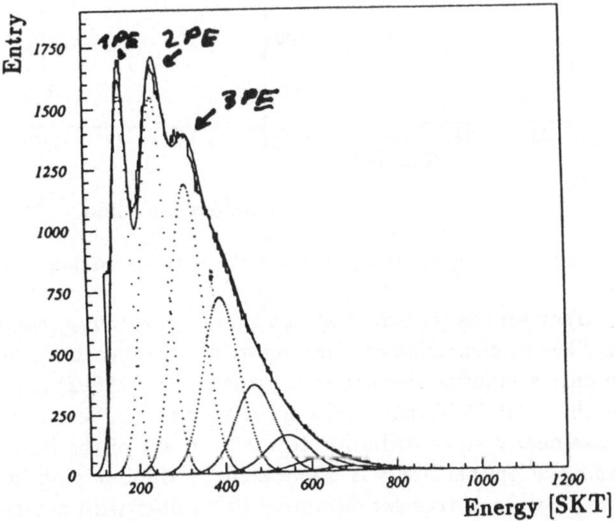

Figure 4: Distribution of integrated charge and a fit with gaussians

has a low deadtime for small signals[5] of about 130*ns* for one PE. Each additional PE increases this number by $\sim 80ns$.

The width of the ECL–outpulse gives the integrated charge and the leading edge the time–information of the phototubepulse. This time–information is obtained by a low threshhold ($\sim 10mV$). A coincidence of a second higher threshhold ($\sim 100mV$) with the 5*ns* delayed low threshhold provides a safe trigger and suppresses noise of the small photomultiplier.

6 Optical Data–transmission

The major feature of the JULIA experiment is the analog transmission of DMQT signals to ship. Each module has its own fiber, so trigger and any other data–processing electronic can be build up with conventional electronic, on board without restrictions on power-consumption and size.

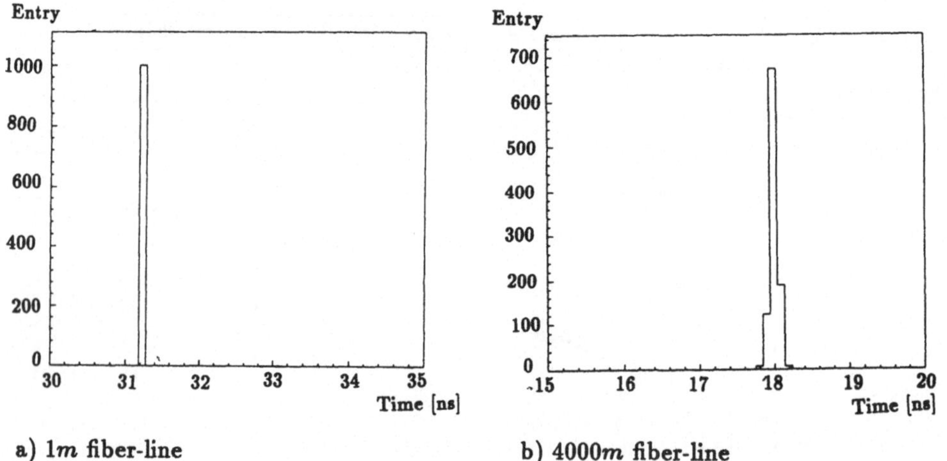

a) 1*m* fiber-line

b) 4000*m* fiber-line

Figure 5: Timing accuracy of the optical transmission line.

The timing–accuracy for the transmission of a short signal (50*ns*) can be seen in figure 6. For a 1*m* optical fiber no dispersion can be seen within the measuring accuracy of 100*ps* (a). Even for 4*km* only a minimal dispersion is measured ($\sim 100ps$) (b).

If the jitter of the DMQT is taken into account, one can see in table 6, that the transmission–line has nearly no contribution to the accuracy of the time–signal.

Since the damping of optical fibers is dominated by the damping in connectors and splices, it should be possible to transfer data over 40*km* fiber with nearly the same accuracy. This might be an important advantage for future underwater experiments as well as any other experiment, where a transmission of data from a detection zone to a laboratory

[5]Keep in mind that most ocean noise is about 1PE

	short NIM–signal	DMQT 1 PE	DMQT 5 PE
$0m$		$\pm 0.21ns$	$\pm 0.42ns$
$1m$	$< 0.1ns$	$\pm 0.27ns$	$\pm 0.45ns$
$4000m$	$\sim 0.1ns$	$\pm 0.31ns$	$\pm 0.45ns$

Table 1: Time accuracy for the transmission of different signals

is desirable. Compared to DUMAND II , where the timing accuracy is limited by the digitisation in the ocean to $2ns$, this method is both cheaper and more accurate.

7 Signal processing on board

The incomming data were processed on board in a conventional way (see figure 6).

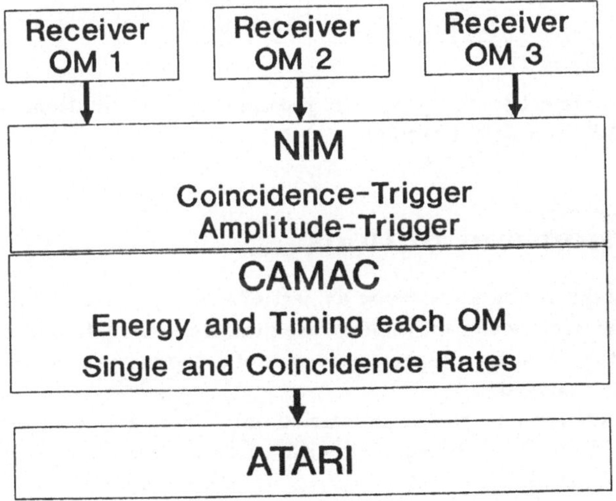

Figure 6: Signal processing on board

The optical receivers for each module are realised as Nim–modules. Different kinds of coincidence conditions allowed the reduction of data. Another kind of trigger, an *amplitudetrigger*, gives the opportunity to set sharp threshholds in energy (number of photoelectrons) [7, 5]. Since the "smart" tube has a very good energy–resolution and the length of the DMQT–signals is proportional to the number of PE, only those signals are accepted which are longer than a certain PE–value. Table 2 shows a table of rejection as an example for the efficiency of an amplitude trigger. Depending on the setting of the amplitude cut, it can be easily calculated, how many percent of the gaussian–distributed[6] 1PE, 2PE, 3PE ...signals are cut out. Due to the good energy–resolution it is easily

[6]Due to the high gain in the first stage of the "smart" tube, each PE-distribution is gaussian.

recognized, that with an amplitude trigger nearly all ocean noise (1PE) can be cut, without loosing signals of higher PE.

After processing with a certain trigger condition, the following values were measured with CAMAC:

1. Energy- and time-information of each module;
2. Single and coincidence countrates (frequently readout).

The CAMAC–crate itself was read out by an Atari Mega ST Computer with a special CAMAC Interface [8]. The maximum readout–rate was $\sim 100Hz$.

cut	1 PE	2 PE	3 PE
1.2 PE	80,4 %	0,8 %	0,0 %
1.3 PE	90,0 %	1,7 %	0,0 %
1.4 PE	95,6 %	3,5 %	0,0 %
1.5 PE	98,4 %	6,5 %	0,0 %
1.6 PE	99,5 %	11,3 %	0,0 %

Table 2: Percentage rejection of equidistant gaussian PE–contributions depending on the cut–threshhold ($\Delta E/E = 55\%$ FWHM)

8 Conclusion and outlook

The technical concept during a testcruise as part of a new project, JULIA , was described. The conclusion are: accuracy improvements are possible, a reduction of costs in nearly all elements of next stage detectors, especially in case of data transmission and the Čerenkov–lightsensors can be achieved

Some future investigations should be carried out. There are two feasibilities.

1. Since the ship "*Sonne*" will be equiped with a new optical fiber cable, tests in dephts down to $5km$ can be done.

2. A german research platform in the Northern Sea could serve as a prototype for a permanent ocean laboratory. The dephth is only $40m$ and the water is not at al clear as ocean-water, but this platform with its good infrastructure can be used for many real–condition tests under nearly laboratory conditions.

Future investigation topis are:

1. Multiplexing of signals from many OMs to one optical fiber.
2. Long term in situ tests, and calibration of OMs.
3. Background studies, especially on K^{40} and bioluminiscence
4. Tests of detector–geometries, and triggers, with the emphasis on the seperation of low energy neutrinos from the background.
5. With a small detector (about 20 OMs) investigations on Muon–, Gamma–, and Neutrino–physics are possible.
6. By placing air–shower–detectors on the sides of the plattform and a Čerenkov–detector at the sea–bottom, Extensive–air–shower–physics is possible.

References

[1] DUMAND Collaboration. DUMAND II — Proposal. *HDC-2-88*, Hawaii DUMAND Center , August 1988.

[2] P.C.Bosetti. Ergebnisse der JULIA Durchführbarkeitsstudie. Internal report, April 1991.

[3] Thomas Mikolajski. Signalwandlung und Übertragung im JULIA Experiment. Internal report. Institut für Hochenergiephysik, Berlin–Zeuthen, Mai 1991.

[4] van Aller et al. A "smart" 35cm Diameter Photomultiplier. *Helvetia Physica Acta*, 59:1119 ff., 1986.

[5] Christopher Wiebusch. Zum Nachweis schwacher Lichtquellen im Ozean mit Hilfe eines neuartigen großflächigen Photomultipliers. *PITHA 91/20* masters thesis, RWTH Aachen, Oktober 1991.

[6] F.Beißel und V.Commichau. A fast charge to time converter V04. Internal report HD04, III.Phys.Inst., Sommerfeldstr., 5100 Aachen, February 1991.

[7] Christoph Ley. Berechnungen über den Nachweis hochenergetischer Neutrinos mit dem DUMAND Detektor. Masters thesis at the Rheinisch–Westfälische Technische Hochschule Aachen, Mai 1990.

[8] F.Beißel, C.Camps, V.Commichau. Atari databox CAMAC-Coupler V01B Internal report HD04, III.Phys.Inst., Sommerfeldstr., 5100 Aachen, February 1991.

Deep-Sea Laboratories for Ocean Sciences

H. Bäcker

GEOMAR Technologie GmbH, Wischhofstraße 1-3, D-2300 Kiel 14

Abstract

A general trend in ocean sciences from single observations and sampling to time series and in situ experiments has led to the development of first small deep-sea monitoring stations operated in free-fall mode or handled by manned submersibles.

There is, however, a rising demand for major benthic laboratories which allow interactive experiments under video control and multidisciplinary approaches. Such stations could be linked by cable to surface vessels or land stations or be left unattended for extended periods of time. They should consist of a basic service unit and various plug-in systems chosen according to the specific research aims.

1. INTRODUCTION

The idea of a deep-sea benthic laboratory for long-term observation and experimentation has been launched already in the early eighties by the Institute of Hydrobiology and Fisheries of Hamburg University. Sponsored by the Westgerman Television Company (WDR) in Cologne, a feasibility study was carried out which was published in 1983. The project could not be realized, partly because of the lack of funds, but mainly because substantial technical components and logistical support could not be secured at that time.

Meanwhile, deep-water technology has advanced considerably, and a number of scientific aims in the deep sea require a new assessment of the matter. In addition to the benthos research there is a rising demand for long-term observations in the fields of sea floor hydrothermalism and ore formation, of microseismicity and tectonics, of astroparticle physics and of fluid and gas exchanges.

In general, there is a trend from "snapshot" observations and sampling to time series and experimentation, and from isolated actions to multi-disciplinary research.

While research fields and objectives clearly have to be defined by the scientific communities, the technological aspects and problems of deep-sea monitoring stations should not be under-estimated. The technical challenges are similar to those related to space research, and the necessary technical and economic expertise clearly exceeds the capabilities of single marine research teams.

For this task pooling of scientific and industrial know-how appears to be necessary.

Small long-term measuring stations are already used for various purposes in the deep sea. They are launched by free-fall techniques or are positioned by manned submersibles or remotely operated vehicles (ROVs).

Any large stations, however, and specially those which should be used for active experimentation, require major efforts and specific logistical approaches.

2. SCIENTIFIC NEEDS AND OBJECTIVES

During the last decades deep-sea investigations have represented an increasing portion of ocean research.

The original belief of a more or less uniform deep-sea environment and the technical and financial restraints had originally limited sampling and observation to widely distributed points and long intervals.

During the seventies and eighties this situation changed due to major national and international programmes related to mineral research, to scientific ocean drilling and to climatic questions.

New sophisticated technical systems were developed which increased the knowledge about the deep sea considerably: manned submersibles, deep-towed acoustical and visual observation systems, multibeam echo sounders, TV-guided sampling devices.

This new technology led to important single discoveries, such as the black smoker metallogenesis and vent communities, but also to the understanding that deep-sea volcanism and fluid exchange through the benthic boundary layer are very important factors in the global environment.

It soon became evident that the usual oceanographic methods only provide snapshot information, while many processes on the deep ocean floor are very dynamic. Consequently scientists were looking after possibilities of continued observation. Submersibles provided a first chance to look at some of these processes. However, submersibles are very expensive, have limited pay load and can be operated only under favourable conditions.

For biological investigations, their use is limited because they influence observations and experiments and lead to selective results. The same is true for certain geophysical experiments.

Free-fall instrument packages, so-called bottom landers or pop-up stations are being used for certain oceanographic, sedimentological and geophysical long-term observations.

A possible solution for many scientific questions are observations, measurements and experimentation from an abyssal benthic laboratory. Teleoperated from a surface vessel and thus serving as a scientific tool not just for one researcher, as usually present in a submersible, but at the same time for multi-disciplinary teams active in front of TV-screens and joysticks.

Presently, biologists, sedimentologists, geochemists, volcanologists, astroparticle physicists, seismologists and oceanographers are eager to make long-term observations on, in and near the deep-sea floor. Other specialists will join as soon as suitable technical means are at hand. Therefore, it is very important to make a thorough survey of the scientific needs and objectives which can be addressed by one or different types of benthic laboratories.

Marine geologists and biologists occupied with processes in the abyssal benthic boundary layer look for accumulation and conversion of matter and sedimentation, but presently lack in situ experiments.

Long-term measurements and in situ experiments will offer new possibilities for understanding the benthic organisms and the chemical and physical processes involved in their metabolism.

Ocean floor hydrothermalism attracts attention of many marine scientists. Enormous mass, heat and fluid fluxes through the sea floor affect the global environment.

Do these dynamic processes depend on short- or long-term magmatic activities, are they coupled to tectonic events or not? Observations in continuous time series would help to understand these spectacular phenomena, give insight into the genesis of ore deposits, and into the hitherto unknown biology of associated communities of animals.

In recent scientific initiatives this was called a key element, for example by the RIDGE research group at Washington University or in the French PNEHO group.

Beyond these hydrothermal phenomena geophysicists demand for data from observation networks in continuous long-time series, if possible over periods of years, measuring microseismic, magnetic, and hydrodynamic parameters. These data should help to understand global water exchange, stress distribution in the crust's plates, and the nature of convection in the earth's mantle below. They will also play an important role in earthquake prediction.

Little is known about the causes, the nature and dynamics of benthic storms and turbidity currents.

Extremely high-sensitive sensors could be used to monitor and analyze weak light sources in the deep sea.

An important question to follow is man's impact on the deep-sea environment, e.g. deep-sea mining, waste disposal and accidential sinking of dangerous freights.

Direct and repeated observation and sampling or even experiments in the abyssal benthic environment could provide the necessary precautionary and accepted limits.

3. TECHNICAL APPROACHES

The term "Deep-Sea Benthic Laboratory" comprises a number of different, partly already existing medium- to long-term deep-sea monitoring stations as well as the idea of a major multi-disciplinary laboratory where interactive research can be carried out using telemetric and robotic methods. The stations presently used are placed and recovered by manned submersibles or constructed as free-fall devices, using the balance of ballast weights and buoyancy as the driving force.

Examples for the latter are ocean bottom seismometers (OBS), particle samplers (sediment traps), respirometers and current meters. While most of these systems are passive, respirometers represent already a small automatic laboratory.

Submersible operated stations are used to observe, monitor and sample fluid sources, hydrothermal vents as well as cold seeps.

The free-fall pop-up method is relatively cheap and timesaving, but there are limitations:

- no exact placement relative to specific sea floor features possible
- payload has to be balanced by (expensive and bulky) floats
- no interactive experiments possible, no realtime observation
- energy problems
- long-term moorings endangered by corrosion and fouling.

Handling by submersibles could be a good solution for certain small stations, but servicing will be weather-dependent and expensive.

A major deep-sea laboratory for multi-disciplinary research requires new technical approaches, which may use a number of components of proved technology, but which certainly will include various R&D fields. System integration will be an important factor: Many single components can easily be found on the markets, but endurance in long-term deep-sea applications and compatibility questions will have to be investigated thoroughly. Various measurements and experiments are sensitive to disturbances from a large station itself and have to be dislocated to safe places. Thus, a deep-sea laboratory might not be a single compact instrument package but a rather complex system of a service station and several satellite experimental units.

The service station should include the components necessary for all measurements and experiments: power supply, data handling, roboters, launching and recovery system. Task-specific instruments, however, should be designed as plug-in systems.

Three main types of application for deep-sea laboratories can be identified, which ideally could be secured by the same station:

- cable link to a land station. This will generally be feasable only if the distance to shore is limited. An interesting new approach for more distant locations is presently discussed: The use of abandoned transocean communication cables which have been replaced by new fibre optic links.

- a second type of station would be connected with a surface vessel via a cable or some other kind of permanent link, to allow major interactive experiments under visual control.

- an untethered version, which will be placed on the ocean floor for a longer period of time to carry out measurements, periodical or event-triggered sampling or visual imaging. Periodical servicing by a research vessel should be envisaged, possibly without moving the station from its position.

Any large or complex deep-sea laboratory will contain a number of technical components which have the character of service units, e.g.:

- support frame, possibly including an emergency buoyancy package
- power supply (electrical, hydraulic)
- pumps
- fixed and steerable video control and documentation cameras, steerable lighting system
- basic hydrographic instrumentation (CTD, current meters)

- (fibre optics) cable for transmission for colour TV, data transmission, commands etc. to and from the surface vessel
- suitable winch for this cable
- links to satellite stations
- working roboter to service these stations and different experiments
- data storage and handling, telemetry.

Concerning the task-specific units which could be plugged in in different combinations, a great number of instruments and systems has been proposed already. Some examples are:

- sediment traps with short sampling intervals (several instruments could be deployed at different distances from the sea floor, using small winches)
- physical and chemical sensor packages (e.g. photo-multipliers or ion-specific sensors)
- biological experimental units, applying in situ incubation methods
- ethological experiments using baits and baited traps
- temperature monitoring probes in sediment and water
- seismometers for monitoring microseismicity
- magnetic reference station for monitoring regional variations of the earth magnetic field
- fluid samplers combined with flow meters.

4. WAYS TO PROCEED

The realization of deep-sea laboratories requires scientific as well as technical and economic input. While interesting scientific questions which could be addressed by deep-sea laboratories will be numerous and manifold, there will certainly appear various technical, logistical and financial restraints.

A growing number of scientists is talking about long-term experimentation and deep-sea stations, but obviously they have quite different facilities in mind. Though multi-purpose equipment carriers with the possibility to carry out changing experiments and to plug in various subsystems should be favoured, we probably will need not just one deep-sea station, but a few quite different approaches.

The planning of a more specific laboratory could be linked to major international scientific programmes, such as INTERRIDGE or OSN (Ocean Seismic Network).

On the other hand, multi-purpose stations require a broader approach. The Commission of the European Communities has recently commissioned studies on benthic laboratories. On the basis of such studies and specific workshops, an interdisciplinary programme could be started which incorporates technical research and development as well as scientific research. One specific problem linked to the use of large sea floor stations is the lack of a suitable research vessel. The existing ones are designed for quick long-distance operations, and deck space and handling facilities are generally not suited for heavy and bulky equipment and containerized laboratories, as required in multi-disciplinary research.

A recent initiative ("NEREIS") launched by IFREMER and sponsored by the Commission of the European Communities and the European Science Foundation has examined, in several workshops, the use and feasibility of a novel research vessel for precise heavy-duty on-station work and profiling. This vessel should be able to handle safely, via a moon pool, complex structures to the size of a 20' container and accomodate various research groups in their own specialized mobile laboratories.

Neutrino Astrophysics

A. Burrows[a,b]

[a]Department of Physics, University of Arizona, Tucson, AZ 85721 USA

[b]Steward Observatory and the Department of Astronomy, University of Arizona, Tucson, AZ 85721 USA

Abstract

Neutrino physics and astrophysics are in a ferment not experienced since the detection of neutral currents and the ascendency of the Electro-Weak theory almost twenty years ago. Whether there is a sea change in our conception of the neutrino in the next few years or merely glorious disappointment in the theoretical ranks will depend in no small measure on the outcome of numerous astronomical and laboratory experiments now underway. With the detection of solar, supernova, and atmospheric neutrinos, neutrino astronomy has come of age. In this brief report, I catalogue the astronomical sites and contexts where neutrinos are important and reflect on the astronomical neutrino's new status as an engine of scientific progress.

1. INTRODUCTION

Traditionally, we studied the neutrino indirectly via the beta decay of isolated isotopes [1] or more directly via the interaciton of accelerator beam dump neutrinos [2] or reactor neutrinos [3] with well-characterized targets. Currently, the crucial questions are whether neutrinos have mass [4], experience vacuum flavor oscillations [5], undergo MSW-like flavor conversions [6], or are of Majorana-type [7]. The putative detection by Simpson [8] and others in beta experiments of a massive neutrino at ~17 keV, new double β-decay measurements and bounds [9], tritium end-point experiments [10], disappearance experiments [11], and Z^o width measurements at LEP [12], are currently pacing the field. Frustratingly, much of what has been derived are upper limits or exclusion regions in Δm_ν^2–$\sin^2 2\theta_\nu$ space. We can not yet say that neutrinos are anything but classic, massless, Dirac particles.

However, many astronomical objects are powerful sources of neutrinos. A useful fraction of these objects can be detected on Earth. Neutrinos from the sun [13], SN1987A [14], and cosmic-rays in the atmosphere [15] have already been detected. Ultra-high energy (UHE) neutrinos from beam dumps thought by some to exist in AGN's or in various galactic X-ray sources (Cyg X-3?), stellar neutrinos, big-bang neutrinos, and the various other neutrino backgrounds may or may not soon be detected, but are of intense theoretical interest. Crucial to the future progress of neutrino astronomy is the extensive underground network of massive detectors that is being established. Facilities such as SNO [16], LVD [17], MACRO [18], Super Kamiokande [19], Baksan [20], Homestake

[21], Frejus [22], ICARUS [23], NUSEX [24], Borexino [25], GRANDE [26], DUMAND [27], and SNBO [28], among others, are already or may soon be the telescopes in this new and exciting astronomical enterprise. This expansion underground is driven by the philosophy that these detectors might illuminate not only astrophysics, but fundamental neutrino physics as well. In what follows, I briefly summarize the role of the neutrino in astrophysics and the larger role of neutrino astronomy in physics.

2. SOLAR NEUTRINOS

Our sun is an average star, but its proximity allows us to test both stellar and neutrino theory. The lion's share of the fusion ν_e's issuing from the $T_c \sim 1.3\,\mathrm{keV}$ solar core are from the pp chain and are below the energy threshold of the pioneering ^{37}Cl experiment of R. Davis [29] and the follow-on Kamiokande (KII) experiment [30]. Rare ($\sim 10^{-4}$) and very temperature sensitive ($\sim T_c^{18}$) side reactions are responsible for what they should see. That they see only 1/2–1/3 of what is predicted is the much-discussed "solar neutrino problem." The two classes of proposed solutions involve either altering the astrophysics (material mixing, opacities, etc.) or altering the neutrino physics (magnetic moments, neutrino oscillation, etc.). It is the latter possibility that has galvanized the particle community. Personally, I find it reassuring and a bit surprising that KII and Homestake are as close as they are to the standard model prediction. The severe core temperature dependence and small branching ratios for the production of the dominating ^{8}B neutrinos argue that factors of 1/2–1/3 are of order unity. Can we really presume to know T_c to 4%? Nevertheless, most of the cognoscenti argue that the discrepancy is real and perplexing. Perhaps it is real, and it is certainly perplexing. Might neutrino propagation effects and/or the MSW conversion of ν_e's into ν_μ's (or ν_τ's) be the answer [31]?

There are two general experimental approaches being pursued to resolve the solar neutrino puzzle: the gallium radiochemical extraction experiments, SAGE [32] and GALLEX [33], that are sensitive to the dominant pp neutrinos, and real-time experiments, such as SNO, Super Kamiokande, and Borex(ino), that are sensitive to the higher energy ($\gtrsim$4 MeV) neutrinos. SNO and Borex(ino), in particular, have good neutral-current and, hence, ν_μ and ν_τ, response. They can detect the neutrinos into which the ν_e's might be oscillating. These are indeed heady times in solar neutrino research. If the gallium experiments unambiguously detect less than $\sim$80 SNU's of the $\sim$132 SNU's of the standard model, then either the sun is not a thermonuclear furnace or new neutrino physics is at work. If the latter, astrophysics will have midwifed a revolution in particle physics. The first SAGE results [34] including data through 15–8–91 imply only $\sim 20\pm 20$ SNU with an upper limit of $\sim$80 SNU's (90% CL). If true, and that is a big IF, new neutrino physics is indicated, as the MSW plots in Figure 1 from Chen and Cherry [35] imply (The large cross in Figure 1 is the Bahcall and Bethe [36] prediction.). One eagerly anticipates the drop of the next shoe (the GALLEX shoe) and the completion of SNO, Borex(ino), and Super Kamiokande.

3. ATMOSPHERIC NEUTRINOS

Cosmic rays, predominantly protons and alpha particles between $\sim$1 GeV and 10^{11} GeV, collide with the Earth's atmosphere and produce showers of particles that are regularly detected around the world by the light they generate in the air, the "pancakes"

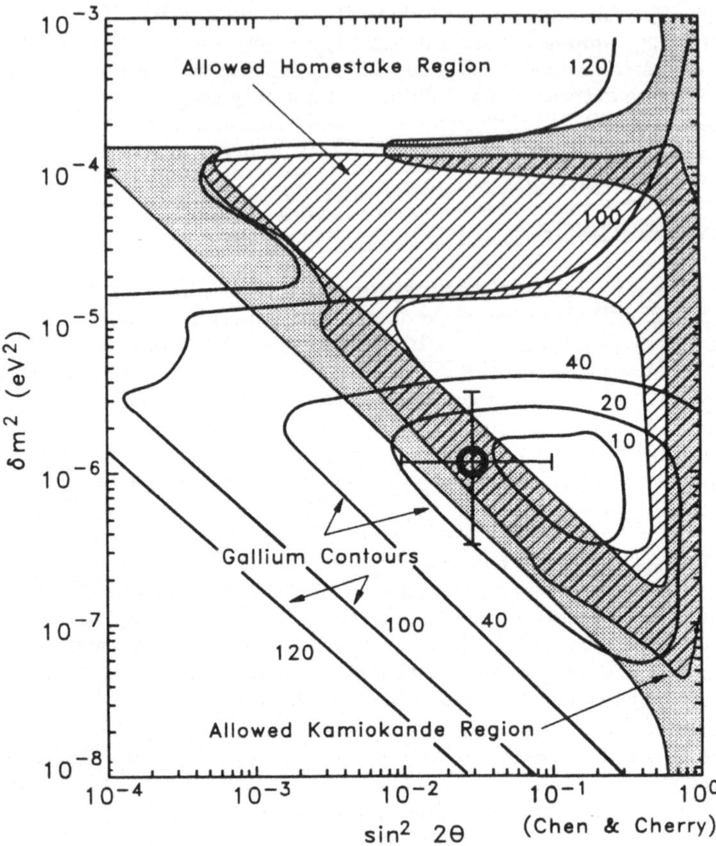

Figure 1. Taken from reference 35, this figure depicts the regions in the MSW $\delta m^2(\mathrm{eV}^3)$ versus $\sin^2 2\theta$ space still allowed by the Kamiokande (stippled) and Homestake (diagonal) solar neutrino data. The lines marked "Gallium Contours" are iso-SNU lines in the gallium detectors. The intersection of these lines with the shaded regions are allowed by the MSW solution. The circle and error bar cross near the center of the plot denote the regions predicted for the gallium experiments in reference 36.

of γe^+e^-'s that reach the Earth's surface, and the muons from the decay of spallation pions and kaons. Before accelerators, cosmic-ray studies led the way in particle physics research. Now, these "beam dump" neutrinos from the decay of the shower mesons and muons provide a means to answer basic questions in neutrino physics. An average of 10–100 atmospheric neutrino events are culled every year in each deep neutrino detector. Such neutrinos are a background for proton-decay studies, show latitude dependences and energy cutoffs that reflect geomagnetic effects on the primary, can be generated by rock interactions ("rock amplification"), and are detected moving both up and down.

About 1000 atmospheric neutrinos events have thus far been identified underground [37]. Their theoretical spectrum is shown in Figure 2 from Totsuka [37] and compared to the spectra from other astrophysical neutrino sources. Figure 2 is a nice summary of the neutrino sky. Note that atmospheric neutrinos should dominate above ~50 MeV.

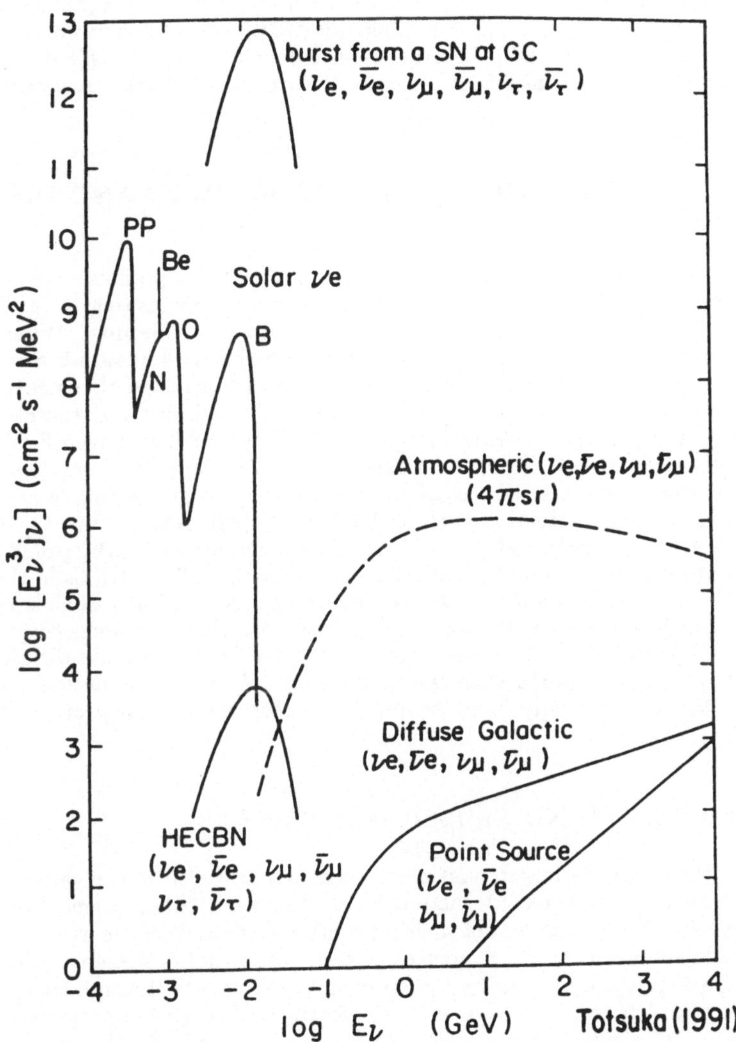

Figure 2. Taken from reference 37, this figure portrays the neutrino sky due to the sun, the atmosphere, the universal diffuse background, an hypothetical UHE point source, the universal supernova background (HECBN) and a supernova at the galactic center (GC). j_ν is in #/(cm^2·s·MeV) and E_ν is in GeV.

However, not all the data are consistent with standard shower calculations, and therein lies another neutrino puzzle. The ratio of $R_{\mu e}$, the ratio of the observed ν flux to the theoretical Monte Carlo flux, for ν_μ's and ν_e's is not one in all the underground detectors. KII saw $0.61^{+0.09}_{-0.08} \pm 0.05$, IMB saw $0.67 \pm 0.09 \pm 0.08$, Frejus saw $1.06^{+0.19}_{-0.16} \pm 0.08$, and NUSEX saw $0.99^{+0.35}_{-0.25}$. That the two largest detectors, IMB and KII, obtained anomalous results is intriguing. Can the ν_μ deficit be real and be a signature of neutrino oscillation? If the solar deficit is explained by MSW conversion from ν_e to ν_μ, the same oscillations can not explain the atmospheric deficit. However, this deficit could be explained by ν_μ–ν_τ oscillations [38–40]. Again, basic particle physics may be illuminated by neutrino astronomy.

4. HIGH-ENERGY NEUTRINOS: POINT SOURCES AND DIFFUSE BACKGROUND

There is not now any evidence for high-energy neutrino point sources in the sky. However, some theories of AGN's and galactic gamma-ray sources (e.g. Cyg X-3, Her X-1, Vela X-1 ??) involve luminous particle beams and beam dumps. When hadronic beams intercept matter, both photon and neutrino beams are generated. Such neutrino beams might be detected on the Earth. It has even been suggested that neutrino beams in quasars can heat the stars in quasars, drive stellar winds, and thereby feed the central quasar engine [41]. Neutrino energies in the TeV, PeV, and even EeV (10^{18} eV) ranges have been discussed (Note that above ~100 TeV, the Earth becomes opaque to neutrinos.). If there are point sources of energetic neutrinos in the galaxy and universe, very massive subsurface detectors such as DUMAND [42], AMANDA [43], GRANDE [44], and LENA [45] are required to see them. The detection of nearby point sources or the general diffuse background of point sources and "Griesen" neutrinos in the universe will speak volumes about fundamental astrophysical processes, but will ve bery difficult. Such work is complementary with ongoing Extensive Air Shower studies (e.g. Fly's Eye, CASA...). A fanciful theoretical spectrum ($<10^4$ GeV) of both the diffuse neutrino background and a point source is shown in Figure 2. Whether the actual spectra look anything like this one can only be determined by observation. I suspect some surprises are in store.

5. STELLAR NEUTRINO EMISSION (PAIRS)

All stars that burn hydrogen emit electron neutrinos because the various hydrogen fusion chains all contain weak interaction links. However, the neutrino losses of such stars amount to only a few percent of their total luminosity and, hence, such losses do not have significant evolutionary consequences. On the other hand, massive stars ($\gtrsim 8\,M_\odot$), whose cores have proceeded from hydrogen to heavy elements through successive thermonuclear burning stages, achieve near their deaths such high temperatures ($>10^9$ K) and densities ($>10^6$ gm/cm^3) that thermal neutrino pair emission dominates the star's luminosity. The most important processes are pair annihilation ($e^+ + e^- \rightarrow \nu\overline{\nu}$), plasmon decay ($\gamma_{\rm pl} \rightarrow \nu\overline{\nu}$), and the photoneutrino process ($\gamma + e^- \rightarrow e^- + \nu\overline{\nu}$). During these later stages, it is the neutrino luminosity that carries away the thermonuclear heat and the surface photon luminosity becomes irrelevant to the subsequent evolution of the star's core. Indeed, just *before* stellar collapse to a protoneutron star and a supernova explosion, the neutrino luminosity can reach $\sim 10^{49}$ ergs/s, which is $\sim 10^{10}$ times the Eddington photon luminosity. Such large neutrino luminosities are responsible for drastically accelerating the evolution of massive stars after carbon burning and for refrigerating their cores into quasi-degenerate "white dwarfs." Without these neutrino emissions, the white dwarf Chandrasekhar mass of $\sim$1.4 $M_\odot$ would not be the relevant mass scale of neutron stars formed from massive stars and we would not expect the observed neutron star masses to cluster as they do near 1.4 $M_\odot$. Indeed, it is possible that black holes, not neutron stars, would be the residue of collapse, because the cores of massive stars might otherwise accumulate masses before collapse ensues in excess of the maximum possible mass of a neutron star. In sum, the very existence of many neutron stars with masses near 1.4 $M_\odot$ may be due to significant neutrino emission before collapse. Unfortunately, even such large neutrino luminosities may be undetectable, since the average energy of the emitted neutrinos lies below $\sim$1 MeV. At such energies, the interaction cross sections are very low and most of the neutrinos are below detector thresholds. Furthermore, even 10^{49} erg/s is not adequate at the $\sim$150 parsec minimum distance from us of a massive star (Betelgeuse).

Another classic stellar site where neutrino emissions dominate a star's evolution is the young ($\lesssim 10^4$ years) neutron star [46]. What we may some day see are the neutron star's surface X-rays, but during the first few thousand years, the evolution of its surface temperature ($\gtrsim 10^6$ K(?)) is tied to the neutrino cooling of its core. The dominant processes are thought to be the modified URCA process (nn$\rightarrow$npe$^-$+$\bar{\nu}_e$: npe$^-$ $\rightarrow$nn+ν_e), neutral current scattering (e.g., nn$\rightarrow$nn+$\nu\bar{\nu}$), and pair bremsstraulung (e$^-$+(Z,A) $\rightarrow$e$^-$ +(Z,A)+$\nu\bar{\nu}$). If there are pions, kaons, or free quarks in "neutron" star cores, similar processes involving them will dominate. Again, these neutrino emissions themselves are detectable through their indirect influence on surface X-ray emissions, but they dominate what can be seen for thousands of years. Similarly, when white dwarfs are formed from stars with masses below $\sim$8 $M_\odot$, they lose their residual heat via neutrinos during the first $\sim 10^6$ years of life. The early evolution of a white dwarf's optical and ultraviolet surface emissions is paced by these neutrino losses.

6. COSMOLOGY

During the last ten years, particle physicists have claimed cosmology as their own. What cosmology can say about particle physics or what particle physics can contribute to our understanding of the universe is too large a subject to review or summarize here. However, from the beginnings of the emerging particle astrophysics symbiosis, the role of the neutrino has been central. If the universe underwent an inflationary epoch, Ω equals one. If the baryonic contribution to the universe is limited to $\Omega_B \lesssim 0.15$ by the standard deuterium nucleosynthesis arguments and what we see amounts to even less, there is non-baryonic dark matter. If the sum of the masses of the neutrinos that we know about is $\sim$100 h^2 eV/c^2 $\sim$20–50 eV/c^2, big bang neutrinos are that dark matter. These arguments are old, but resilient. Furthermore, the gravity of such neutrinos would affect the power spectrum of density inhomogeneities by smoothing out perturbations on small scales (the so-called "Hot" dark matter scenario, that is not currently in favor). Their effect on Large-Scale Structure and galaxy clustering is measurable (even if they themselves are not). In addition, it has been shown [47] that the primordial ^{4}He abundance is sensitive to the number of light neutrinos and limited to be below 3.4 (i.e., = 3). This has recently been verified by LEP, which finds that $N_\nu = 2.99 \pm 0.05$ [12].

Cosmological arguments can also constrain the properties of Simpson's 17 keV beast [48]. That $\Omega \sim 1$ forces it to be unstable ($\tau < 10^{12}$ s) into much lighter species. Observations of density fluctuations on 5 Mpc/h scales and the theory that galaxy clustering requires a period of matter domination brings the decay time to below 10^6 s. Curiously, the upper limit to the mass of a Dirac neutrino derived from the SN1987A $\overline{\nu}_e$ signal is on the order of, but greater than, 20 keV [49] and this fails to eliminate Simpson's neutrino. However, if pion cooling is significant in protoneutron stars, this limit might be decreased substantially.

Cosmology is an arena in which the dialogue between neutrino physics and astrophysics has been most profitable and prolonged. This will no doubt fruitfully continue to be the case for the foreseeable future.

7. SUPERNOVA NEUTRINO BURSTS

Theory supposed that fabulous neutrino bursts accompany the death of massive stars, the birth of neutron stars, and most supernova explosions. The detection [50,51] of just such a burst from SN1987A seems to confirm this theory, though the residual neutron star has yet to be seen. Little can be said here about the neutrinos from 87A that hasn't been said in the one thousand or so readable papers on this subject generated over the last five years. The total radiated energy, average event energy, and duration fit the standard model developed during the eighties [52,53]. Limits have been obtained on the ν_e mass, the ν_e magnetic moment, the ν_e lifetime, the properties of axions, the weak principle of equivalence, and a variety of putative particle parameters [54]. On the whole, SN1987A has proven useful.

However, there is an agonizingly long list of questions concerning supernova and protoneutron star physics that were left unresolved. What is the mechanism of the explosion? Are ν_μ's and ν_τ's in fact produced in generous amounts? Is there a shock break-out flash of ν_e's? Are protoneutron stars convective? How do the neutrino spectra evolve? How are the emissions in the various neutrino channels correlated? Does the neutrino luminosity fluctuate before explosion? What might be the role of neutrino oscillations and the MSW effect in supernovae? How long is the entire burst? What is the nuclear equation of state? These are only a few of the interesting and germane questions on which SN1987A was almost entirely mute. Fortunately, our ignorance needn't be permanent. The nineties will see the establishment of an international network of massive underground detectors with exquisite sensitivity to galactic neutrino bursts. The thousands of events anticipated in this collection of neutrino telescopes from a core collapse anywhere in our galaxy will provide not only a quantitatively, but a qualitatively, better view of supernova physics. Table 1 lists the facilities that are now either built, being built, or planned, along with some of their pertinent characteristics and estimates of the total number of events anticipated at 10 kiloparsecs. The latter should be contrasted with the total of 19 events from both KII and IMB.

I will not here repeat the detailed discussion of the potential of this network for supernova and neutrino science that is found in the recent paper by Burrows, Klein, and Gandhi [55]. Rather, I will illustrate only one small, but exotic, signature of a finite ν_e mass in the super massive "Super" Kamiokande (SK). Figure 3 depicts the early evolution of the spectrum of energies that could be detected by SK at 10 kpc if the electron neutrino mass were 8 eV/c^2. A generic model of the neutrino emissions constructed in reference 55 was employed, but the precise details of that model are not important. What are important are the striking characteristics of the spectra and their evolution from 50 milliseconds to 600 milliseconds. Not only is there a dramatic sweep

Supernova Neutrino Telescope Characteristics

Detector	Total Mass (tonnes) (Fiducial Mass (tonnes))	Composition	Depth (mwe)	Threshold (MeV)	# Events at 10kpc	*Resolution (e,t,a)
ČERENKOV:						
IMB3	8000 (6800)	H_2O	1570	10	940	e,t,a
KII	3000 (2140)	H_2O	2700	7	370	e,t,a
Super KII	40,000 (32,000)	H_2O	2700	7	5510	e,t,a,
SNO	1600/1000	H_2O/D_2O	6000	5	781	e,t,a (t for NC)
SCINTILLATION:						
LVD	1800 (1200)	Kerosene	3100–4000	5–7	375	e,t, (a?)
MACRO	1000	"CH_2"	3100–4000	10	241	e,t
Baksan	330 (200)	"White Spirits"	850	10	70	e,t
LSND	200	"CH_2"	-	5	70	e,t
Borex[ino]	1760[300]	$(BO)_3(OCH_3)_3$	3100–4000	4.7	~325[33]	e,t
CalTech	1000	-	?	2.8	290	e,t
DRIFT CHAMBER:						
ICARUS	3600	^{40}Ar	3100–4000	5	120	e,t,a
RADIOCHEMICAL:						
Homestake ^{37}Cl	610	C_2Cl_4	4200	0.814	4	
Homestake ^{127}I	1000	NaI	4200	0.664	~25	
Baksan ^{37}Cl	3000	C_2Cl_4	800	0.814	22	
EXTRAGALACTIC:						
SNBO	100,000	$CaCO_3$	?	-	10,000	t
JULIA	40,000	H_2O	?	-	10,000	e,t,a

*e, t, and a mean that the detector has or will have energy, time, or angle resolution.

with time from high average and minimum energies, but the predicted fluctuations in the *luminosity* versus time are echoed in the individual *spectra* at a given time. These remarkable signatures of a finite ν_e mass are only a few of the rich features and diagnostics that the new array of neutrino telescopes promises. We will just need a little luck.

CONCLUSION

I hope that this short article adequately conveys the variety of astrophysical contexts in which neutrinos play a central role and the usefulness to fundamental neutrino physics of an ongoing dialogue with astrophysics. Much has already been achieved at the interface between these disciplines. With the establishment of massive underground and underwater neutrino telescopes, this dialogue can only accelerate to mutual benefit.

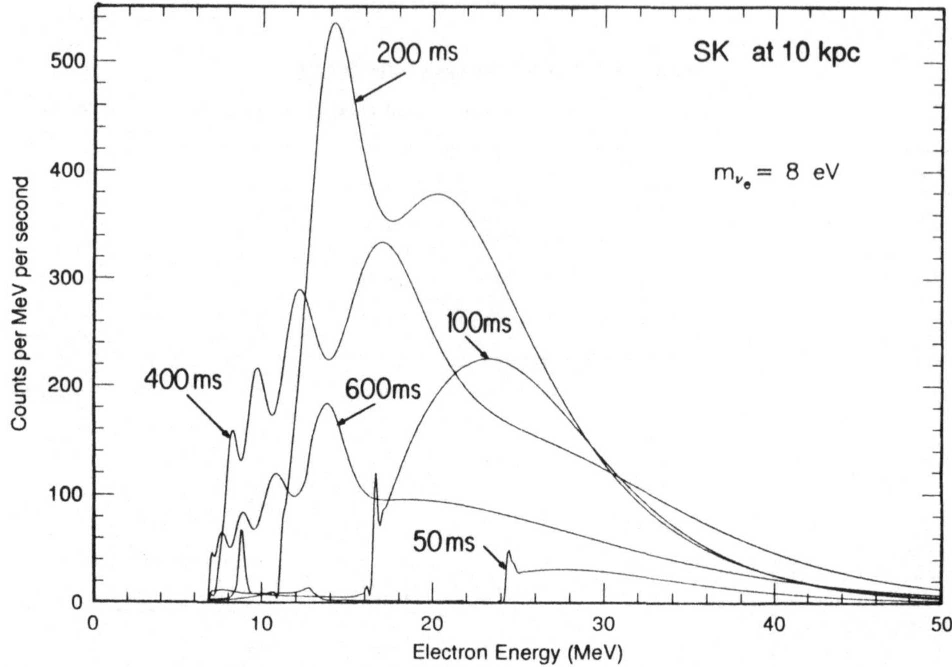

Figure 3. The number of counts per MeV per second in Super Kamiokande at 10 kpc versus the event electron energy in MeV for the generic neutrino burst model in reference 55 at 50, 100, 200, 400, and 600 milliseconds after core bounce. For illustration, an electron neutrino mass of 8 eV/c^2 and a detection threshold of 7 MeV was assumed. See text for discussion.

Acknowledgments

The author would like to thank Raj Gandhi and David Klein for useful discussions and both the NSF and NASA, under grants AST89-14346 and NAGW-2145, respectively, for their generous financial support. Thanks are also extended to the organizers, especially Peter Bosetti, for arranging such a stimulating workshop.

9. REFERENCES

1 O. Kofoed-Hansen and C. J. Christensen, Handbuch der Physik 42, part 2, p. 1 (1962).
2 H. Kendall, Nobel Lecture 1990.
3 P. O. Lagage, Nature, 316 (1985) 420.
4 H. Robertson, in Neutrino '88, ed. J. Schneps et al., World Scientific, p. 10, 1988.
5 L. Wolfenstein, Phys. Rev., D17 (1990) 2369.

6 S. P. Mikheyev and A. Yu. Smirnov, Sov. J. Nucl. Phys., 42 (1985) 913; L. Wolfenstein, Phys. Rev,. D17 (1979) 2369.
7 G. B. Gelmini and M. Roncadelli, Phys. Lett., B99 (1981) 411.
8 J. J. Simpson, Phys. Rev. Lett., 54 (1985) 1891.
9 D. O. Caldwell et al., Nucl. Phys., B13 (1990) 547.
10 R. G. H. Robertson et al., Phys. Rev. Lett., 67 (1991) 957.
11 J. Bouchez, in Neutrino '88, ed. J. Schneps et al., World Scientific, p. 28, 1988.
12 E. Amaldi, in the proceedings of the Texas/ESO/CERN Symposium held in Brighton, England, 1990, in press.
13 K. S. Hirata et al., Phys. Rev. Lett., 63 (1989) 16; R. Davis, Jr. in ICOBAN '86, ed. J. Arafune, World Scientific, p. 237, 1987.
14 W. D. Arnett, J. N. Bahcall, R. P. Kirshner, and S. E. Woosley, Ann. Rev. Astron. Astrophys., 7 (1989) 629.
15 T. K. Gaisser, T. Stanev, and G. Barr, Phys. Rev., D38 (1988) 85.
16 H. B. Mak, in the Proceedings of the XXIIIth Recontre de Moriond, eds. O. Fackler and J. Tran Thanh Van, Editions Frontières, p. 321, 1988.
17 A. I. Pless, ibid., p. 309.
18 B. Barish, in the proceedings of the Supernova Watch Workshop, ed. D. Cline, 1990, in press.
19 H. Minakata and H. Nunokawa, Phys. Rev., D41 (1990) 2976.
20 E. N. Alexeyev et al. Phys. Lett., B205 (1988) 209.
21 K. Lande, in Neutrino '90, ed. K. Winter, in press.
22 C. Berger et al., Phys. Lett., B245 (1990) 305.
23 D. Cline, in the proceedings of the Supernova Watch Workshop, ed. D. Cline, 1990, in press.
24 M. Aglietta et al. Europhys. Lett., 8 (1989) 611
25 R. S. Raghavan and S. Pakvasa, Phys. Rev., D37 (1988) 849.
26 A. Adams et al., UC Irvine preprint GRANDE 90–005 (1990).
27 P. Bosetti et al., proposal July 27, 1988 (Univ. Hawaii); J. Learned, this conference.
28 D. Cline et al., preprint (1991).
29 R. Davis, Jr., Proc. of the Thirteenth International Conference on Neutrino Physics and Astrophysics, Neutrino 88, ed. J. Schneps et al., World Scientific, 1989, p. 518.
30 K. S. Hirata et al., Phys. Rev. Lett., 65 (1990) 1297.
31 S. P. Rosen and J. M. Gelb, Phys. Rev., D34 (1986) 969.
32 V. N. Gavrin, in Neutrino '90, ed. K. Winter, in press.
33 T. Kirsten, in Neutrino '90, ed. K. Winter, in press; G. Heusser, this workshop.
34 A. I. Abazov et al., Phys. Rev. Lett., 67 (1991) 3332.
35 C. Chen and M. L. Cherry, Ap. J. Letters, 377 (1991) L105.
36 J. N. Bahcall and H. A. Bethe, Phys. Rev. Lett., 65 (1990) 2233.
37 Y. Totsuka, 1991, to be published in Reports on Progress in Physics.
38 G. Barr, T. K. Gaisser, and T. Stanev, Phys. Rev., D39 (1989) 3532.
39 H. Lee and S. A. Bludman, Phys. Rev., D37 (1988) 122.
40 J. Learned et al., preprint (1991).
41 J. McDonald, T. Stanev, and P. Biermann, Ap. J., 378 (1991) 30.
42 P. R. F. Grieder, Nuovo Cimento, C9 (1986) 222.
43 F. Halzen, this workshop.
44 R. C. Svoboda, in Neutrino '88, ed. J. Schneps et al., World Scientific, 1988, p. 358.
45 M. Sasaki et al., in the Proc. of the 2nd Workshop on the Elementary Particle Picture of the Universe, ed. M. Yoshimura et al., KEK, 1988, p. 181.
46 K. Nomoto and S. Tsuruta, Ap. J., 312 (1987) 711.

47 K. Olive et al. Phys. Lett., B236 (1990) 454.
48 L. M. Krauss, Phys. Lett., B263 (1991) 441.
49 R. Gandhi and A. Burrows, Phys. Lett., B246 (1990) 149.
50 R. M. Bionta et al. Phys. Rev. Lett., 58 (1987) 1494.
51 K. Hirata et al., Phys. Rev. Lett., 58 (1987) 1490.
52 A. Burrows, Ap. J., 334 (1988) 891.
53 A. Burrows, Ann. Rev. Nucl. Part. Sci., 40 (1990) 181.
54 see, for example, L. Krauss, Nature, 329 (1987), p. 689.
55 A. Burrows, D. Klein, and R. Gandhi, Phys. Rev. D, 1991, in press.

DARK MATTER AND HIGH ENERGY NEUTRINOS

V.S. Berezinsky

INFN, Laboratori Nazionali del Gran Sasso, Assergi, Italy
and
Institute for Nuclear Research, Moscow

ABSTRACT

The production of high energy neutrinos by dark matter particles (DMP) is reviewed. We confine ourselves to two types of DMP: supersymmetric relics and exotic relics. Two supersymmetric particles can be considered nowdays as the candidates for DMP: neutralino and gravitino. Neutrinos are the best tracers for very heavy neutralino. High energy neutrinos can be produced due to the capture of neutralinos in the earth and the sun with the subsequent annihilation of these particles there. Neutralino decay is another source of neutrinos if R-parity is weakly violated. The gravitino as DMP is unobservable directly, unless R-parity is violated and the gravitino decays. Neutrinos can serve also as a tracer for the exotic superheavy (up to $\sim 10^{18}$ GeV) metastable relics.

1. INTRODUCTION

The indirect search for DMP includes neutrinos as a tracer. Usually,the heavier the dark matter particle, the more effective the neutrino search is.

High energy neutrino astronomy started and developed as a science connected with the accelerated particles (for a review see Ref.[1] - [3]). The most interesting "accelerator" sources of high-energy neutrinos are the galactic supernovae, hidden sources and AGN's. Much interest was recently shown in the AGN model suggested in Ref.[4].

There can also be the sources not connected with accelerated particles.

In Ref. [5] evaporating mini-black holes were suggested as a source of ultra-high energy neutrinos with energies up to 10^{22} eV.

However, as was shown in [2], high energy neutrinos are strongly self-absorbed ($\nu + \bar{\nu} \rightarrow e^{+} + e^{-}$) due to the collisions in the neighborhood of the black hole.

Another hypothetical source of high energy neutrinos is a superconductive cosmic string [6]. However, (see Ref. [7]) all high energy particles including neutrinos are catastrophically degraded in energy due to the processes in the strong magnetic field around a string.

Cusps in the usual (non-superconducting) cosmic strings can be the source of high energy neutrinos [8].

In the last several years we began to realize that big bang particles are a promising source of high energy neutrinos. For the first time this possibility was considered in Ref. [9] and then [10].

In this paper two subjects will be reviewed:

(i) High energy neutrinos from supersymmetric dark matter particles and

(ii) High energy neutrinos from exotic big bang relics.

2. HIGH ENERGY NEUTRINOS FROM SUPERSYMMETRIC DARK MATTER PARTICLES

Within the Minimal Supersymmetric Standard Model (MSSM) there are three candidates for the lightest supersymmetric particle (LSP) and dark matter particle (DMP): neutralino, gravitino and sneutrino. The last particle is now excluded by the combination of LEP data and the direct search for the sneutrino.

The gravitino meets several difficulties as DMP.

First of all in most of the models it is not LSP. However, this objection is not general, it is connected with the particular mechanisms of supersymmetry breaking and of generation of particle masses. Actually, the idea that gravitino is a DMP was left in the 80's due to some other arguments.

The first one was connected with the requirement of too low a reheating temperature [11, 12] after inflation. It must be $T_R \sim 10^{13}$ GeV, while the mechanisms of baryogenesis, known at that time, required $T_R \sim 10^{15} - 10^{16}$ GeV. This argument is now irrelevant in the light of nonperturbative baryon number violation in electroweak interaction [13]: in any case this process washes away any baryon excess, which was produced earlier than electroweak phase transition $T \sim 10^4$ GeV.

The second argument is concerned with the second lightest supersymmetric particle (SLSP). If the SLSP is photino (or more generally neutralino), its decay to gravitino and photon results in an overproduction of D and ^{3}He-nuclei [15,16]. This problem can be resolved assuming SLSP is sneutrino $\tilde{\nu}$. The decay $\tilde{\nu} \rightarrow G + \nu$ is harmless for nucleosynthesis.

If stable, the gravitino as a DMP is unobservable. In the case of R-parity violation, the gravitino decay results in observable phenomena, including high energy neutrinos [16].

The most probable candidate for DMP is the neutralino. The neutralino is a linear superposition of the four neutral, spin 1/2 fields in the MSSM:

$$\chi = Z_{11}\tilde{W}_3 + Z_{12}\tilde{B} + Z_{13}\tilde{H}_1 + Z_{14}\tilde{H}_2 \tag{1}$$

where $\tilde{W}_3$ and $\tilde{B}$ are wino and bino and $\tilde{H}_1$ and $\tilde{H}_2$ are two higgsinos. The neutralino is the lightest mass eingenstate, four of which diagonalize the mass matrix. It is a Majorana particle.

The space of possible neutralino states is conventionally described (instead of three independent values of Z in Eq. (1)) by mass of wino, M_2, by mixing parameter μ of two higgsinos $\mu\, \tilde{H}_1 \tilde{H}_2$, and by $\tan\beta \equiv V_2/V_1$, where V_2 and V_1 are the two VEV's of the model.

2.1. *The neutralino as a dark matter particle*

The relic mass density, ρ_χ, of neutralinos is almost entirely determined by the cross-section of $\chi\chi$-annihilation

$$<\sigma v>_{ann} = \tilde{a} + \tilde{b}x \tag{2}$$

where conventionally $x = v^2/3$ and v is the relative velocity.

More specifically one obtains

$$\Omega_\chi h^2 = 3.4 \cdot 10^{-11}\, N_f^{-1/2} \frac{\mathrm{GeV}^{-2}}{\tilde{a}x_f + \tilde{b}x_f^2/2} \tag{3}$$

where $\Omega_\chi = \rho_\chi/\rho_c$, $\rho_c = 1.88 \cdot 10^{-29}\, h^2$ g/cm^3 is the critical density, $h = H_o/100$ km s^{-1} Mpc^{-1}, H_o is the Hubble constant, $x_f = T_f/m_\chi$, T_f is the freezing temperature and N_f is the number of unfrozen degrees of freedom at the freezing temperature T_f (N_f here is defined according to $\rho = N_f T^4 \pi^2/30$, while in [14] $\rho = N'_f T^4 \pi^2/15$ is used) the typical freezing values are $N_f \approx 100$ and $x_f \approx 1/20$.

In the general case the calculations of cross-section have to be performed numerically. However, following [17] we shall give here the analytical expressions for several particular compositions of neutralino. We confine ourselves to the case $m_b << m_\chi < m_t$, where m_b and m_t are the masses of b-quark and top-quark, respectively.

In case of gaugino-dominated neutralino (Z_{13} and Z_{14} are small):

$$\tilde{a} = \frac{\pi}{216}\, \alpha_{em}^2 \frac{m_b^2}{\tilde{m}^4} \left(\frac{Z_{11}}{\sin\theta_W}\right)^4 \left(9 - 6y + 5y^2\right)^2 \tag{4}$$

$$\tilde{b} = \frac{\pi}{54}\,\alpha_{em}^2\,\frac{m_\chi^2}{\tilde{m}^4}\left(\frac{Z_{11}}{\sin\theta_W}\right)^4\left(567-108y+1242y^2-12y^3+2023y^4\right) \tag{5}$$

where $\tilde{m}$ is the mass of a squark, θ_w is the Weinberg angle and

$$y = (Z_{12}/Z_{11})\tan\theta_W \tag{6}$$

In particular, for the photino ($Z_{11} = \sin\theta_w$, $y = 1$) we have

$$\tilde{a} = \frac{8}{27}\,\pi\alpha_{em}^2\,\frac{m_b^2}{\tilde{m}^4} \tag{7}$$

$$\tilde{b} = \frac{1856}{27}\,\pi\alpha_{em}^2\,\frac{m_\chi^2}{\tilde{m}^4} \tag{8}$$

For the zino ($Z_{11} = \cos\theta_w$, $Z_{12} = -\sin\theta_w$):

$$\tilde{a} = 6.55\pi\alpha_{em}^2\,\frac{m_b^2}{\tilde{m}^4} \tag{9}$$

$$\tilde{b} = 150\pi\alpha_{em}^2\,\frac{m_\chi^2}{\tilde{m}^4} \tag{10}$$

For a higgsino-dominated neutralino (Z_{11} and Z_{12} are small):

$$\tilde{a} = \frac{3\pi}{8}\,\alpha_{em}^2\,\frac{m_b^2}{\left(m_Z^2-4m_\chi^2\right)^2}\,\frac{\left(Z_{13}^2-Z_{14}^2\right)^2}{\cos^4\theta_W\sin^4\theta_W} \tag{11}$$

$$\tilde{b} = 2\pi\alpha_{em}^2\,\frac{m_\chi^2}{\left(m_Z^2-4m_\chi^2\right)^2}\,\frac{\left(Z_{13}^2-Z_{14}^2\right)^2}{\cos^4\theta_W\sin^4\theta_W}\left(\frac{40}{3}\sin^4\theta_W-10\sin^2\theta_W+\frac{21}{4}\right) \tag{12}$$

In the case $\chi\chi$-annihilation is dominated by Higgs exchange (it occurs when annihilation involves the gaugino part from one neutralino and the higgsino part from the other one):

$$\tilde{a} = 6\pi\alpha_{em}^2\,\frac{\tan^2\beta}{\sin^4\theta_W}\,\frac{m_b^2 m_\chi^2}{m_W^2\left(m_A^2-4m_\chi^2\right)^2}\left(Z_{11}-Z_{12}\tan\theta_W\right)^2\left(Z_{13}\cos\beta-Z_{14}\sin\beta\right)^2, \tag{13}$$

$$\tilde{b} = 72\pi\alpha_{em}^2 \frac{\tan^2\beta}{\sin^4\theta_W} \frac{m_b^2 m_\chi^4}{m_W^2\left(m_A^2-4m_\chi^2\right)^3}\left(1+\frac{1}{16}\frac{\left(m_A^2-4m_\chi^2\right)m_b^2}{m_\chi^4}\right)\cdot$$
$$\cdot\left(Z_{11}-Z_{12}\tan\theta_W\right)^2\left(Z_{13}\cos\beta-Z_{14}\sin\beta\right)^2 \tag{14}$$

where m_A is the mass of the Higgs pseudoscalar.

For $x_f \approx 1/20$ in all cases $\tilde{b}x_f/2 > \tilde{a}$ or $\tilde{b}x_f/2 >> \tilde{a}$ and the term $\tilde{a}x_f$ in Eq. (3) can be neglected. Eq.s (3) – (14) allow one to estimate $\Omega_\chi h^2$ for the various parts of plot in Fig. 1.

The basic observation is that $\tilde{b}$ is governed by the term $\sim\alpha_{em}^2 m_\chi^2/\tilde{m}^4$ for the case of a gaugino-dominated neutralino, by the term $\sim\alpha_{em}^2 m_\chi^2/(m_Z^2-4m_\chi^2)^2$ for a higgsino-dominated neutralino and by the term $\sim\alpha_{em}^2 m_b^2 m_\chi^4/\left(m_W^4(m_A^2-4m_\chi^2)^3\right)$ for a strongly mixed neutralino. A neutralino composition plot is given in Fig. 1 according to the calculations of Roszkowski [18]. It corresponds to $\tan\beta = 2$ and $\mu < 0$. The diagonal full line separates the neutralino states with more than 50% contribution of gaugino (the area under the line) from neutralinos with more than 50% contribution of higgsino (the area above the line). The area of almost pure photino states (95%) is marked by $\tilde{\gamma}$ and limited by the dashed line. The areas excluded by LEP and CDF experiments are explicitly shown. The CDF limit is connected with the lower limit for gluino mass $m_{\tilde{g}} \geq 150$ GeV [19]. By using the GUT relation

$$M_2 = (\alpha_2/\alpha_3)\, m_{\tilde{g}} \tag{15}$$

(where α_2 and α_3 are coupling constants for SU(2) and SU(3) groups), one obtains $M_2 \geq 45$ GeV), shown on the graph.

Most probably the allowed region is limited from above by the thick horizontal dashed line. It is the theoretical limit obtained using the fine-tuning arguments [18, 20]: if the gluino mass $m_{\tilde{g}} \leq 1$ TeV, then according to Eq. (15) $M_2 \leq 300$ GeV, as shown in Fig. 1. Another rather extreme option is to neglect fine tuning arguments. In this case the neutralino can be a DMP if $m_\chi \leq 1$ TeV and the annihilation to the gauge bosons and higgses becomes the main process which regulates the relic abundance of neutralinos [18, 19]. I think that the fine-tuning limit $m_{\tilde{g}} \leq 1$ TeV is too rigid. Some compensation in the mass graphs can occur and a more liberal limit (*e.g.* $M_2 < 1000$ GeV, see Fig. 1) can take place.

Following neutralino-isomass curves in Fig. 1 (see the solid lines marked by the values of neutralino masses 25, 60, 150 and 300 GeV), one observes that the curve for $m_\chi = 25$ GeV passes only marginally through the allowed area and the lighter neutralino is excluded (more accurately the lower limit $m_\chi \geq 20$ GeV [18, 23]). Similarly, the fine-tuning upper limit for neutralino mass is 150 GeV (see the corresponding isomass curve in Fig. 1).

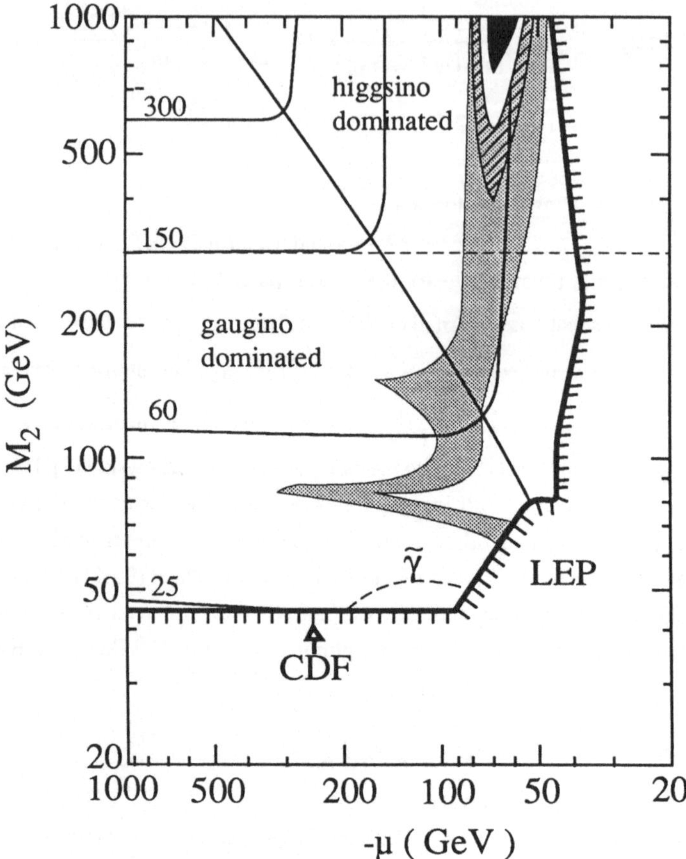

Fig. 1. Neutralino composition plot in M_2-μ parameter space for $\tan\beta = 2$ and for other parameters given in the text (the calculations of Ref. [18]).
The regions excluded by LEP and CDF data are shown by the corresponding curves. The fine-tuning upper limit on M_2 is shown by the dashed horizontal line. The full diagonal line separates gaugino dominated states (below the line) from higgsino dominated states (above the line). The full curves marked 50, 60, 150 and 300 are neutralino isomass curves with the corresponding masses in GeV. The cosmologically excluded region $\Omega_\chi h^2 > 1$ is shown by black, the hatched area corresponds to $0.25 \leq \Omega_\chi h^2 \leq 0.5$ (the most favourite case) and in the dotted area $0.025 \leq \Omega_\chi h^2 \leq 0.1$ which is marginally allowed (see the text for the further discussion).

The relic neutralino density is also displayed in Fig. 1 according to the calculations of Ref. [18] for the following set of parameters: tan $\beta = 2$, the pseudoscalar Higgs mass $m_A = 200$ GeV, the selectron mass $\tilde{m}_e = 80$ GeV and all other sfermion masses $\tilde{m}_f = 1$ TeV. The cosmologically excluded region $\Omega_\chi h^2 > 1$ is shown by black. The hatched area corresponds to $0.25 \leq \Omega_\chi h^2 \leq 0.5$, *i.e.* $\rho_\chi \sim \rho_c$ at $h \sim 0.75$. Within the dotted area $0.025 \leq \Omega_\chi h^2 \leq 0.1$ which is marginally allowed for $h = 0.5$ and for dark matter as a mixture of neutralinos and the hot dark matter (*e.g.* neutrino).

The neutralino relic abundance displayed in Fig. 1 is parameter dependent. One can obtain the analytical expressions for some cases from Eq.'s (3) – (14). In particular for the case of almost pure photino (see area $\tilde{\gamma}$ in Fig. 1),

$$\Omega_\chi h^2 = 2.4 \cdot 10^{-7}\, \tilde{m}^4 / m_\chi^2 \tag{16}$$

and therefore photino-dominated neutralino marginally can be DMP for $m_\chi \approx 40$ GeV and $\tilde{m} \approx 200 - 300$ GeV. In Eq. (16) the masses are in GeV.

More generally, for the gaugino-dominated neutralino one has

$$\Omega_\chi h^2 \approx 8.9 \cdot 10^{-4} \left(\tilde{m}^4 / m_\chi^2\right)\left(Z_{11} / \sin\theta_W\right)^{-4} \cdot \cdot \left(567 - 108y + 1242y^2 - 12y^3 + 2023y^4\right)^{-1} \tag{17}$$

with the masses measured in GeV. Eq. (17) demonstrates the crucial dependence of $\Omega_\chi h^2$ on $\tilde{m}^4 / m_\chi^2$.

As far as galaxy formation is concerned the neutralino, as well as any other cold dark matter particle (CDMP), cannot explain all structures in the universe (galaxies, clusters and superclusters). The computation of momenta of spatial distribution of galaxies [24] is in excellent agreement with the distribution of galaxies according to IRAS observations on the scale of ~ 5 Mpc. However, on a scale of ~10 Mpc the agreement is marginal and on scale of ~ 20 Mpc the observations strongly contradict the calculations.

On the other hand hot dark matter (neutrinos) naturally explains large-scale structures such as superclusters, but meets serious difficulties in formation of galaxies.

The best compromise is given by the mixed DM composition [25] with ≤ 30% of HDM (neutrinos) and ≥ 70% CDM (presumably neutralinos).

2.2. *Accumulation of neutralinos in the earth and in the sun*

The capture of neutralinos by the earth and the sun was first noticed by Press and Spergel [26]. The detailed calculations were performed by Gould [27] (see also Ref. [28]).

The physical essence of this phenomenon is as follows.

If the neutralino velocity relative to the earth (sun) is v_i, then its velocity at the surface is $v_i + v_{esc}$, where v_{esc} is the escape velocity equal to 618 km/s for the sun and to 11 km/s for the earth. Scattering off a nucleus, a neutralino can lose its velocity down to a value $v < v_{esc}$. In this case a neutralino cannot leave the region of gravitational attraction: it comes back to the earth (sun) scatters again and in the end will be trapped in the central region of this celestial body.

For the masses $m_\chi > 20$ GeV one can neglect the evaporation of neutralinos and therefore the capture rate is equal to the rate of neutralino annihilation $\chi + \chi \rightarrow f + \bar{f}$.

The capture rate can be written down as [27, 28]

$$C = c\frac{\rho_{0.3}}{v_{300}} \sum_i f_i \varphi_i S_i \frac{\sigma_i / 10^{-36}\,\mathrm{cm}^2}{m_\chi m_i / \mathrm{GeV}^2} \tag{18}$$

where $c = 5.88 \cdot 10^{28}\ s^{-1}$ for the sun and $c = 5.80 \cdot 10^{19}\ s^{-1}$ for the earth, $\rho_{0.3}$ is the local neutralino mass density in units $0.3\ \mathrm{GeV/cm^3}$ and v_{300} is a mean halo velocity of neutralinos, σ_i is a cross-section for neutralino scattering on the nucleus with atomic number A_i, $m_i \approx A_i m_N$ is the mass of a nucleus, f_i is the mass fraction of the nuclei i in the earth (sun) and φ_i is the dimensionless gravitational potential equal on average to 3.2 for the sun and 1.4 for the earth [27,28]. And finally S_i is so-called the kinematic suppression factor (see Ref. [27] and also [29]). We shall shortly discuss it.

Kinematics is the most important factor in the capture of neutralinos. A neutralino loses a velocity most effectively if $m_\chi \sim m_i$ (mass of a nucleus). When $m_\chi << m_i$ or $m_\chi >> m_i$ the velocity loss $\Delta v/v << 1$ and only a small fraction of neutralinos is captured. The peak in the neutralino capture rate at $m_\chi \sim m_i$ is often referred to as a "resonance". The energy (velocity) loss can be adequatelly described [27, 29] by the parameter.

$$A = \frac{3}{2} \frac{m_\chi m_i}{\left(m_\chi - m_i\right)^2} \left(\frac{v_{esc}}{\bar{v}}\right)^2 \varphi_i, \tag{19}$$

where $\bar{v} \approx 300$ km/s is a mean velocity of neutralinos.

The factor S_i in Eq. (18) is defined in such a way that in the maximum of resonance, when $A >> 1$, $S_i \approx 1$. The regions $m_\chi << m_i$ and $m_\chi >> m_i$ are kinematically suppressed: for $A << 1$, $S_i \approx A << 1$.

The purpose of the discussion below is to find a range for m_χ where the neutrino flux from the earth is higher than from the sun. In principle, the sun captures neutralinos more effectively than the earth because of the higher escape velocity. However, smaller distance from a detector to the center of the earth and the "resonant" scattering on the heavy nuclei in the earth (mostly on iron) compensate the difference in the escape velocities.

The ratio of neutrino fluxes in a detector is equal to C_{earth}/C_{sun} multiplied by the ratio of distances squared, $5.5 \cdot 10^8$. Then from Eq. (18) we find

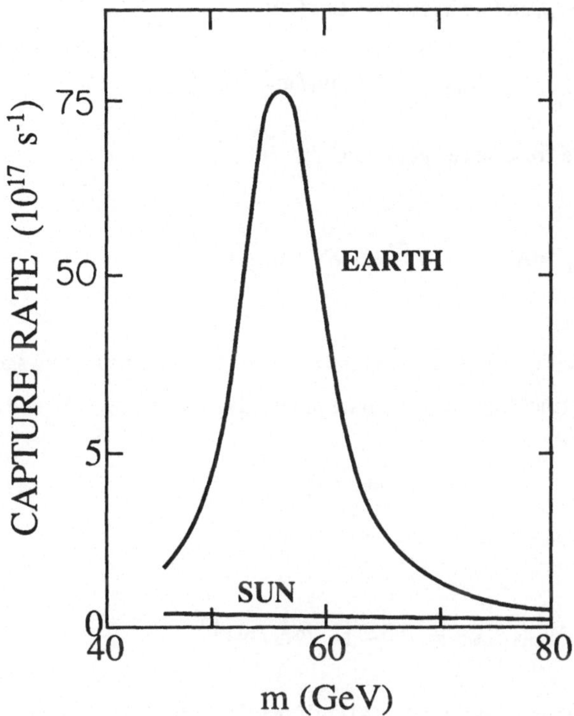

Fig. 2. The capture rate for heavy Dirac neutrinos (according to the calculations of Ref. [27]. The solar rate is scaled down by a factor $(r_{sun}/R_{earth})^2 = 5.5 \cdot 10^8$ so that the rates are proportional to neutrino fluxes from the earth and the sun, respectively. The figure illustrates the dominance of the neutrino flux from the earth within the "resonance" peak.

$$j_{earth} / j_{sun} = 0.24 \sum_i \frac{\sigma_i}{\sigma_H} \frac{f_i}{A_i} \frac{S_i}{S_{sun}} \tag{20}$$

where we assumed the pure hydrogen (H) composition for the sun.

For the sun $m_\chi >> m_i \equiv m_H$, and we have from Eq. (19)

$$S_{sun} \approx A \approx 20\, m_H / m_\chi \tag{21}$$

For the earth $S_i \approx 1$ in the "resonance" peak and

$$S_{earth} \approx A \approx 2.1 \cdot 10^{-3} \left(70 \text{GeV} / m_\chi\right) \left(1 - \frac{50}{m_\chi}\right)^{-2} \tag{22}$$

outside the "resonance" peak. Eq. (22) is given for the iron nuclei. For the neutralino states with the coherent scattering on nuclei (the Higgs exchange) we put into Eq. (20) $\sigma_i / \sigma_H \approx A_i^2$ and thus obtain:

$$j_{earth} / j_{sun} \sim 10 \tag{23}$$

in the "resonance" peak and

$$j_{earth} / j_{sun} = 0.025 \left(1 - m_{Fe} / m_\chi\right)^{-2} \tag{24}$$

outside the peak. It is easy to verity that at $m_\chi \geq 70$ GeV the flux from the sun dominates even for the case of the coherent scattering.

These calculations can be qualitatively illustrated by Fig. 2 for the heavy Dirac neutrino [27]. Neutrino flux from the sun dominates at $m_\nu > 80$ GeV.

As far as a signal in an underground detector is concerned, one must take into account the duration of observation of each source (the sun is invisible in the daytime) and the properties of a detector. For some compositions of the neutralino, both sources can be simultaneously detected, which is the most clear signature of the neutralino.

2.3. *Search for the neutralino in the earth and the sun*

Neutrino flux is produced due to annihilation of neutralinos in the earth and the sun. The signature of the neutralino is the neutrino flux from the core of the earth and from the sun. Neutrinos are assumed to be detected due to the accompanying muon flux.

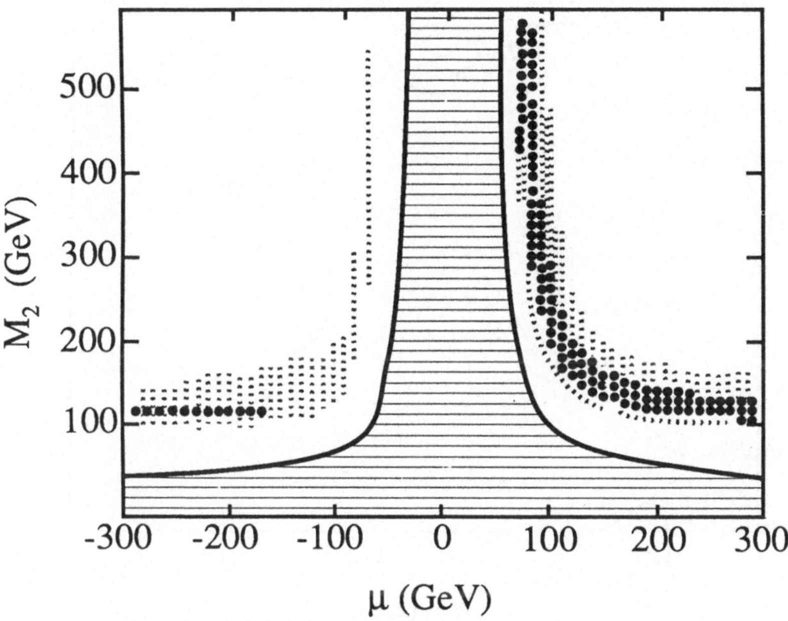

Fig. 3. Exclusion plot ($\tan\beta = 8$ and the Higgs mass $m_h = 80$ GeV) for the neutralino according to the calculations of Bottino et al [30].

The hatched area is excluded by the LEP data. By black filled circles is shown the area excluded by the Kamiokande data on up-going muons, which could originate due to the neutralino annihilation in the center of the earth. In the dotted region high energy neutrinos from the center of the earth can be registered by 10^5 m^2-detector during two years of observation as 4σ excess over the background due to atmospheric neutrinos.

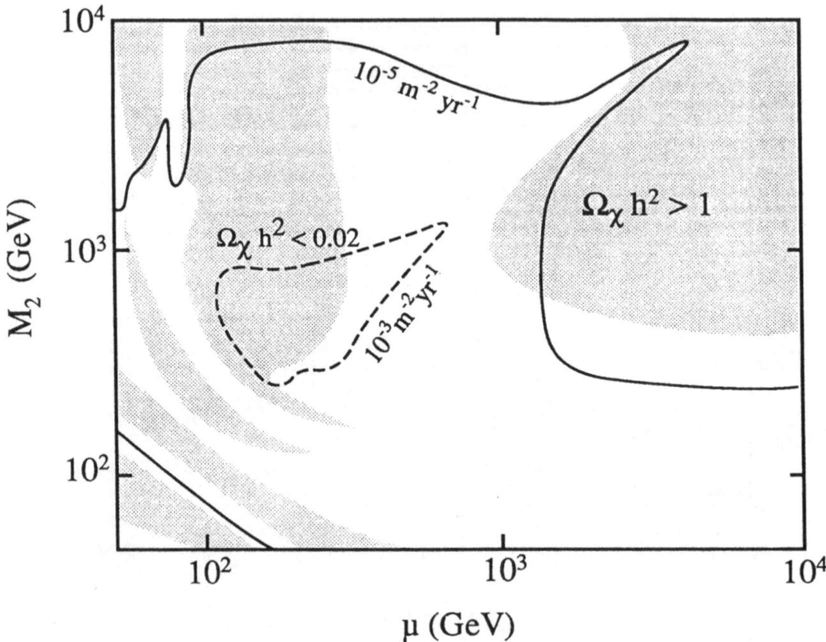

Fig. 4. High energy neutrinos from the annihilation of heavy neutralinos in the sun, according to the calculations of Ref. [31].

The shaded area shows where the neutralino is not DMP: at the right region its density is too high $\Omega_\chi h^2 > 1$) at the left – too small ($\Omega_\chi h^2 < 0.02$). The full curve limits the area where the predicted high energy muon flux produced by high energy neutrinos underground is equal or higher than 10^{-5} m^{-2} yr^{-1}. The dashed line corresponds to the muon flux 10^{-3} m^{-2} yr^{-1}. The parameters used in the calculations [31]: tan $\beta = 2$, the Higgs mass $m_H = 50$ GeV and the squark mass $\tilde{m} = 2.5\, m_\chi$.

A possibility to search for the neutralinos in the earth is illustrated by Fig. 3, according to the calculations of Ref. [30]. The region in neutralino space M_2, μ ($\tan\beta = 8$) covered by the filled circles is excluded by the Kamiokande observations of up-going muons.

The dotted region shows where the neutralino can be found as the DMP by a future underground detector with area $S = 1 \cdot 10^5$ m^2. The detectability condition is 4σ excess of the signal over the atmospheric neutrino background in two years. (In the calculations the Higgs mass was assumed $m_h = 50$ GeV and $\tan\beta = 8$).

For the heavy neutralino the neutrino flux from the sun dominates. The possibility to search for the neutralino in the sun is illustrated by Fig. 4, taken from Ref. [31]. The annihilation of the neutralinos, with a composition described by the points along the full line in Fig. 4, results in the underground muon flux 10^{-5} m^{-2} yr^{-1}. This flux can be marginally registrated by the detector with area 1 km^2. The dashed line corresponds to the muon flux 10^{-3} m^{-2} yr^{-1}.

In the dotted area the neutralino is excluded as DMP. One can see from Fig. 4 that in the case of heavy neutralino the underground detector with area ~ 1 km^2 controls almost the whole M_2-μ area, where the neutralino is DMP.

Therefore, high energy neutrino astronomy gives us the unique possibility of searching for the neutralino as DMP in the mass interval between 40 and 1000 GeV. The signature is the directionality of the signal (the earth, the sun or both).

The underground detector with an area S ~1 km^2 controls, for all neutralino masses, the large region in M_2-μ plot, where the neutralino can be DMP.

2.4. *The neutralino decay*

LSP is stable if R-parity is strictly conserved.

R-parity is a discrete symmetry which is postulated *ad hoc* in supersymmetric theories to avoid the catastrophically fast proton decay. There is no deep theoretical justification for this symmetry and most probably it must be broken at some level.

In this case the neutralino decays to the usual particles. If the neutralino is DMP, its lifetime is larger (or much larger) than the age of the Universe t_o. It implies a very weak R-parity violation.

Phenomenologically, one can consider the neutralino decay to the various channels, characterising each one by the branching ratio. However, there can be the specific mechanisms of R-parity violation which result in a strong dominance of the decay into neutrinos. This is the case of the spontaneous R-parity breaking with the massless Goldstone boson remaining the physical particles, the Majoron [32].

For this mechanism of R-parity violation, the decay $\chi \rightarrow \nu + J$ (where J is the Majoron) strongly dominates over all other channels [33]. The neutrino flux (mostly of extragalactic origin) is determined by the value t_o/t_χ:

$$I_\nu = \frac{c}{4\pi}\frac{\rho_c}{m_\chi}\frac{t_o}{\tau_\chi} \approx 2.5\cdot 10^3 \frac{t_o}{\tau_\chi}\frac{10\,\mathrm{GeV}}{m_\chi}\ \mathrm{cm^{-2}\,s^{-1}\,sr^{-1}}, \tag{25}$$

where t_o is the age of the Universe and τ_χ is the neutralino lifetime. The ratio t_o/τ_χ can be considered as a parameter of R-parity violation.

In the decay $\chi \rightarrow \nu + J$ the monoenergetic neutrinos with energy $E_\nu = m_\chi/2$ are produced. The width of this line is determined by the red-shifted neutrinos produced at large distances from the earth.

The clear signature of the events is given by the electron neutrinos and antineutrinos, which produce the monoenergetic showers inside a detector due to the reactions:

$$\nu_e + N \rightarrow e + \mathrm{hadrons} \tag{26}$$

The lower limits on the neutralino lifetime can be obtained [33] from the observations of the diffuse X- and gamma-radiation and the positrons in our Galaxy. Recently, the Kamiokande group [34] further improved these limits using the up-going muons.

The rate of monoenergetic events inside a detector, $N_\nu = 4\pi I_\nu \sigma_{\nu N} N_N$, does not depend on neutralino mass (see Ref. [33]) and is given by

$$N_\nu = \frac{1}{2}c\rho_c\tilde{\sigma}N_N(t_o/\tau_\chi) = 2(10^7\,t_o/\tau_\chi)h^2\,\mathrm{kt^{-1}\,yr^{-1}} \tag{27}$$

where N_N is the number of nucleons in the detector, $\sigma_{\nu N} = \tilde{\sigma}\, m_\chi/2$ is the νN-cross-section ($\tilde{\sigma} \approx 0.8\cdot 10^{-38}\ \mathrm{cm^2/GeV}$) and h is the Hubble constant in the units 100 km/s Mpc. Therefore, future detectors with the mass 10^4 kt will be able to search for the parameter of R-parity violation t_o/τ_χ as small as 10^{-11}.

3. HIGH ENERGY NEUTRINOS FROM EXOTIC BIG-BANG RELICS

Frampton and Glashow [9] considered very heavy long-lived particles decaying as $X \rightarrow \nu + \mathrm{all}$. The neutral question arises, whether these decays can occur at large red shifts z ($\tau_X \ll t_o$), so that all products of the decay are absorbed by the relic radiation and only neutrinos, with their unique penetrating ability, reach us.

The answer to this question is negative. In the end of this section I hope to convince a patient reader that detectable neutrino flux can be produced only by long-lived particles with the life time $\tau_X > t_o$ (or $\tau_X \gg t_o$).

3.1. *Neutrino absorption*

This is the most straightforward argument which limits the horizon for HE-neutrino observations. The relevant processes are due to neutrino annihilation with the black-body neutrinos ν_{bb}: $\nu + \bar{\nu}_{bb} \rightarrow e^+ + e^-$, $\nu + \bar{\nu}_{bb} \rightarrow \mu^+ + \mu^-$, $\nu + \bar{\nu}_{bb} \rightarrow$ pions etc. To a less extent the neutrino flux diminishes due to the elastic scattering $\nu + \nu_{bb} \rightarrow \nu + \nu$. The neutrino absorption was studied in Ref.s [2, 35-37].

The condition for the absorption of the neutrino with the present-day (red shift $z = 0$) energy $E_{\nu o}$ can be written down as

$$\int_{t_a}^{t_o} dt\, n_\nu(t)\sigma(E_{\nu o},t)c = 1 \tag{28}$$

where n_ν (t) is the spatial density of relic neutrinos, σ ($E_{\nu o}$, t) is the cross-section of annihilation $\nu + \bar{\nu}_{bb} \rightarrow$ all at time t (red shift z) and t_a is the age of the Universe when the absorption sets in for a neutrino with the present-day energy $E_{\nu o}$.

The details of the calculations can be found in Ref.s [2, 35-37]. Here we just give the approximate formulae for t_a and z_a:

$$z_a \approx 8 \cdot 10^4 \,(E_{\nu o} / 1\ \mathrm{TeV})^{-1/3} \tag{29}$$

$$t_a \approx 5 \cdot 10^9 \,(E_{\nu o} / 1\ \mathrm{TeV})^{2/3}\ \mathrm{sec} \tag{30}$$

Actually these expressions contain another factor weakly dependent on $E_{\nu o}$.

Eq.s (29) – (30) show that the cosmological horizon for high energy neutrino astronomy is limited by the relatively recent epochs ($z \leq 10^5$).

3.2. *High energy neutrinos from thermal relics in standard cosmology*

In this subsection we shall confine ourselves to the thermal relics, *i.e.* to the particles which were once in the thermal equilibrium in the Universe. In the calculations [36] the following assumptions are used:

i) *Standard cosmology*. We assume that apart from the exotic relic X, all other particles belong to the GUT extension of the MSSM model.

ii) *The annihilation cross-section* $X + \bar{X} \rightarrow$ all. If the annihilation is mediated by the gauge bosons of SU(3) x SU(2) x U(1) group, then the cross-section is given by

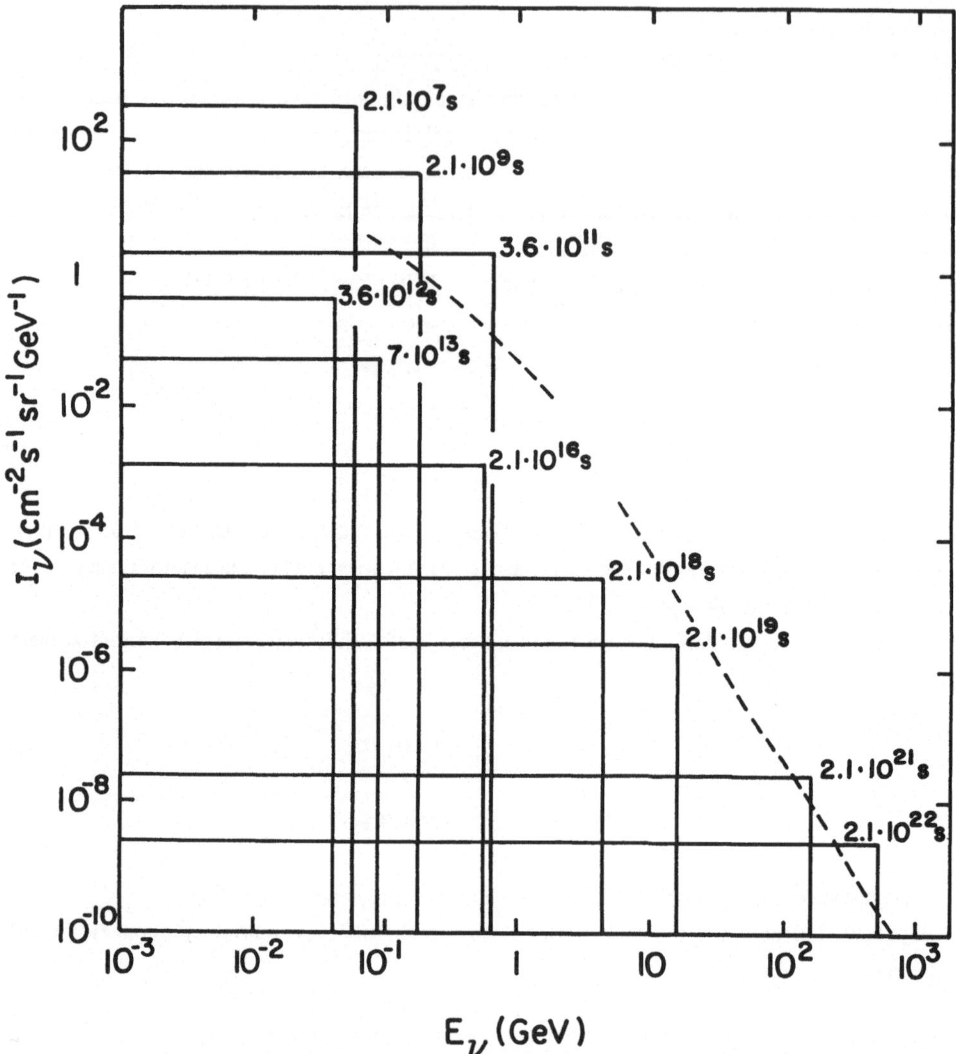

Fig. 5. Differential neutrino fluxes $I_\nu(E)$ for different X-particle lifetimes τ_X (shown at the right corner of each box). The vertical line shows the upper limit of E_ν for a given τ_X. The horizontal lines give the differential neutrino flux. Note that it is not a spectrum (see explanation in the text). The dashed line shows the spectrum of atmospheric neutrinos. The flux from decay of X–particles can exceed the atmospheric flux at $E \sim 0.1 - 10\,TeV$.

$$\sigma v = A\alpha^2 / m_X^2 \tag{31}$$

with $\alpha \equiv 1 \cdot 10^{-2}$ and $A \sim 1$ (see the accurate expressions in Ref. [36]).

iii) *Decay* modes. Each decay mode, $X \rightarrow \nu + \text{all}$, $X \rightarrow e + \text{all}$ etc, is characterised by the branching ratio b_ν, b_e etc. However, in numerical calculations all branching ratios are assumed to be of the same order of magnitude.

Actually, only assumption (i) is essential for the conclusions listed below. We can state even more: the results obtained in Ref. [36] qualitatively hold if

no violent entropy production occurs after the decoupling of X-particles.

As a result of the straightforward calculations we arrive at the following conclusions.

1. Predicted neutrino flux strongly depends on the lifetime τ_X, but does not depend on m_X.
2. Various physical phenomena, such as nucleosynthesis, distortion of 2.7 K microwave radiation, positron production and, most notably, diffuse X- and gamma-radiation constrain neutrino energy E_ν, but not the neutrino flux I_ν, (E_ν).
3. The absolute upper limit for the neutrino energy is imposed by the condition $\rho_X(t_o) < \rho_c$ and is equal to ~ 10 TeV.
4. The detectable HE neutrino flux can be produced only by the long-lived ($\tau_X > t_o$) massive particles.
5. Predicted flux is marginally detectable at $E_\nu \geq 0.3 - 1$ TeV.

The calculated fluxes are plotted in Fig. 5 together with the upper limits for neutrino energy E_ν shown by the vertical lines. The corresponding lifetimes τ_X are shown at the right corner of each box. Note that the graph gives the mean differential flux for the mean neutrino energy $\overline{E}_\nu$. As m_X diminishes, $\overline{E}_\nu$ becomes smaller, but the mean value of flux remains the same. The flux of atmospheric neutrinos is shown by the dashed line. As one can see from Fig. 5, neutrino flux from relic particles can exceed the flux of atmospheric neutrinos at 0.1 – 10 TeV. The signal reveals itself as a bump in continuous spectrum of atmospheric neutrinos.

3.3. *Superheavy metastable relics*

In the works [38, 39] very heavy relic particles (with masses up to 10^{18} GeV) were considered as DMP. The theoretical motivation for such particles is given by examples of technibaryons [40] and "cryptons" [38]. Cryptons were suggested as a solution of the fractional electric-charge problem in the superstring models. They are the confined states of the fractionally charged particles. The mass of the lightest crypton can reach 10^{18} GeV, while its lifetime can be much larger than the age of the Universe.

At first glance, this assumption contradicts the results of Section 3.2. Physically this contradiction comes from the too small annihilation cross-section for very heavy particles, $\sigma_{ann} \sim 1 / m_X^2$. After decoupling these particles practically do not annihilate and later on their relic density decreases as

a^{-3} (t), while the mass density of the photons decreases faster, as a^{-4} (t) (here a(t) is a scaling factor of the expanding Universe). As a result one obtains at present too large a ratio ρ_X / ρ_γ, above the value for critical density.

This contradiction arises because in Section 3.2 we assumed that no burst of entropy production takes place after the decoupling of X-particles. However, such a burst can occur if one or several particles Y_i, decouple later and annihilate producing photons. Then the ratio n_X / n_γ decreases by a jump after the annihilation of each pair $Y_i \overline{Y}_i$ and now the ratio n_X / n_γ can be as low as needed for the case of the critical density. This possibility is exploited in the works [37, 39].

The results of the calculations are presented in Fig. 6. The area on m_X-τ_X-plot, excluded by HE neutrino data of Frejus, IMB and Fly's Eye detectors, is explicitly shown. Beyond this limit the neutrino flux can be registered by the larger detectors. In the approach of Ref. [37] the accompaning e.-m. radiation is not considered. In fact it is tacitly assumed that X-particles decay only into neutrinos. Obviously, this is a very exotic case: even if the branching ratio for the neutrino channel $b_\nu = 1$, the other decay partner(s) can start the electromagnetic cascade and thus give the upper limit for the neutrino flux.

We shall discuss here this upper limit. For $\tau_X > t_o$ energy density of the electromagnetic cascade radiation is limited, due to 100 MeV SAS-2 data, by $\omega_{cas} \leq 5 \cdot 10^{-6}$ eV / cm^3 [36]. If the X-particles constitute the dark matter, then one finds for the energy density of cascade photons

$$\omega_{cas} = \rho_c \, r_{em} \, t_o / \tau_X \tag{32}$$

where $r_{em} = b_{em} \cdot f_{em}$, b_{em} is the branching ratio for X-decay into a particle which can start an e-m cascade (*e.g.* electron, photon and hadron) and f_{em} is a fraction of energy transferred to such a particle. Using the limit $\omega_{cas} \leq 5 \cdot 10^{-6}$ eV / cm^3, one obtains the lower limit for $\tau_X / (t_o \, r_{em})$ shown in Fig. 6.

The neutrino flux from the decays $X \to \nu$ + al can be readily estimated as

$$I_\nu = \frac{c}{4\pi} \frac{\rho_c}{m_X} b_\nu \frac{t_o}{\tau_X}. \tag{33}$$

Using Eq. (32) one obtains

$$I_\nu = \frac{c}{4\pi} \frac{b_\nu}{r_{em}} \frac{\omega_{cas}}{m_X} < 1 \cdot 10^{-11} \frac{b_\nu}{r_{em}} \frac{10^{-6}\,\text{GeV}}{m_X} \text{cm}^{-2}\text{s}^{-1}\text{sr}^{-1}, \tag{34}$$

where the flux (34) corresponds to the average neutrino energy $\overline{E}_\nu = f_\nu \, m_X$.

For $b_\nu / r_{em} \geq 1$ this is a detectable flux for large underground detectors with an area of $S \geq 10^5$ m^2. In particular the flux (34) is higher than the upper limit for the model of Ref. [4].

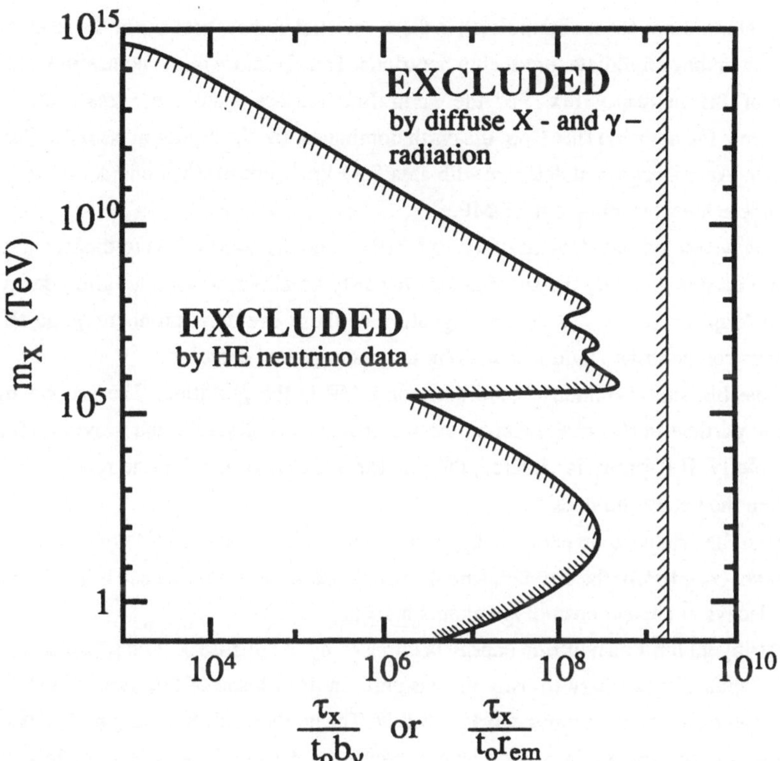

EXCLUDED
by diffuse X - and γ - radiation
EXCLUDED
by HE neutrino data
10^{15}
10^{10}
10^{5}
1
m_X (TeV)
10^{4}
10^{6}
10^{8}
10^{10}
$\frac{\tau_x}{t_o b_\nu}$ or $\frac{\tau_x}{t_o r_{em}}$

4. CONCLUSIONS

High energy neutrinos are very effective tool to search for the supersymmetric candidates for DM. In particular, the neutralino can reveal itself due to high energy neutrino radiation from the earth, the sun or both. Neutralinos are accummulated in these celestial bodies, because of the scattering off the nuclei, and then they annihilate, producing neutrinos. The signature of the neutralino thus is the directionality of the neutrino flux. For the neutralino masses inside the "resonance" peak $40 < m_{\chi} < 60$ GeV the neutrino flux from the earth dominates, for the higher masses the flux from the sun is higher. An underground detector with area S ~1 km^2 controls the considerable region in M_2-μ-space, where the neutralino is the DMP.

If R-parity is weakly violated, neutralinos as DMP can be discovered due to their decays. The most interesting case is given by the spontaneous R-parity breaking, when neutralino decays to a neutrino and a Majoron ($\chi \rightarrow \nu + J$). The signature of these events is monoenergetic showers produced by electron neutrinos (antineutrinos) in the underground detectors.

The last possible supersymmetric candidate for DMP is the gravitino. The second lightest supersymmetric particle in this case must be the scalar neutrino. If stable, the gravitino does not reveal itself directly. If R-parity is violated, the gravitino decays producing the reach observable phenomena, like the neutralino does.

Neutrinos can be produced by exotic heavy relic particles from Big Bang. Inspite of the strong penetration power ($z_a \sim 10^5$ for the Tev-neutrinos), the detectable neutrino flux can be produced only by X-particle decays at present cosmological epoch $t = t_o$.

The most stringent limit on neutrino energy is imposed by the diffuse X- and gamma-radiation. In the case of standard cosmology (mostly the assumption of absence of the burst of the entropy production) the neutrino energy is limited by $E_{\nu} < 8$ TeV. The predicted flux is marginally detectable.

The superheavy long-lived DM particles can produce the detectable neutrino flux in scenarios with the burst production of entropy after decoupling of X-particles. In this case the energy of neutrino can reach $\sim 3 \cdot 10^5$ TeV.

REFERENCES

1. T.K. Gaisser, Proc. II Workshop "Neutrino Telescopes" (Venezia, 1990) 397.
2. V.S. Berezinsky, S.V. Bulanov, V.A. Dogiel, V.L. Ginzburg and V.S. Ptuskin, Astrophysics of Cosmic Rays, North-Holland, 1990.
3. F. Halzen, Neutrino Telescopes (ed. M. Baldo Ceolin), 437, 1991.
4. F.W. Stecker, C. Done, M.H. Salamon and P. Sommers, Phys. Rev. Lett. 66 (1991) 2697.
5. M.A. Markov and I.M. Zheleznykh in Proc. 1979 DUMAND workshop (ed. J. Learned, Khabarovsk and Baikal) 177, 1979.
6. E. Witten, Nucl. Phys. B 249 (1986) 557.
   J.P. Ostriker, C. Thompson and E. Witten, Phys. Lett. 180 B (1986) 231.
   C.T. Hill, D.N. Schramm and T.P. Walker, Phys. Rev. D 36 (1987) 1007.
7. V.S. Berezinsky and H.R. Rubinstein, Nucl. Phys. B 323 (1989) 95.
8. J.H. Mac Gibbon and R.H. Brandenberger, NP B 331 (1990) 153.
9. P.H. Frampton and S.L. Glashow, Phys. Rev. Lett. 44 (1980) 1481.
10. M.Yu. Khlopov and V.M. Chechetking, preprint IPM, Moscow, N 41, 1985.
11. J. Ellis et al, Nucl. Phys. B 238 (1984) 453.
12. J. Ellis, J.E. Kim and D.V. Nanopoulos, Phys. Lett. B 145 (1984) 181.
13. V.A. Kuzmin, V.A. Rubakov and M.E. Shaposhnikov, Phys. Lett. B 155 (1985) 36.
14. J. Ellis et al, Nucl. Phys. B 238 (1984) 453.
15. J. Ellis, D.V. Nanopoulos and S. Sarkar, Nucl. Phys. B 259 (1985) 175.
16. V.S. Berezinsky, Phys. Lett. 261 (1991) 71.
17. V.S. Berezinsky, A. Bottino and V. de Alfaro, Phys. Lett. B 247 (1992) 122.
18. L. Roszkowski, Phys. Lett. B 278 (1992) 147.
19. L.G. Pondom (CDF collaboration) talk at Int. Conf. on High Physics, Singapore, 1990.
20. R. Barbieri and G.F. Giudice, Nucl. Phys. B 306 (1988) 63.
21. K. Griest, M. Kamionkowski and M. Turner, Phys. Rev. D 41 (1990) 3565.
22. M. Kamionkowski and M. Turner, Phys. Rev. D 42 (1990) 3310.
23. K. Hidaka, Phys. Rev. D 44 (1991) 927.
24. W. Saunders et al (Q DOT collaboration), Nature 349 (1991) 32.
25. A. van Dalen and R.K. Shaefer, preprint BA-91-67.
26. W.H. Press and D.N. Spergel, Ap. J. 296 (1985) 679.
27. A. Gould, Ap. J. 321 (1987) 571.
28. G. Gelmini, D. Gondolo and E. Roulet, Nucl. Phys. B 351 (1991) 623.
29. M. Kamionkowski, Phys. Rev. D 44 (1991) 3021.
30. A. Bottino, V. de Alfaro, N. Fornengo, G. Mignola and M. Pignone, Phys. Lett. B.
31. F. Halzen, T. Stelzer and M. Kamionkowski, preprint MAD/PH/685.
32. A. Masiero and J. Valle, Phys. Lett. B 251 (1990) 273.

33. V.S. Berezinsky, A. Masiero and J.W.F. Valle, Phys. Lett. B 266 (1991) 382.
34. M. Mori et al, Phys. Lett. B 287 (1992) 217.
35. V.S. Berezinsky, preprint Gran-Sasso Lab., LNGS-91/02, 1991 (to be published in Nucl. Phys. B).
36. P. Gondolo, G. Gelmini and S. Sarkar, preprint UCLA/91/TEP/31, 1991.
38. J. Ellis, J. Lopez and D.V. Nanopoulos, Phys. Lett. B 247 (1990) 257.
39. J. Ellis et al, Nucl. Phys. B 373 (1992) 399.
40. R.K. Kaul, Rev. Mod. Phys. 55 (1983) 449.

2nd International Conference on Trends in Astroparticle Physics; Aachen, October 10 – 12, 1991

Outlook and Conclusions

D.H. Perkins

Nuclear Physics Laboratory, University of Oxford

Earlier today we heard four excellent reviews on neutrino astrophysics, on high energy gamma rays, on cosmology and particle physics and on new detector concepts and new experimental projects. From these, you will have formed an excellent view of the prospects and future of this field. There is little for me to add: it seems to me that astroparticle physics is in good shape and that all I can do is to encourage people to keep going. I am simply going to limit myself to a few comments on some aspects of the subject.

Peter Bosetti asked me to speak particularly about neutrino properties, including oscillation experiments at accelerators, on astronomical point sources and on the experimental situation regarding a possible heavy (17 keV) neutrino. This last topic has been discussed at length but not, I feel, very constructively, at the recent summer conferences. If such a heavy neutrino existed, it would have profound effects in both particle physics and cosmology, neither of which would be the same again: so it is worth some discussion.

1. NEUTRINO PROPERTIES

The number of *neutrino flavours* is frequently quoted from the value of the total Z° width measured at LEP:

$$N_\nu = 2.99 \pm .05 \tag{1}$$

This number however refers to the number of types of neutrino of $M_\nu < M_{Z^\circ}/2$ with the standard model (SM) coupling where neutrinos are assigned weak isospin $I_3 = +\frac{1}{2}$, $g_V = g_A = I_3$ and couple to $Z^\circ(I_Z = 1)$. For example, one could conceive of isosinglet neutrinos which would have zero coupling to Z°, and about which LEP says *nothing*. This restriction to SM coupling does not apply to the limit quoted from Big Bang nucleosynthesis (BBN), where isosinglet neutrinos would contribute to the radiation density at nucleosynthesis via reactions such as $e^+ + e^- \to \nu + \bar{\nu}$. The limit quoted is [1]

$$N_\nu \lesssim 3.4 \tag{2a}$$

This limit comes from reconciling the helium/hydrogen mass ratio (~ 0.24) with the *inferred* baryon/photon ratio $N_B/N_\gamma \simeq 3.10^{-10}$ coming from Li abundances as measured today in ten population II stars (the individual values vary by an order of magnitude and the relevance to the universal Li abundance 12 billion years ago is far from clear). The *directly measured* $N_B/N_\gamma \simeq 6.10^{-11}$ can be regarded as a safe lower limit, which results in the less restrictive flavour number

$$N_\nu \lesssim 5 \tag{2b}$$

The bottom line is to be very cautious about cosmologist's limits which imply that there cannot be any extra neutrinos beyond those prescribed by LEP.

The *nature* of neutrinos — whether Dirac or Majorana particles — has arisen again because of the "17 keV phenomenon". Neutrinoless double β-decay is in principle possible for Majorana neutrinos, and its non-observation yields the limit

$$|\sum_i \eta_i V_i^2 m_i| < 3 \text{ eV} \tag{3}$$

where m_i is the mass of a Majorana neutrino of type 'i', U_i is its coupling to nucleons (quarks) and $\eta_i = \pm 1$ is its CP parity. Eqno(3) can be satisfied for large m_i (e.g. 17 keV) only if there exist two states almost degenerate in mass, with opposite CP parities. Such a doublet is however really indistinguishable from a massive Dirac neutrino, with one "active" helicity state for the particle and the other for the antiparticle (so, it is sometimes termed a "pseudo-Dirac" particle).

There is no evidence that light (< 1 GeV) Majorana particles exist, but very heavy Majorana neutrinos have been postulated in connection with "majoron" decay of heavy Dirac neutrinos, as discussed later.

2. NEUTRINO OSCILLATIONS, MASSES, MAGNETIC MOMENTS

The direct mass limits on the three established flavours of neutrino are

$$\begin{aligned} m_{\nu_e} &< 9.4 \text{ eV} \\ m_{\nu_\mu} &< 270 \text{ keV} \\ m_{\nu_\tau} &< 35 \text{ MeV} \end{aligned} \tag{4}$$

If neutrino flavour is not conserved, mixing may occur and the limits on the mixing angles from reactor and accelerator experiments are given in Table 1, which covers differences in the squared masses of the mass eigenstates $\Delta m^2 \gtrsim 1 \text{ eV}^2$.

In the SM, the predicted value of magnetic moment is [2]

$$\mu_{SM} = 3.10^{-19}(m_\nu/1 \text{ eV})\mu_B \tag{5}$$

Table 1
Present and expected 90% CL limits on Mixing Angles

	$\sin^2 2\theta$ (present)	$\sin^2 2\theta$ (1995 expected)
$\nu_\mu \to \nu_e$	<0.0034	—
$\nu_\mu \to \nu_\tau$	<0.004	< .0004
$\bar{\nu}_e \to \bar{\nu}_x$	<0.16	—
$\nu_e \to \nu_x$	<0.14	—
$\nu_e \to \nu_\tau$	<0.12	<0.03

where μ_B is the Bohr magneton ($= e\hbar/2m_e c$). As expected, μ vanishes for zero neutrino mass.

The best laboratory limit on μ comes from comparing the cross-section for $\nu - e$ scattering with the SM prediction. $\sigma(\text{SM}) \propto \text{E}_\nu$ for $E_\nu >> m_e c^2$. Any excess may be ascribed to magnetic moment scattering, where $\sigma(\mathrm{MM}) \propto 1/E_\nu$. Thus, reactor experiments are the most sensitive, and Kyuldjiev [3] quotes, from the reactor experiment of Reines, Gurr and Sobel [4]

$$\mu(\bar{\nu}_e) < 1.5.10^{-10}\mu_B \qquad (6)$$

However this limit uses cross-sections calculated from the reactor neutrino fluxes estimated by Avignone [5]. A more exact spectrum calculation, by Vogel and Engel [6] increases the cross-section to 1.5–2 times the SM prediction and actually suggests a finite moment! A more conservative limit than (6) is therefore

$$\mu(\bar{\nu}_e) < 4.10^{-10}\mu_B \qquad (7)$$

In accelerator experiments, the limits to $\mu(\nu_\mu)$ are about one order of magnitude larger than (7). Less direct astrophysical limits have been deduced from the cooling rate of red giant stars and from SN1987A and give

$$\mu\ (\mathrm{ASTR.}) < 10^{-12}\mu_B \qquad (8)$$

The possibility [7] of a large ($\mu \sim 10^{-10}\mu_B$) moment has arisen over the last few years, because of an apparent anticorrelation of the solar neutrino rate with sunspot number in the ^{37}Cl experiment, but the evidence was not strong and is not supported by the results of the Kamioka experiment.

Figure 1(a) and Table 1 show existing limits on mixing angle for $\nu_e \to \nu_\mu$ oscillations, Figure 1(b) the limits for $\nu_e \to \nu_\tau$ and $\nu_\mu \to \nu_\tau$ from the Fermilab emulsion experiment E531 by Ushida *et al* [8]. As indicated, the proposed P261 and P254 experiments at CERN, and P803 at Fermilab, should reduce these limits by factors of 10 and 4 respectively. The goal on $\nu_e \to \nu_\tau$ mixing is important as it lies below the value $\sin^2 2\theta \sim 0.04$ required if the 17 keV neutrino exists and is predominantly ν_τ. All these experiments involve neutrinos in the few GeV energy range, path lengths of order 1 km and hence, sensitivity to mass differences of order $\Delta m^2 \geq 1$ eV^2.

Very long baseline experiments have also been proposed (P805, P822 at Fermilab). The former utilises a neutrino beam aimed at the IMB 8 kiloton water Cerenkov detector in the Moreton Salt Mine, Ohio, with L=570 km, while P822 is for a baseline L=800 km pointing at the Soudan 2 one kiloton drift chamber detector, in North Minnesota. The sensitivities of these proposed experiments are given in Figure 2.

For the IMB experiment, one relies on the ratio $R_{\mu/\nu}$ of entering muons (from ν_μ CC interactions in the surrounding rock) to total number of neutrino vertices in the detector. It should also be possible to differentiate, among the vertex events, CC interactions due to ν_μ, giving a penetrating muon and those due to ν_e, giving a showering electron. The limits on $\sin^2 2\theta$ for $\nu_\mu \to \nu_e$ and $\nu_\mu \to \nu_\tau$ transitions are as indicated. A weakness of this proposal is that interpretation of the value of $R_{\mu/\nu}$ relies heavily on a Monte Carlo simulation: there is no possibility of an experimental check through a short-baseline experiment.

The Soudan 2 experiment has a smaller sensitive mass and is at a greater distance, so is statistically a factor 5 weaker than the IMB experiment. On the other hand, since it is a fine-grain device, it can measure both $R_{\mu/\nu}$ and the NC/CC ratio. Most important, the detector is modular and some modules can be sited at a "near station" ($L \sim 1$ km) with the tremendous advantage that one is comparing events in identical detectors at different distances.

The goal of the long baseline proposals is to try to probe the Δm^2 region down to 10^{-2} – 10^{-3} (eV)^2, which is also the mass region to which atmospherically-produced neutrinos are sensitive, and where indeed Kamioka have found apparent evidence for oscillations, as shown

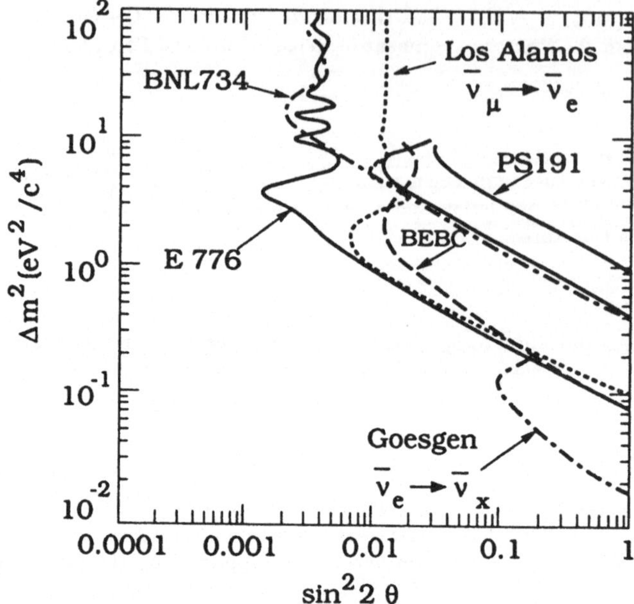

Figure 1(a). Existing limits on $\nu_e \to \nu_\mu$ oscillations (after Moscoso [9]).

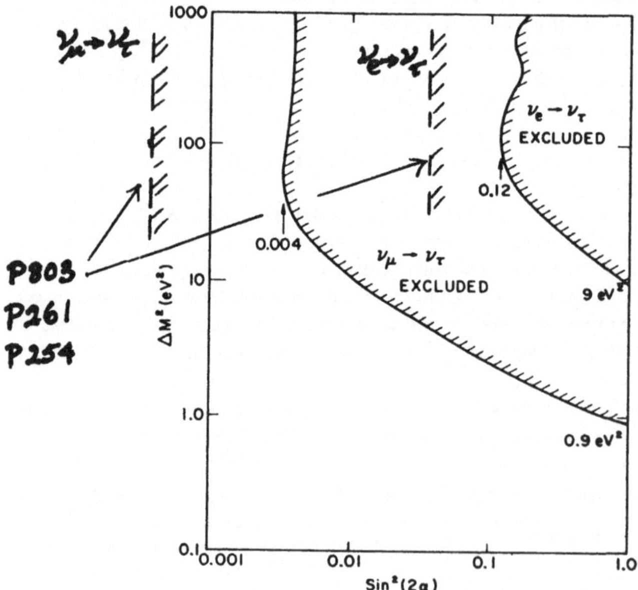

Figure 1(b). Limits for $\nu_e \to \nu_\tau$ and $\nu_\mu \to \nu_\tau$ (Ushida *et al* [8]) together with those fron proposals.

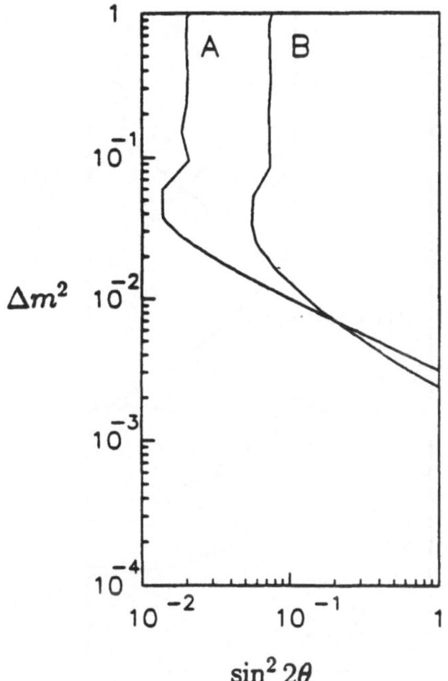

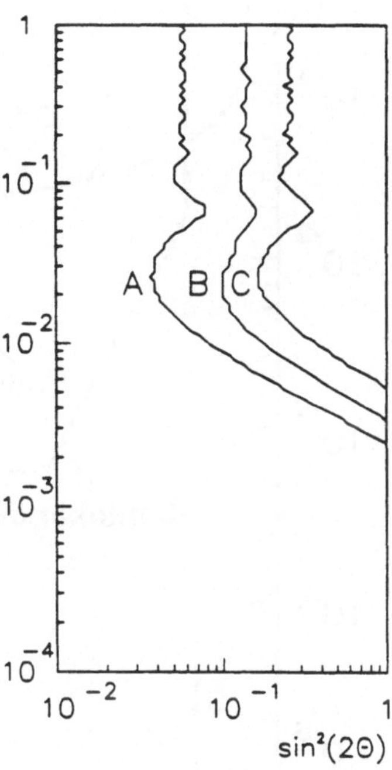

Figure 2. On left, the proposed limits for P805 (IMB detector) based on ratio $R_{\mu/\nu}$ of entering muons to vertex events, for (A) $\nu_\mu \to \nu_e$ and (B) $\nu_\mu \to \nu_\tau$ oscillations. On right are shown limits for $\nu_\mu \to \nu_\tau$ of P822 (Soudan 2 detector) for (C) $R_{\mu/\nu}$ method, (B) NC/CC ratio and (A) two detector measurements.

in Figure 3. This gives the 90% CL LOWER limits on $\sin^2 2\theta$ from Kamioka and the 90% CL UPPER limits from Frejus, obtained from comparing the number of events identified as due to ν_e and ν_μ respectively [9]. At the GeV energies considered here, all atmospheric pions, and most muons, decay in flight. In these two decays, $\pi^+ \to \mu^+ + \nu_\mu$ and $\mu^+ \to e^+ + \nu_e + \bar{\nu}_\mu$, and the corresponding decays for antiparticles, one produces just twice as many $(\nu_\mu + \bar{\nu}_\mu)$ as $(\nu_e + \bar{\nu}_e)$, and it turns out that all of them have on average, a similar fraction (20–25)% of the original pion energy. Hence, the expected flux ratio $\phi(\nu_e)/\phi(\nu_\mu) \simeq 0.50$ (actually a few per cent more because of polarisation effects). Instead, the Kamioka experiment finds comparable fluxes of ν_e and ν_μ which, if genuine, could indicate oscillations with a large mixing angle ($\sin^2 2\theta > 0.2$). The Frejus result is however consistent with no oscillations.

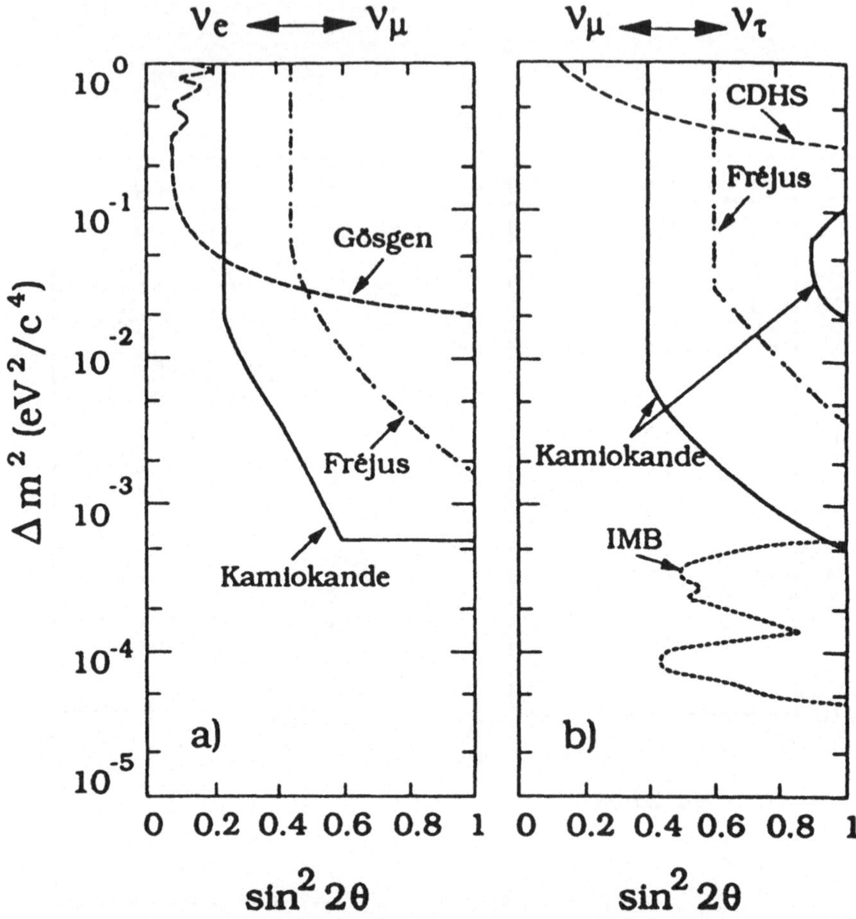

Figure 3. Results on atmospheric neutrinos from the Kamioka and Frejus experiments.

It is clear that a long baseline accelerator experiment would be very useful. However, the: will require the new Fermilab injector and new, downward-inclined neutrino beams. They coul be completed by 1995 if the funding is forthcoming.

3. POINT SOURCES OF HIGH ENERGY COSMIC RAYS

Over the past years, several reports have appeared of signals from air-shower arrays or from underground (muon) detectors, apparently correlated with known point sources (Crab, Cygnus X3, Hercules X1 ...) and involving particle energies in the TeV or PeV region. While there is no question that DC point sources of γ-rays of about 1 GeV energy are clearly established from satellite observations, there is at present only one ground experiment claiming a steady signal, based on observation of atmospheric Cerenkov radiation at the Whipple Observatory [10]. By measuring an event 'shape', they were able to select preferentially γ-induced showers, and obtained, when pointing at the Crab, a (source on-source off) signal at the level of about 20σ significance. These results refer to energies $E_\gamma > 0.4$ TeV and are averaged over a period of several years. They have not yet been confirmed in other experiments.

The question arises whether sources may give rise to non-steady, that is episodic, signals. For example, the binary Cygnus X3 occasionally emits intense radioflares which, for a few days, are the brightest radio sources in the sky. Figure 4 shows results from the Soudan 2 underground experiment [11], sensitive to muons of energy > 0.6 TeV at ground level, during January 1991.

The histogram shows the observed number of muons per day, the dots the expected number based on the background muon rate from nearby sky-directions. The variations in the expected rate are associated with the fact that new detector modules were being brought into operation and the entire detector was turned off for some periods. There is a clear excess on January 20 and 23. During the period Jan 19 — Jan 24 inclusive, 51 muons were recorded, compared with an expectation of 28. The significance of this result is best judged by comparing with expectations, the integrated muon rate during the last $2\frac{1}{2}$ years, during which time there have been 6 flares of intensity > 5Jy. The January 1991 flare is the only one showing an excess; the chance probability of such an excess during any of the 6 flares is just under 1%. This is sufficiently unlikely as to warrant future investigation and also comparison with results of other experiments. These results, [13] for the January 1991 flare only, are shown in Figure 5. The first 4 plots are of underground muons, the last 3 are from airshower arrays. The statistical weight of the KGF, Agasa and Macro data is very weak. The significance of the IMB data is dubious, since the angular resolution (= cone semi-angle) is 7° and thus the signal/noise ratio is about 1/25 that of Soudan 2.

Of the remaining experiments — CASA and HEGRA — CASA observe no significant effect, while the HEGRA array [12] shows a 3.8σ excess on January 20 and 1.5σ on January 23. (Unfortunately, the actual numbers of events and comparison with the background rate are not given). If taken seriously, the last result indicates, presumably, that Cygnus X3 emits γ-rays of energy 50 TeV or more. An underground muon excess, if genuine, would indicate completely new physics. At least, these results are sufficiently intriguing that it is necessary for the experiments to continue, and, most importantly, to operate with a better duty cycle than is indicated in Figure 5.

4. THE 17 KEV NEUTRINO PHENOMENON

Evidence for a neutrino of mass $\sim$ 17 keV emitted as an admixture in beta decay with the normal light neutrino (mass < 9.3 eV) was first presented by Simpson [14] in 1985, on the basis of deviations in the Kurie plot in tritium decay. Subsequent experiments in other laboratories, with both solid-state counters, like those of Simpson, and magnetic spectrometers, and using ^{35}S and ^{63}Si sources, found no evidence for the Simpson effect. During the last 2 years however, several experiments have reported positive results, and the matter is open.

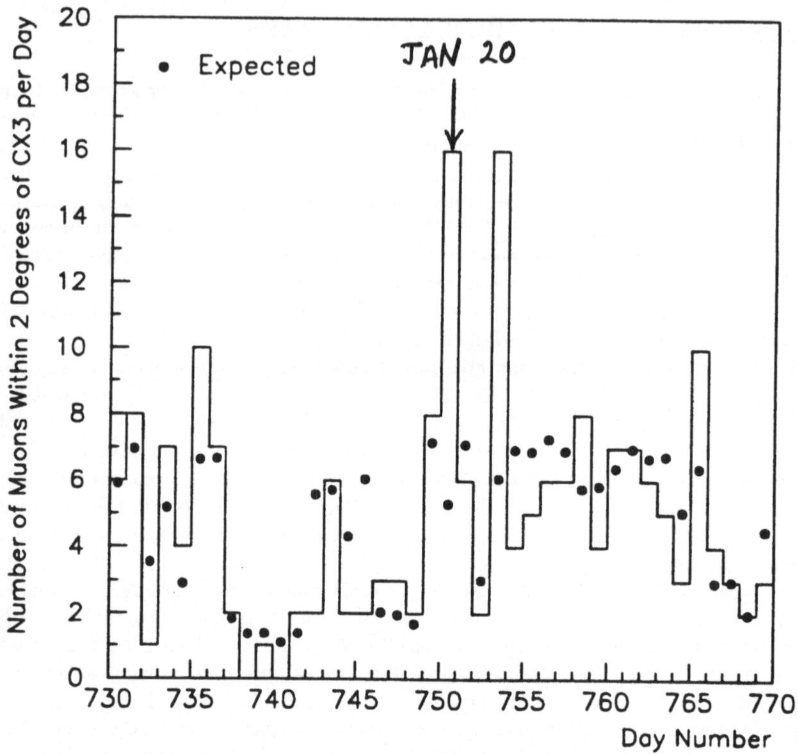

Figure 4. Muon flux in Soudan 2 as a function of time, inside a cone of semi-angle 2° pointing at Cygnus X3, during the radioflare of Jan 20–23 1991. The cone angle was chosen *before* the test, and based on the measured angular resolution to optimize signal/noise.

Assuming that the ν_e is a superposition of mass eigenstates ν_1 and ν_2, with $M_1 < 9$ eV and $M_2 = 17$ keV, then

$$|\nu_e >= \cos\theta|\nu_1 > + \sin\theta|\nu_2 > \tag{9}$$

where the positive experiments suggest $\sin^2\theta \simeq 0.010$ and the negative ones place limits $\sin^2\theta <$ .003. Figure 6 shows how a massive neutrino will affect the Kurie plot. The top picture shows a Kurie plot for $m_\nu = 0$, cutting the axis at electron kinetic energy $T = Q$, together with a second Kurie plot, for $m_\nu = M_2$, which cuts the axis at $T = Q - M_2$. Together they give a plot with a change in slope and a 'kink' at $T = Q - M_2$. The lower 2 diagrams in Figure 6 show (a) the effect of normalizing to the best straight line fit between $T = Q - M_2$, and (b) the ratio of data to the best straight line fit to all the data points.

During the summer 1991 conferences, reviewers have concluded that the 17 keV phenomenon is dead, apparently on the basis of a majority vote. For example Morrison [15] concludes that

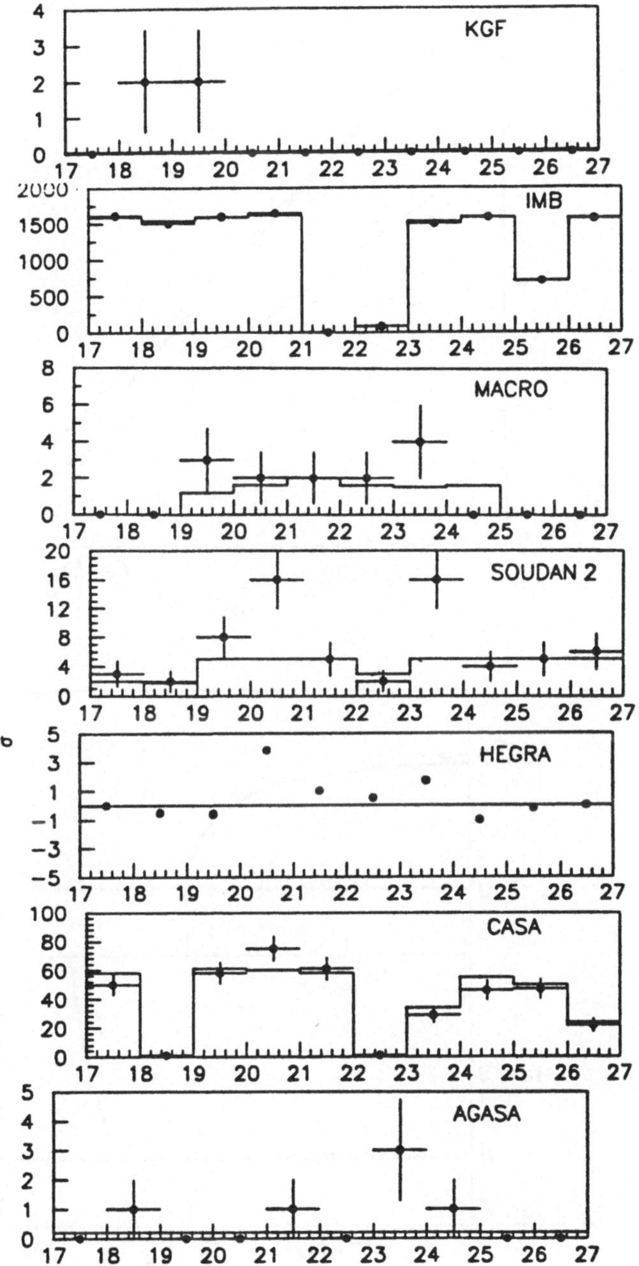

Figure 5. Event rates during period Jan 17–27 1991 associated with the direction of Cygnus X3, for various underground and surface detectors. Expected and observed numbers indicated by histogram and dots respectively [13].

4
K
2
0
Q - 3M2
Q - M2
Q
T
a
tan²θ
0
(S − 1)
b
tan²θ / 2
0
~ − tan²θ / 2
Q − 2M2
Q − M2
Q
ENERGY

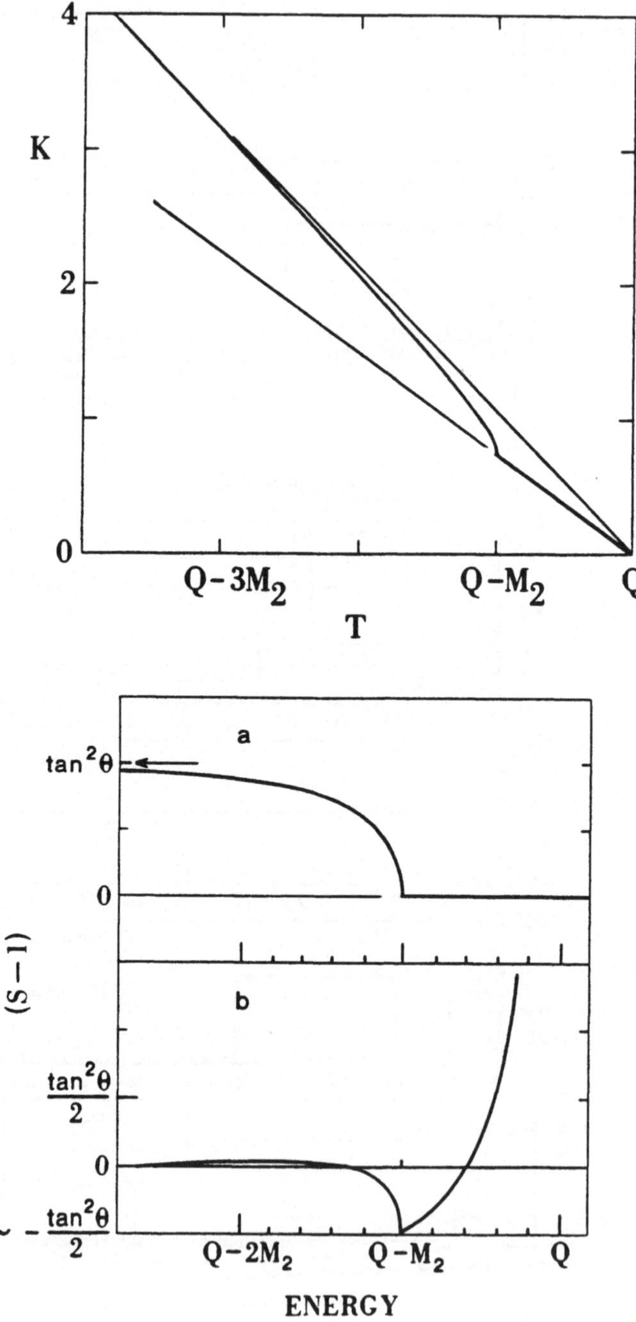

Figure 6.

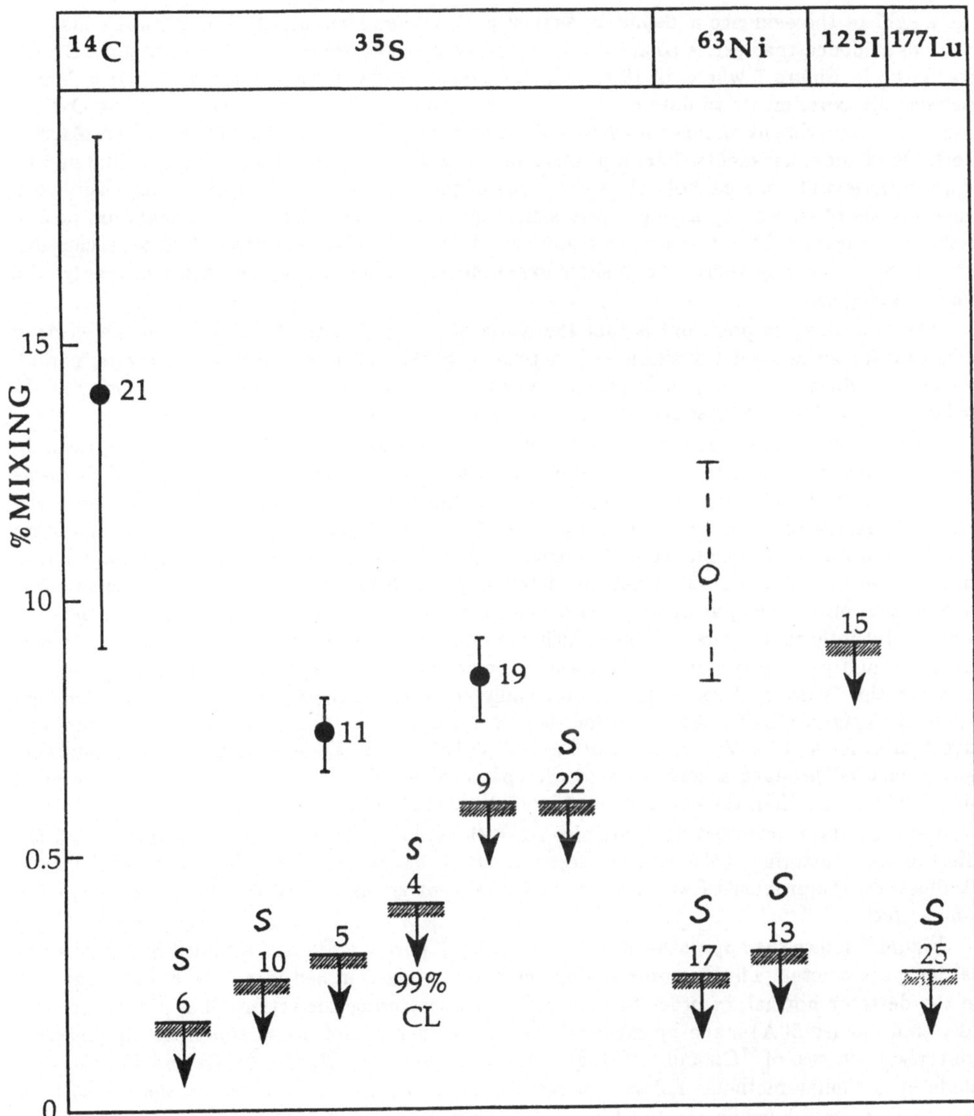

Figure 7. Experimental results on percentage mixing, $\sin^2\theta$, for various experiments with various beta sources, after Morrison [15]. The numbers attached refer to references in Morrison's paper. I have marked the magnetic spectrometer experiment with an 'S'. A recent ^{63}Ni result from Oxford [16] has been inserted as an open circle with dashed errors.

the weight of the evidence is definitely against a heavy neutrino, and further remarks that the positive evidence that exists comes from just 3 people. The reasons for Morrison's conclusion are shown in Figure 7 where, of 13 experiments, only 3 show a positive effect. Morrison's plot includes all experiments to date except those on tritium, a recent ^{63}Ni experiment at Oxford, and a ^{71}Ge experiment on inner bremsstrahlung following electron capture (IBEC) from Zagreb. Both the latter experiments claim a positive signal. However, whether the odds are 13:3 or 13:5 is quite irrelevant: one cannot assess the value of physics results by voting. I can easily turn the odds the other way by asking a very simple question: which of the experiments did not, in fitting their results, have to resort to empirical 'shape factors' in order to obtain an acceptable χ^2? If one keeps only those, the positive experiments number 4 and the negative, 1 only! So, what is going on?

The first thing to point out is that the Kurie plot — a straight line version of the electron spectrum for an allowed transition — is a **purely textbook phenomenon**. *No one*, period, has ever produced a Kurie plot in practice which is an exact straight line to an accuracy of 1% or better. *All* observed plots show deviations. Where do they come from?

To measure a beta spectrum to better than 1% precision is actually a formidable task. In magnetic spectrometers, there are small effects distorting the spectrum: scattering of electrons by the momentum-defining slits; energy losses in the source, if it is thick; energy losses and scattering in the counter window (for a gas counter) or backscattering effects (for a solid state counter); variation of acceptance with momentum arising from the potential gradient applied to an extended source; temporal variations in the magnetic field, etc. etc. All of these effects are, in principal, measurable, using calibration line (internal conversion) sources in place of the beta source. But internal conversion lines (which one relies on to measure the momentum resolution) are often multiple and few and far between, so one has to use interpolations and extrapolations. However this is done, it seems there are usually small extra corrections necessary. Although small, the magnitude of these corrections is such that, over the fitted part of the spectrum one investigates for a '17 keV effect' — say the last 40 keV before the end point — these empirical 'shape factors' produce a correction to the spectrum which can be as much as one order of magnitude bigger than the effect of the 17 keV neutrino itself.

For solid-state detectors there are also corrections, for example thick source effects and the effect of backscattering of the incident electrons out of the crystal when using an external source. To illustrate the problems I will discuss one experiment giving a positive effect, and one giving a null effect.

Figure 8 shows the apparatus of Hime and Jelley [7] using a ^{35}S and a cooled Si(Li) detector as the spectrometer. The electrons hitting the detector are confined to a cone of half angle 10° to the detector normal, in order to reduce the backscattering correction. The ^{35}S source is a very thin one (< 50Å) made by chemical deposition on a 2.8μm mylar substrate: the internal conversion sources, of ^{57}Co and ^{109}Cd, giving electron lines at 128, 122, 85, 81 and 60 keV, were made by a similar method. Figure 9 shows the total backscatter fraction for silicon, with its very weak energy dependence, and Figure 10 the resolution of the 128 keV ^{57}Co line.

Figure 10(a) shows the 'tail' on the low energy side of the peak, due to backscattering, while Figure 10(b) shows the distribution of recorded energy (= that left in crystal) divided by incident energy, E/E_0. Because of confusion with a Compton edge, this distribution can be measured only down to $E/E_0 = 0.55$. The histogram is the expected distribution from a polynomial fit to previous extensive data in the distribution of backscattering (from which Figure 8 was obtained). The procedure followed was to use the data from the internal conversion lines to parameterize the resolution function (including backscattering and any target/collimator absorption or scattering effects) and fold this with the theoretical spectrum for a massless neutrino and finally divide

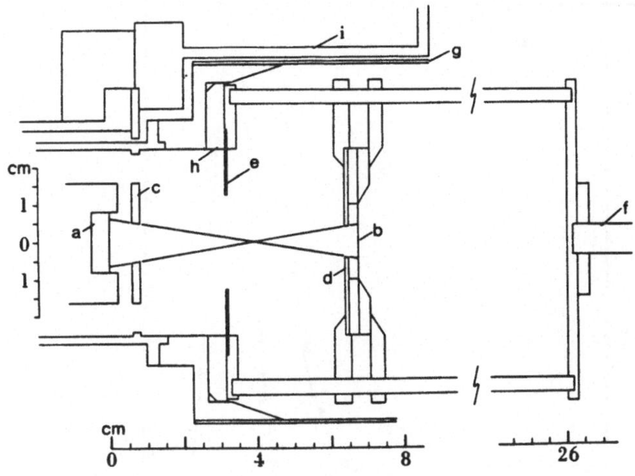

Figure 8. Apparatus of Hime and Jelley [17].
a: Si(Li) detector; b: source; d,c: baffles of copper and aluminium.

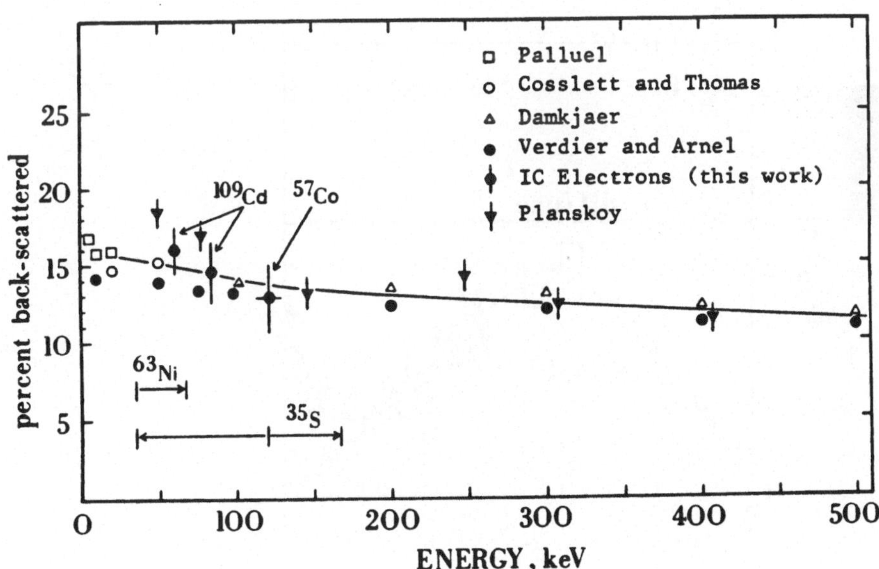

Figure 9. Backscatter fraction for normal incidence in silicon, as a function of electron energy.

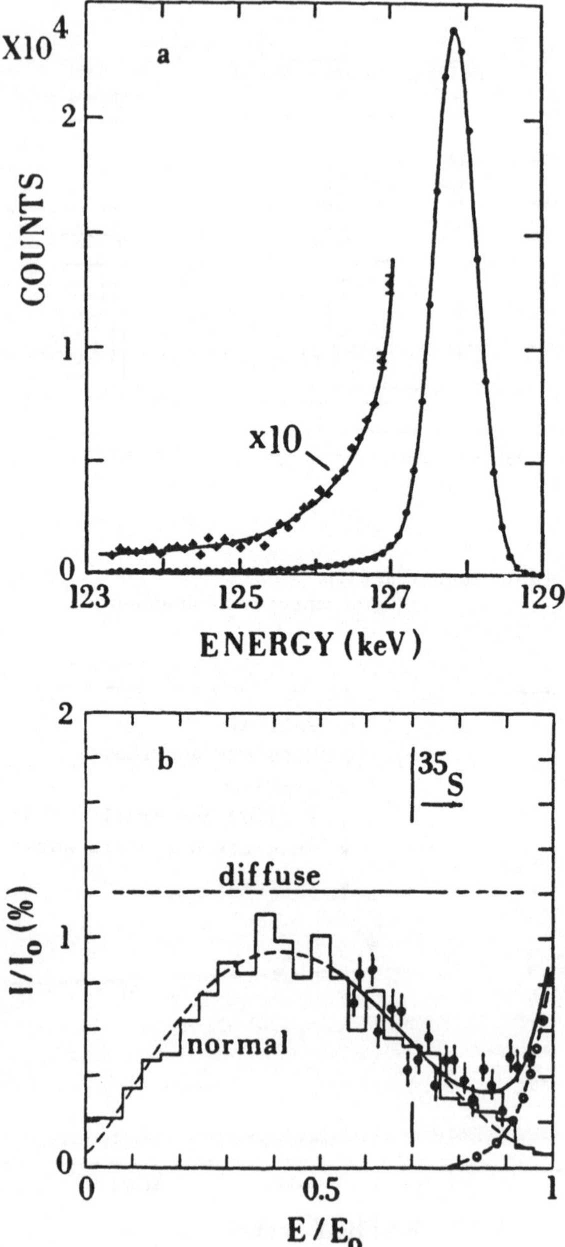

Figure 10(a). Shape of 128 keV line, with backscattering tail.
Figure 10(b). Fractional energy distribution of backscattering.

the data points by this expected spectrum (shown as a straight horizontal line). This gave Figures 11(a) and (b), which correspond to two runs with different collimator geometry. The data undershoots the Kurie plot for $m_\nu = 0$ some 17 keV below the end-point, and overshoots near the end point, just as in Figure 6(b). Figure 11(c) shows all the data, normalized to a straight line Kurie plot over the last 17 keV of electron energy, as in Figure 6(a). These plots are consistent with the existence of a heavy neutrino of $M_2 = 17 \pm 0.4$ keV, and mixing angle $\sin^2\theta = 0.0084 \pm 0.001$.

Of course, it is *also* possible to fit the data with empirical shape factors and $m_\nu = 0$ only. To do this, one has to go to a polynomial in T involving terms up to T^3, and with, of course, 3 arbitrary coefficients. The result of this experiment depends on knowing the resolution function (backscattering effects) at different electron energies and therefore some interpolation: however, it should be emphasized that while the total backscattering fraction is $\sim 12\%$, that part affecting the spectrum between the end point and 40 keV from the end-point (the region of the fit) is less than 1%, with an estimated uncertainty of less than 0.1%.

The second experiment I discuss is that of Hetherington *et al* [18] at Chalk River in 1987, which used a $\pi\sqrt{2}$ double-focussing magnetic spectrometer, as shown in Figure 12. The layout of the apparatus included an extended, 19-strip source, and 22 gas counters. Applying a suitable P.D. across the strips, one can arrange that all electrons of given momentum will reach the same focus regardless of where they originate in the source. The use of a multistrip source gives higher counting rates and the use of 22 counters allows one to observe several momenta simultaneously. There are penalties however. The source was produced by evaporating ^{63}Ni on to a thick plexiglass backing: the total backscattering from this source was 18%. The gas counters were separated from the spectrometer vacuum by a window of polypropylene, gold and chromium ($\sim 170\mu g/cm^2$), which drastically cuts the transmission of electrons below 30 keV. Clearly, the counter efficiency will depend on electron energy, and the 22 counters have to be cross-calibrated.

Despite the great care used in calibrating the spectrometer, there were unfortunately residual uncertainties, for example in the window thickness, the effects of the edges of the detector entrance slits, discriminator losses, etc. These were parameterized by introducing a 'shape factor', of the form $S = 1 + \alpha T$ where T is the electron kinetic energy. A typical value of $\alpha = 0.0006$ keV^{-1} for the fitted range $T = 50 - 67$ keV, which corresponds to a 1% variation in the intensity over the fitted interval, somewhat larger than the 0.5% effect predicted by the Hime-Jelley results for a 17 keV neutrino. Figure 13 shows the results obtained for a narrow scan spectrum ($T = 46 - 54$ keV). The dashed curve is for a heavy (17 keV) neutrino with 0.8% mixing, which is clearly a poorer fit to the data. In fact, the authors set an upper 90% CL of 0.3% for the mixing.

As far as one can see, this experiment was done very carefully and, like that of Hime and Jelley, it is not easy to see where things could have gone wrong. While the postulate of a shape factor to give acceptable fits clearly indicates that some features of the technique were not completely understood, it is difficult to see how a smoothly-varying 1% intensity correction could hide a 'kink' in the spectrum. As stated before, the question of the 17 keV neutrino is open: if it does not exist, an explanation has to be found for the 'kink' in their spectrum: and if it does exist at the level of 0.8% why did Hetherington *et al* not see it? The matter can only be settled by further experiments, and investigations may well continue for some years before the problem is resolved.

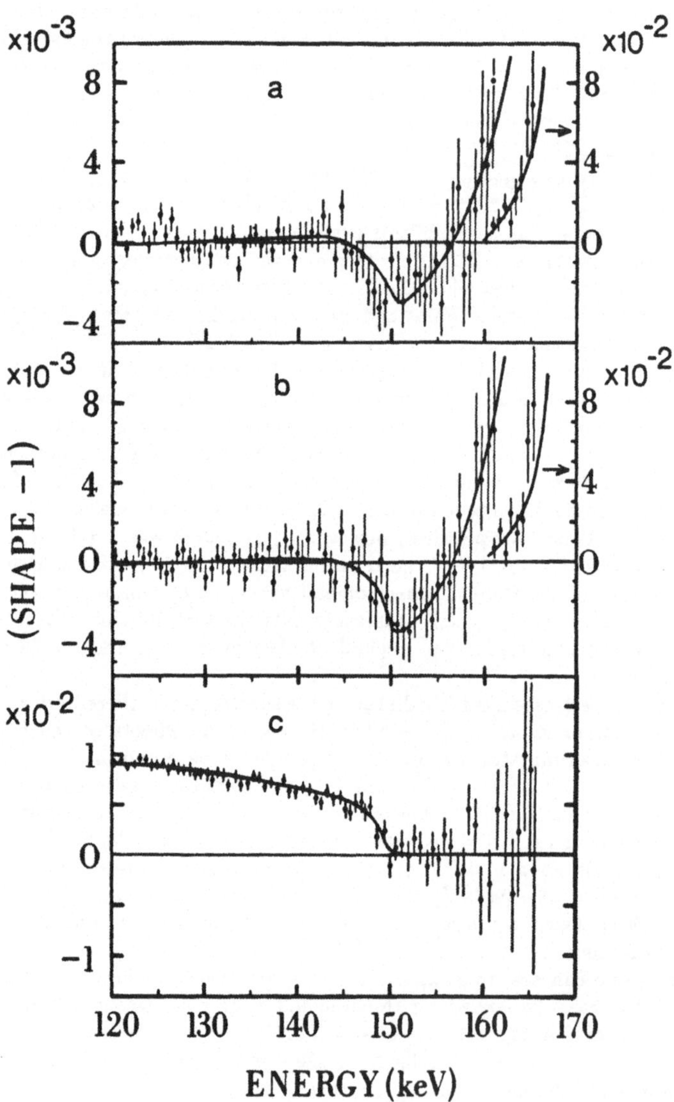

Figure 11. ^{35}S data of Hime and Jelley [17].

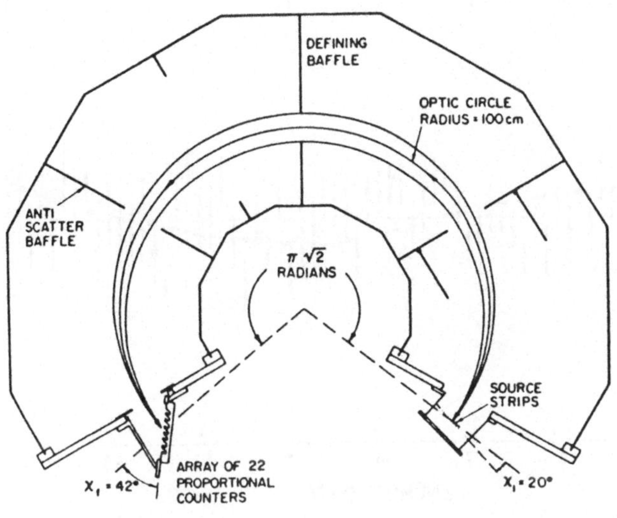

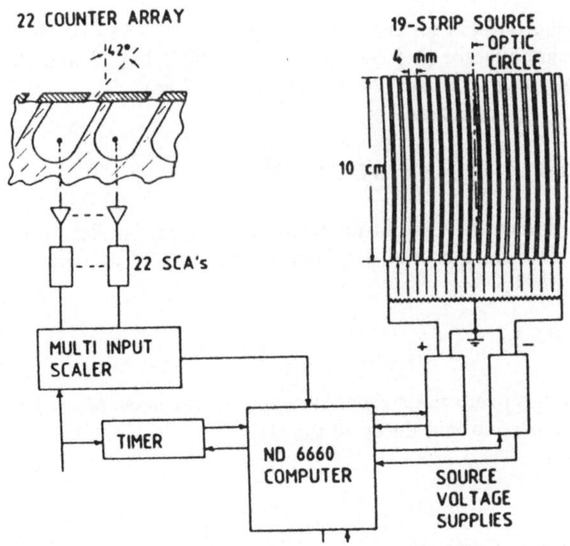

Figure 12 (top) Spectrometer of Hetherington *et al* [18] used for the beta spectrum of ^{63}Ni. (bottom) Multiple strip source and counter layout.

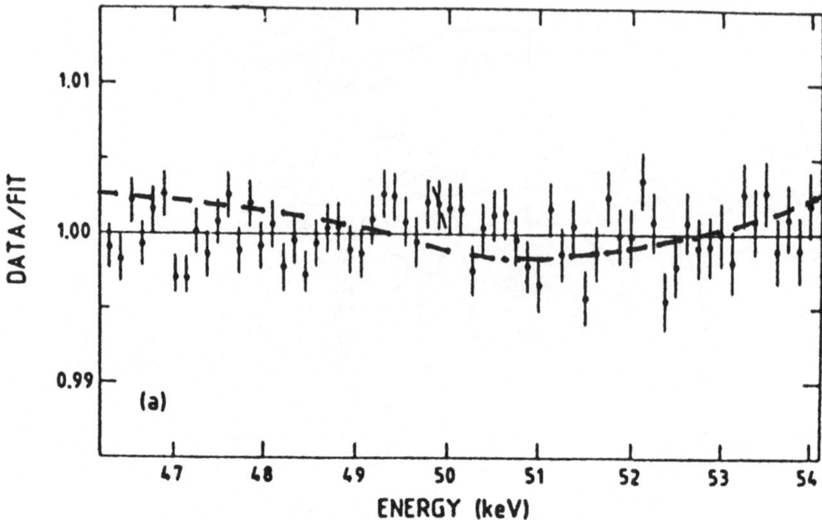

Figure 13. Fits of Hetherington *et al* for the energy range $T = 46 - 54$ keV. Solid horizontal line ratio of data to fit for a shape factor $(1 + \alpha T)$ where $\alpha = .00075$ keV^{-1} and there is no mixing ($\sin^2 2\theta = 0$). The dashed curve is for $\sin^2 2\theta = .008$, $M_2 = 17$ keV.

5. IMPLICATIONS OF A 17 KEV NEUTRINO

In terms of the three known flavours of neutrino, limits from oscillation experiments indicate that a 17 keV neutrino must be the dominant component of ν_τ, that is the orthogonal combination to (9):

$$|\nu_H> \equiv |\nu_\tau> = |\nu_2> \cos\theta - |\nu_1> \sin\theta \tag{10}$$

where ν_1 is an (approximately) zero mass eigenstate and ν_2 has mass $M_2 = 17$ keV.

Then the heavy neutrino ν_H would decay to ν_e:

$$\nu_H \to \nu_e + \nu_x + \bar{\nu}_x \tag{11}$$

where ν_x is any light neutrino. Comparing with

$$\mu^+ \to e^+ + \nu_e + \bar{\nu}_\mu$$

we get for the lifetime from the Sargent rule, the very large value

$$\tau_H = (\tau_\mu / \sin^2\theta)(m_\mu/M_H)^5 = 2.10^{15}\text{sec} \tag{12}$$

taking $m_H = 17$ keV, $\sin^2\theta = .01$. Radiative decay $\nu_H \to \nu_e\gamma$ is even slower. These lifetimes for ν_H are embarrassingly long from the viewpoint of cosmology. Since the critical mass for neutrinos surviving from the Big Bang is about 30 eV (the mass which would make the density of the universe equal to the critical density for closure), it is obvious that a stable neutrino of $m_H = 17$ keV would be a disaster: the Big Crunch would have followed the Big Bang billions of years ago.

If ν_H undergoes rapid decay to light particles which get the same red-shift as photons, as the universe expands and cools, the above difficulty can be avoided. For $kT < 17$ keV ($t > 1$ hour after the Big Bang), the heavy neutrinos become non-relativistic and the density varies as R^{-3} or T^3 (since the radius parameter $R \propto T^{-1}$), instead of as T^4 for relativistic particles. So the massive neutrino should decay to relativistic particles before the temperature falls to $T \simeq (17\,\text{keV}/30\,\text{eV})\,T_0 \approx 600T_0$, where $T_0 = 2.7°$K is the present background radiation temperature. Since $t \propto T^{-2}$, this implies decay before a time $t = t_0/(600)^2 \simeq 10^{12}$ secs, where $t_0 = 3.10^{17}$ secs is the present age of the universe. This estimate

$$\tau_H < 10^{12}\ \text{secs}\ (\approx 10^5\ \text{years}) \tag{13}$$

clearly contradicts (12) and implies some extra interactions to hasten the decay (11). So, the existence of ν_H (if proven) not only entails an extension of the Higgs sector to accommodate a heavy neutrino, but new couplings in addition. For example Chikashige, Mohapatra and Peccei [19] invoke massive Majorana neutrinos associated with a mass scale M of lepton non-conservation, with a Majoron χ as the massless Goldstone boson of this broken global symmetry. Provided M is not too large ($\lesssim 10^5$ GeV), the lifetime τ_H for the decay $\nu_H \to \nu_L + \chi$ can fulfil the limit (13). For further details, see a recent review by Peccei [20].

In summary, massive stable neutrinos appear to cause unacceptable problems for cosmology. However it does seem possible to fulfil most of the constraints from cosmology, SN1987A etc. by invoking new interactions in the lepton sector, which will allow such a neutrino to decay quickly. Hence, the experimental verification or disproof of the 17 keV neutrino is a matter of great importance.

5. FUTURE PROSPECTS

The experiments in particle astrophysics without accelerators, which form the main subject of this meeting, seem to me to offer considerable promise of totally new physics, outside the realm of the Standard Model. For example, the recent furore over the 17 keV neutrino — which may or may not exist — is evidence of that. The position in high energy physics today bears some parallels with that of 45 years ago, when I started research. I was forcefully reminded of this when Shapiro asked me to contribute a talk on the discovery of the pion, at a special session commemorating Carl Anderson at the Dublin Cosmic Ray Conference in August 1991, and entitled "Physics with Nature's Accelerators".

In 1946, the Holy Grail of nuclear physics was the Yukawa quantum, the fundamental carrier of the strong interactions which, if found, was expected to lead to final understanding of nuclear forces. The pion, with apparently the right properties, was indeed found in May 1947, by Powell and his colleagues. But already, by December 1947, the uniqueness of the pion, as *the* unique quantum of Yukawa, was in question: V-particles were observed at Manchester, and these seemed to complicate the picture. Looking back now, we see that the discovery of the pion was very important in promoting the building of larger and larger accelerators and marked the blossoming of particle physics, but that the pion itself was not fundamental, but just the lightest of many possible combinations of up and down quarks.

Indeed, today we realise that the muon is far more fundamental than the pion — even if the Nobel prize was awarded for the pion rather than the muon. In 1947, we were not only getting the wrong answers, but not even asking the right questions! Of course it is easy to be wise years after the event. But the point I am trying to make is that we have sometimes been wrong and can be wrong again.

Today, the Holy Grail of particle physics is the Higgs boson, to the discovery of which the SSC and LHC are dedicated. If it is found, the Higgs is supposed to reveal the secrets of mass generation and be the final missing piece to complete the vindication of the Standard Model (assuming, that is, that the top quark has already been found). For me, however — based on my own experience over the past 45 years — there appears to be a distinct possibility that we might be asking the wrong question. It seems to me that to do experiments which are not locked into, or motivated by, any particular model, but allow even a modest exploration of high energy physics disconnected from the main stream, may be very well worthwhile. The questions themselves — the nature of neutrino masses and mixings, the solar neutrino discrepancy (even the atmospheric neutrino discrepancy), the nature of dark matter, etc. etc. are very real questions, not motivated by any theory but by experimental fact, and are crying out for answers.

REFERENCES

1 K.A. Olive, D.N. Schramm, G. Steigman and T.P. Walker, Phys. Lett. 236B (1990) 454.
2 K. Fujikawa and R.E. Schrock, Phys. Rev. Lett. 45 (1980) 963.
3 A.V. Kyuldjiev, Nucl. Phys. B243 (1984) 387.
4 F. Reines, H.S. Gurr and H.W. Sobel, Phys. Rev. Lett. 37 (1976) 315.
5 F.T. Avignone, Phys. Rev. D2 (1970) 2609.
6 P. Vogel and J. Engel, Phys. Rev. D39 (1989) 3378.
7 M.B. Voloshin, M.I. Vysotsky and L. Okun, Sov. Phys. JETP 64 (1986) 446.
8 N. Ushida *et al.* Phys. Rev. Lett. 57 (1986) 2897.
9 L. Moscoso, Proc. 14th Int. Conf. on Neutrino Phys. and Astrophys. CERN: Nucl. Phys. B (Proc. Suppl.) 19 (1991) 147.
10 T.C. Weekes *et al*, Astro. Journal 342 (1989) 379.
11 M.A. Thompson *et al* (Soudan 2 collaboration), Phys. Lett. B269 (1991) 220.
12 M. Merck *et al*, Proc. Cosmic Ray Conf., Dublin (August 1991).
13 M.A. Thomson, private communication.
14 J.J. Simpson, Phys. Rev. Lett. 54 (1985) 1891.
15 D.R.O. Morrison, CERN-PPE/91-140 (1991).
16 A. Hime and N. Jelley, OUNP-91-21 (1991).
17 A. Hime and N. Jelley, Phys. Lett. B257 (1991) 441.
18 D.W. Hetherington *et al* Phys. Rev. C36 (1987) 1504.
19 J. Chikashige, R.N. Mohapatra and R.D. Peccei, Phys. Lett. 98B (1981) 265; 45 (1981) 1926.
20 R.D. Peccei, “Constraints and Model Considerations for a 17 keV Neutrino’, UCLA/91/TEP/42 (1991).